I0027465

Brain Tumor Drug Development: Current Advances and Strategies

(Part 1)

Edited by

Prashant Tiwari
Department of Pharmacology
College of Pharmaceutical Sciences
Dayananda Sagar University
Bengaluru, Karnataka, India

Pankaj Kumar Singh
Department of Pharmaceutics
National Institute of Pharmaceutical Education
and Research (NIPER), Hyderabad, Telangana, India

&

Sunil Kumar Kadiri
Department of Pharmacology
College of Pharmaceutical Sciences
Dayananda Sagar University
Bengaluru, Karnataka, India

Brain Tumor Drug Development: Current Advances and Strategies *(Part 1)*

Editors: Prashant Tiwari, Pankaj Kumar Singh and Sunil Kumar Kadiri

ISBN (Online): 979-8-89881-171-6

ISBN (Print): 979-8-89881-172-3

ISBN (Paperback): 979-8-89881-173-0

© 2025, Bentham Books imprint.

Published by Bentham Science Publishers Pte. Ltd. Singapore,

in collaboration with Eureka Conferences, USA. All Rights Reserved.

First published in 2025.

BENTHAM SCIENCE PUBLISHERS LTD.
End User License Agreement (for non-institutional, personal use)

This is an agreement between you and Bentham Science Publishers Ltd. Please read this License Agreement carefully before using the ebook/echapter/ejournal (**"Work"**). Your use of the Work constitutes your agreement to the terms and conditions set forth in this License Agreement. If you do not agree to these terms and conditions then you should not use the Work.

Bentham Science Publishers agrees to grant you a non-exclusive, non-transferable limited license to use the Work subject to and in accordance with the following terms and conditions. This License Agreement is for non-library, personal use only. For a library / institutional / multi user license in respect of the Work, please contact: permission@benthamscience.org.

Usage Rules:

1. All rights reserved: The Work is the subject of copyright and Bentham Science Publishers either owns the Work (and the copyright in it) or is licensed to distribute the Work. You shall not copy, reproduce, modify, remove, delete, augment, add to, publish, transmit, sell, resell, create derivative works from, or in any way exploit the Work or make the Work available for others to do any of the same, in any form or by any means, in whole or in part, in each case without the prior written permission of Bentham Science Publishers, unless stated otherwise in this License Agreement.
2. You may download a copy of the Work on one occasion to one personal computer (including tablet, laptop, desktop, or other such devices). You may make one back-up copy of the Work to avoid losing it.
3. The unauthorised use or distribution of copyrighted or other proprietary content is illegal and could subject you to liability for substantial money damages. You will be liable for any damage resulting from your misuse of the Work or any violation of this License Agreement, including any infringement by you of copyrights or proprietary rights.

Disclaimer:

Bentham Science Publishers does not guarantee that the information in the Work is error-free, or warrant that it will meet your requirements or that access to the Work will be uninterrupted or error-free. The Work is provided "as is" without warranty of any kind, either express or implied or statutory, including, without limitation, implied warranties of merchantability and fitness for a particular purpose. The entire risk as to the results and performance of the Work is assumed by you. No responsibility is assumed by Bentham Science Publishers, its staff, editors and/or authors for any injury and/or damage to persons or property as a matter of products liability, negligence or otherwise, or from any use or operation of any methods, products instruction, advertisements or ideas contained in the Work.

Limitation of Liability:

In no event will Bentham Science Publishers, its staff, editors and/or authors, be liable for any damages, including, without limitation, special, incidental and/or consequential damages and/or damages for lost data and/or profits arising out of (whether directly or indirectly) the use or inability to use the Work. The entire liability of Bentham Science Publishers shall be limited to the amount actually paid by you for the Work.

General:

1. Any dispute or claim arising out of or in connection with this License Agreement or the Work (including non-contractual disputes or claims) will be governed by and construed in accordance with the laws of Singapore. Each party agrees that the courts of the state of Singapore shall have exclusive jurisdiction to settle any dispute or claim arising out of or in connection with this License Agreement or the Work (including non-contractual disputes or claims).
2. Your rights under this License Agreement will automatically terminate without notice and without the

need for a court order if at any point you breach any terms of this License Agreement. In no event will any delay or failure by Bentham Science Publishers in enforcing your compliance with this License Agreement constitute a waiver of any of its rights.

3. You acknowledge that you have read this License Agreement, and agree to be bound by its terms and conditions. To the extent that any other terms and conditions presented on any website of Bentham Science Publishers conflict with, or are inconsistent with, the terms and conditions set out in this License Agreement, you acknowledge that the terms and conditions set out in this License Agreement shall prevail.

Bentham Science Publishers Pte. Ltd.
No. 9 Raffles Place
Office No. 26-01
Singapore 048619
Singapore
Email: subscriptions@benthamscience.net

CONTENTS

FOREWORD

In recent years, brain tumor research has made remarkable strides, bringing to light new pathways, mechanisms, and potential therapeutic strategies. However, the complexity of the brain and the heterogeneity of brain tumors, particularly glioblastomas, continue to present formidable challenges in developing effective treatments. Against this backdrop, the book "Brain Tumor Drug Development: Current Advances and Strategies (Part 1)," edited by Dr. Prashant Tiwari, Dr. Pankaj Kumar Singh, and Dr. Sunil Kumar Kadiri, emerges as a critical resource for both experienced researchers and new entrants into the field.

The editors, each with extensive expertise in oncology, pharmacology, and drug development, have compiled a remarkable collection of insights that encompass the full spectrum of current advances in brain tumor therapeutics. From the latest molecular and genomic approaches to emerging targeted therapies, this volume encapsulates a dynamic and evolving landscape. The emphasis on precision medicine, immunotherapy, and the repurposing of existing drugs highlights the cutting-edge strategies that are driving the next generation of brain tumor treatments.

In particular, the book addresses some of the most pressing questions facing the field: How can we overcome the blood-brain barrier to deliver effective drugs? What role do molecular biomarkers play in predicting patient outcomes? And how can we harness the immune system to combat tumors that have historically been resistant to conventional treatments?

The editors have ensured that this book not only reflects the state-of-the-art in brain tumor drug development but also points toward the future. By exploring the potential of novel therapeutic agents, innovative delivery systems, and combination treatments, this book provides readers with a comprehensive understanding of the strategies that are being employed to tackle one of the most aggressive forms of cancer.

I hope that this book will inspire and inform the next wave of discoveries, bridging the gap between fundamental research and clinical application. Whether you are a researcher, clinician, or industry professional, Brain Tumor Drug Development: Current Advances and Strategies is an invaluable resource that provides both a broad overview and in-depth discussions of the ongoing efforts to develop effective therapies for brain tumors.

I congratulate the editors and contributors for compiling such a timely and impactful book, and I am confident that it will make a lasting contribution to the field of brain tumor research.

Shashi Alok
Department of Pharmacognosy
Institute of Pharmacy
Bundelkhand University, Jhansi (U.P.)
India

PREFACE

The development of effective treatments for brain tumors represents one of the most formidable challenges in modern oncology. Brain tumors, particularly glioblastomas, are characterized by their aggressive nature and resistance to conventional therapies. Despite advancements in surgery, radiotherapy, and chemotherapy, the prognosis for patients with brain tumors has remained poor, necessitating novel approaches and a deeper understanding of the underlying biology of these malignancies.

In recent years, research in brain tumor drug development has progressed significantly, driven by advancements in molecular biology, genomics, immunology, and drug delivery systems. New insights into the genetic and molecular makeup of brain tumors, combined with breakthroughs in drug design and delivery, have opened exciting new possibilities for targeting tumors more effectively. However, the journey from bench to bedside remains long and complex, with many scientific, clinical, and regulatory hurdles still to be overcome.

This book, titled Brain Tumor Drug Development: Current Advances and Strategies (Part 1), aims to provide a comprehensive overview of the latest developments in the field, from the discovery of new therapeutic targets to innovative drug delivery systems. It brings together leading experts from various disciplines, each contributing their knowledge and experience in the pursuit of better treatment options for brain tumors. By focusing on current advances and emerging strategies, we hope to shed light on the future directions of brain tumor drug development and inspire further research and collaboration.

This book seeks to highlight both the opportunities and challenges in brain tumor drug development. While we have witnessed remarkable progress in understanding the biology of brain tumors and developing potential therapies, much work remains to be done. The translation of laboratory findings into clinically effective treatments requires ongoing collaboration between researchers, clinicians, pharmaceutical companies, and regulatory bodies.

As editors, we are deeply grateful to the contributors whose expertise and dedication have made this book possible. Their cutting-edge research and forward-thinking approaches represent the best of what the scientific community has to offer in the fight against brain cancer. We also wish to express our appreciation to Bentham Science for supporting this project and facilitating its publication.

We hope that Brain Tumor Drug Development: Current Advances and Strategies (Part 1) will serve as an indispensable resource for researchers, clinicians, and pharmaceutical professionals who are committed to advancing the treatment of brain tumors. By fostering greater understanding and innovation, we believe this book will play a significant role in shaping the future of brain tumor therapeutics and ultimately improve the lives of patients worldwide.

Prashant Tiwari
Department of Pharmacology
College of Pharmaceutical Sciences
Dayananda Sagar University
Bengaluru, Karnataka, India

Pankaj Kumar Singh
Department of Pharmaceutics
National Institute of Pharmaceutical Education and Research (NIPER)
Hyderabad, Telangana, India

&

Sunil Kumar Kadiri
Department of Pharmacology
College of Pharmaceutical Sciences
Dayananda Sagar University
Bengaluru, Karnataka, India

List of Contributors

Aarti Tiwari	Department of Pharmacy, Guru Ghasidas Vishwavidyalaya (Central University), Bilaspur, Chhattisgarh, India
Asha Raghav	Department of Pharmaceutics, School of Health Sciences, Sushant University, Gurugram, Haryana, India
Bhaskar H. Vaidhun	LSHGCT's Gahlot Institute of Pharmacy, Kopar Khairane, Navi Mumbai, Maharashtra-400709, India
Deepa Mandlik	Department of Pharmacology, Bharati Vidyapeeth (Deemed to be University), Poona College of Pharmacy, Pune-411038, Maharashtra, India
Dipika Pawar	LSHGCT's Gahlot Institute of Pharmacy, Kopar Khairane, Navi Mumbai, Maharashtra-400709, India
Ekta Singh	Aditya Bangalore Institute of Pharmacy Education and Research, Bengaluru, Karnataka, India
Meghraj Suryawanshi	Department of Pharmaceutics, Sandip Institute of Pharmaceutical Sciences (SIPS), Affiliated to Savitribai Phule Pune University (SPPU, Pune), Nashik, Maharashtra-422213, India AllWell Nutritech LLP Dharangaon, Maharashtra-425105, India
Meenakshi Attri	School of Medical & Allied Sciences, K. R. Mangalam University, Gurugram, Haryana-122103, India
Mohit Agrawal	School of Medical & Allied Sciences, K. R. Mangalam University, Gurugram, Haryana-122103, India
Nagaraj Sreeharsha	Department of Pharmaceutics, Vidya Siri College of Pharmacy, Bengaluru-560035, Karnataka, India
Pankaj Kumar Singh	Department of Pharmaceutics, National Institute of Pharmaceutical Education and Research (NIPER), Hyderabad-500037, India
Pradeep Kumar Samal	Department of Pharmacy, Guru Ghasidas Vishwavidyalaya (Central University), Bilaspur, Chhattisgarh, India
Prakash Goudanavar	Department of Pharmaceutics, Sri Adichucnhanagiri College of Pharmacy, Adichunchanagiri University-571448, Karnataka, India
Pranjal Gujarathi	Department of Pharmacology, Vidhyadeep Institute of Pharmacy, Vidhyadeep University, Surat, Gujarat-394110, India
Prashant Tiwari	College of Pharmaceutical Sciences, Dayananda Sagar University, Bengaluru-562112, India
Rahul Kumar	Department of Biological Science, National Institute of Pharmaceutical Education and Research (NIPER), Hyderabad-500037, India
Sakshi Soni	Department of Pharmaceutical Sciences, Dr. Harisingh Gour Central University, Sagar, Madhya Pradesh-470003, India
Sandeep Waghulde	LSHGCT's Gahlot Institute of Pharmacy, Kopar Khairane, Navi Mumbai, Maharashtra-400709, India
Santosh Kumar Guru	Department of Biological Science, National Institute of Pharmaceutical Education and Research (NIPER), Hyderabad–500037, India

Sonal Dubey   College of Pharmaceutical Science, Dayananda Sagar University, Bengaluru-562112, Karnataka, India

Sushil K. Kashaw   Department of Pharmaceutical Sciences, Dr. Harisingh Gour Central University, Sagar, Madhya Pradesh-470003, India

Thippeswamy Mallamma   Department of Pharmaceutics, Sri Adichucnhanagiri College of Pharmacy, Adichunchanagiri University-571448, Karnataka, India

Vandana Soni   Department of Pharmaceutical Sciences, Dr. Harisingh Gour Central University, Sagar, Madhya Pradesh-470003, India

CHAPTER 1

Current Advances in Drug Development, Design, and Strategies for Brain Tumors

Sandeep Waghulde[1,*], Dipika Pawar[1] and Bhaskar H. Vaidhun[1]

[1] LSHGCT's Gahlot Institute of Pharmacy, Kopar Khairane, Navi Mumbai, Maharashtra-400709, India

Abstract: Cancer manifests itself differently in each patient due to various genetic abnormalities that allow cancer to develop in vulnerable cells. The most common method of treating brain tumours is surgery; however, complete removal is challenging due to the tumor's invasiveness and lack of clear boundaries. Effective brain tumor-targeted drug delivery requires careful consideration of numerous factors, including the tumour microenvironment, tumour cells, and the obstacles involved in the process, as brain tumours differ significantly from peripheral tumours owing to their complex oncogenesis. Physiological barriers like the Blood-Brain Tumour Barrier (BBTB) and overexpressed efflux pumps prevent the drugs from penetrating tumours. Optimising the medication distribution volume allows for effective intraventricular infusion by preventing backflow. Research suggests that during interstitial infusion, fluid convection, rather than simple diffusion, maintains a pressure gradient that enhances the distribution of both large and small molecules in cancerous and brain tissues. As nanoparticles can cross the porous blood-brain barrier, this is one potential method of drug delivery to brain tumours. When treating many tumour antigens at once, a vaccine is often the most effective approach. Instead of designing several CAR structures, it is far more practical to include multiple peptides into a vaccine formulation. In recent times, there has been an unexpected rise in the appeal of cell treatments, which now rank as the third most promising experimental treatment strategy for cancer.

Keywords: Brain tumour, Cerebrospinal fluid, Clinical trials, Glioma, Intra-arterial, Intracerebroventricular, Microdialysis, Monoclonal antibody, Nanocarriers, Nanoparticles, Oncogenesis, P-glycoprotein, Prodrugs, Sonoporation, Tumour barrier, Vaccine, Viruses.

INTRODUCTION

Cancer manifests itself differently in each patient due to the numerous genetic abnormalities that enable cancer to develop in vulnerable cells. Over the last

---

* Corresponding author Sandeep Waghulde: LSHGCT's Gahlot Institute of Pharmacy, Kopar Khairane, Navi Mumbai, Maharashtra-400709, India; E-mail: sandeepwaghulde@yahoo.com

Prashant Tiwari, Pankaj Kumar Singh & Sunil Kumar Kadiri (Eds.)
All rights reserved-© 2025 Bentham Science Publishers

several decades, growing research indicates that a subtype of cancer cells mimicking normal stem cell characteristics may be responsible for tumor development and heterogeneity, treatment resistance, and recurrence [1].

Many people in the West are concerned about developing a brain tumor, one of the most challenging illnesses to treat. This issue affects about 25% of the population [2]. Despite the fact that there is currently no definitive cure for central nervous system diseases, such as HIV encephalopathy, neurodegenerative disorders, epilepsy, or cerebrovascular disease, research has increasingly focused on brain-targeted therapies [3].

According to the most recent worldwide cancer statistics published by the World Health Organisation (WHO) in 2020 [4], brain tumours are responsible for around 1.6% of tumour incidence and 2.5% of tumour deaths. Gliomas account for 30% of all brain tumours and are the most prevalent and invasive kind of the disease. They are characterised by significant invasion, a high recurrence rate, and a poor prognosis [5].

The most common method of treating brain tumours is surgery; however, full removal is challenging due to the tumor's invasiveness and lack of clear boundaries. More than 90% of recurrences occur after surgery [6 - 10]. There has been some encouraging but modest success with more modern glioma therapies, including gene therapy, immunotherapy, and angiogenesis inhibition [11]. Therefore, it is crucial to create drugs for brain tumours that are targeted, effective, and less hazardous.

Glioma infiltration makes it difficult to fully eliminate tissues damaged by pathogens or tumors without impairing normal brain functioning [12, 13], which is one reason why this conventional treatment is associated with a poor prognosis and rapid recurrence.

For a long time, scientists have been working to develop methods for delivering therapeutic drugs to tumor sites while minimizing side effects on healthy brain and peripheral tissues. Over the last few decades, there has been a significant focus on currently available and important drug delivery systems for brain tumors. Effective brain tumor-targeted drug delivery requires careful consideration of numerous factors, including the tumour microenvironment, tumour cells, and the obstacles involved in the process, as brain tumours differ significantly from peripheral tumours owing to their complex oncogenesis. To achieve the targeted treatment using nanocarriers, various targets have been employed. In this chapter, we provide a concise overview of many potential targeted delivery techniques for brain tumours.

CHALLENGES IN DEVELOPING DRUGS FOR BRAIN CANCER

Many obstacles exist in therapy for brain cancers compared to peripheral malignancies (Fig. **1**). A counterargument suggests that medications are unable to penetrate tumors due to physiological barriers, such as the blood-brain tumor barrier (BBTB) and overexpressed efflux pumps. There is a significant failure and recurrence rate in brain cancer treatments due to the tumour microenvironment (TME) and cancer stem cells (CSC)-induced heterogeneity, immune evasion, drug resistance, invasion, and infiltration [9]. The majority of tumors exhibit heterogeneity, which is one of the most challenging behaviors in cancer ecosystems. Heterogeneity contributes to tumor resistance, more aggressive metastasis, and recurrence, which are key factors that hinder the long-term effectiveness of solid tumor therapies. Cellular interactions drive heterogeneity, and understanding the mechanisms governing the tumor microenvironment (TME), as well as how distinct cellular subtypes relate to clinical outcomes, will significantly improve current therapeutic strategies. Patients with brain tumors receiving standard treatment have a median survival of approximately 20 months, with survival rates of only 27% at 2 years and 10% at 5 years [14].

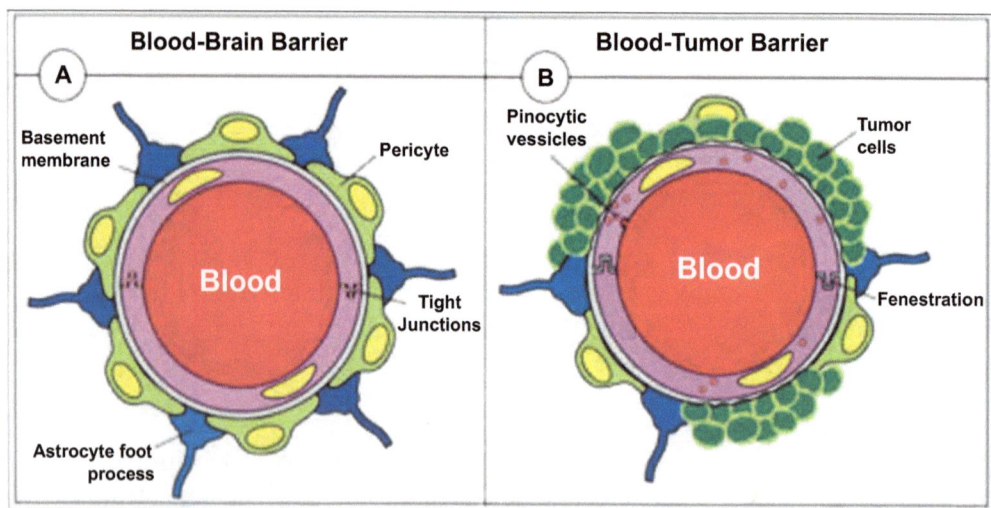

Fig. (1). Diagram showing how BBB and BBTB differ from one another. Blood-brain tumour barrier (BBTB) and blood-brain barrier (BBB).

Barriers to Targeted Drug Delivery Strategies and their Role

Types of gliomas include astrocytomas, oligodendrogliomas, and ependymomas. Tumours that form outside of the blood-brain barrier are called meningiomas. Tumours that expand to cover various areas of the body are called metastases. They typically originate from malignancies of the colon, lung, melanoma, or melanoma [15]. Extensive infiltration of neighbouring tissues and substantial endothelial growth of the aglomeruloid multilayer vasculature are the most aggressive hallmarks of high-grade brain tumours (glioblastoma multiforme). Despite optimal medication administration, glioblastoma develops resistance due to factors, such as high intratumoral interstitial pressure (ITIS), partial blood-brain barrier (BBB) integrity, inadequate blood flow, and the emergence of new drug resistance mechanisms. Fig. (**1**) shows a functional and structural difference between the BBB and the choroid plexus, another barrier that separates the blood from the cerebrospinal fluid (CSF) [16]. Treatment of brain illnesses frequently involves the administration of the prescribed medications either orally or intravenously. Bioavailability is inadequate for optimal therapeutic effect due to poor blood-brain barrier (BBB) permeation associated with oral drug delivery [17]. The fundamental causes of this ineffectiveness include a large molecular weight, undesirable physicochemical characteristics (polar functional groups), enzymatic breakdown before reaching the target organ (the brain), or extrusion at the cell level of the cerebrovascular endothelium prior to reaching the neurons in the brain. We must correctly diagnose the issue before addressing the carrier systems, as BBB is a governing factor [18, 19].

Chemicals are unable to pass through the brain due to the blood-brain barrier (BBB), which consists of neurons, astrocytes, pericytes, and brain capillary endothelial cells (BCEC). More than 98% of small-molecule medications and all macromolecular pharmaceuticals are blocked from accessing the CNS by the blood-brain barrier (BBB). Aside from small hydrophilic molecules and lipophilic compounds, the paracellular barrier prevents the passive diffusion of all other substances into the CNS. The close connections that exist between BCEC cells maintain this barrier. As BCEC cells have lower endocytosis activity than other brain cells, medicinal transcellular transport across the blood-brain barrier is severely limited [20]. Moreover, beta cell death receptors (BCECs) express a variety of enzymes involved in the metabolic process of pharmaceuticals, including peptidase, phosphatase, esterase, nucleotidase, and cytochrome P450 [21]. This has resulted in the BBB cells gaining a substantial metabolic capability. The immune system creates a barrier composed of mast cells, macrophages, and microglia to facilitate the removal of drugs. Actively pumping out drugs and reducing permeability, efflux proteins are overexpressed in the BBB [22]. ATP-binding cassette transporters and solute carrier transporters are two examples of these proteins (P-gp, BCRP, and MRPs). Additionally, they are a major factor in why brain tumours might become resistant to some drugs.

When a brain tumor exceeds a volume of approximately 2 mm^3, the blood–brain tumor barrier (BBTB) begins to form due to the disruption of the blood–brain barrier (BBB) integrity, primarily driven by tumor-induced angiogenesis [23].

This pathological transformation compromises the BBB's normal function, creating a new barrier that presents unique challenges for therapeutic delivery. Historically, the most widely used strategy for nanoparticle accumulation in tumors has been passive targeting *via* the enhanced permeability and retention (EPR) effect. However, in brain tumors, the EPR effect is significantly diminished. Brain cancers possess much smaller vascular pores, typically ranging from 7 to 100 nm, and exhibit weaker vascular permeability, which severely limits the effectiveness of EPR-based drug delivery [24]. As a result, the BBTB is now recognized as a major obstacle in brain tumor therapy, impeding the efficient transport of therapeutic agents to tumor tissues. Overcoming this barrier is crucial for enhancing drug delivery and improving treatment efficacy in patients with brain cancer.

TYPES OF BRAIN TUMORS

A brain tumor, also referred to as an intracranial tumor, is an abnormal mass of tissue resulting from the unregulated growth and proliferation of brain cells, bypassing the normal mechanisms that control cellular division and death. These tumors can disrupt neurological function and pose serious health risks depending on their location, size, and growth rate. There are more than 150 distinct types of brain tumors, broadly classified into two major categories: Primary brain tumors – originating within the brain or its surrounding structures and metastatic (secondary) brain tumors – arising from cancers elsewhere in the body that have spread to the brain *via* the bloodstream or direct invasion. Fig. (**2**) illustrates the classification and key distinctions between primary and metastatic brain tumors.

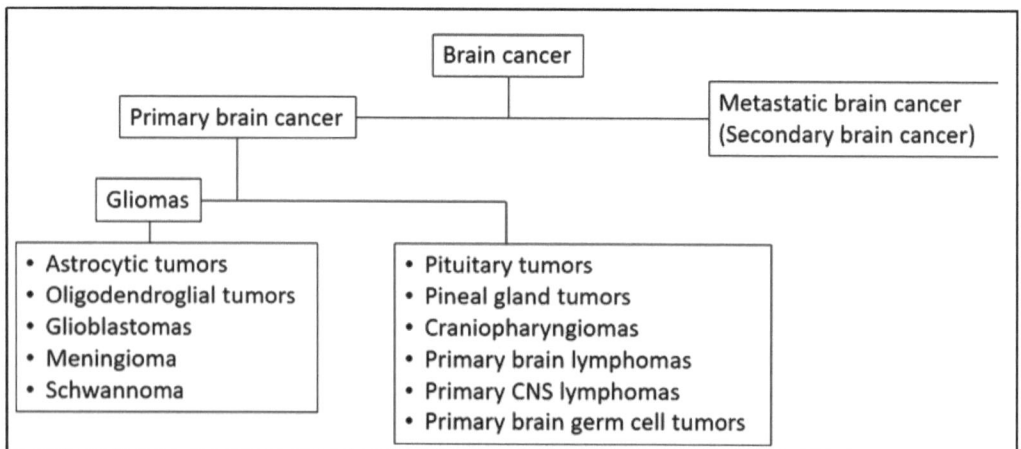

Fig. (2). Common types of brain tumors [6-8].

Glioblastoma multiforme (GBM) is the most common and aggressive malignant brain tumor, accounting for approximately 81% of all malignant central nervous system (CNS) tumors [25, 26]. GBM is classified as a subtype of astrocytoma and is designated as a World Health Organization (WHO) Grade IV tumor, indicating its high malignancy and poor prognosis [27]. Treatment protocols for Grade III brain tumors and GBM are often similar, involving a combination of surgery, radiotherapy, and chemotherapy. Regardless of malignancy level, intracranial tumors pose significant risks due to their potential to invade or displace critical brain regions, thereby compromising neurological function [28]. Common clinical manifestations include seizures, peritumoral edema, venous thrombosis, profound fatigue, and cognitive and mental impairment [29].

GBM is clinically categorized into two major subtypes: Primary GBM, which represents approximately 95% of cases, typically arises *de novo* in older adults and progresses rapidly within 3 to 6 months, and secondary GBM, which develops in younger individuals from pre-existing low-grade astrocytomas, with a slower progression over 10 to 15 years [30]. Although both subtypes are treated using similar therapeutic approaches, their molecular profiles, clinical trajectories, and patient demographics differ, which may influence treatment response and prognosis [31, 32].

STRATEGY OF BRAIN TUMOR TARGETING

Traditional Treatment Approach: Preventing the Growth of Brain Tumours

The role of tyrosine kinases in cancer has been extensively investigated. Aberrant expression of tyrosine kinases, which account for over 50% of proto-oncogenes and oncogenic products, disrupts the regulation of cell proliferation and ultimately contributes to cancer progression [33]. Dysregulated tyrosine kinase activity is associated with tumor invasion, metastasis, neovascularization, and chemoresistance [34]. Consequently, antitumor therapies are being developed to target more than twenty distinct families of receptor and non-receptor tyrosine kinases, including EGFR, VEGFR, PDGFR, FGFR, among others [35]. Nimotuzumab, an anti-EGFR drug (Fig. **3**), is commonly used in the treatment of brain cancer. However, when administered to pediatric patients with high-grade gliomas, it has shown only a marginal improvement in overall survival rates [36].

Neovascularization and tumour growth are both aided by the VEFG/VEGFR signalling pathway. To inhibit tumour neovascularization, it is crucial to discover anticancer drugs that target this signalling mechanism [37]. Many nations have now approved the use of bevacizumab, a monoclonal antibody that targets VEGF, to treat glioblastoma [38].

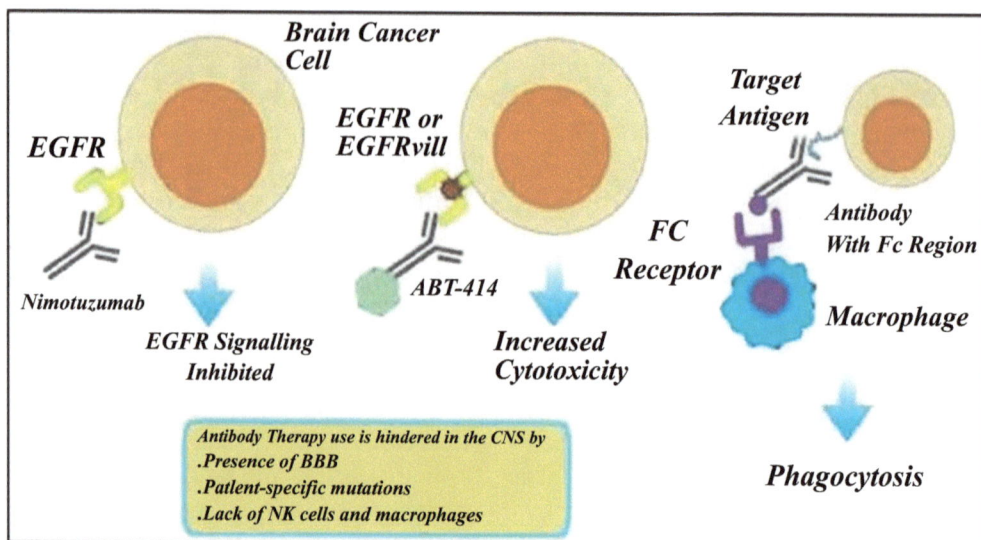

Fig. (3). The therapeutic approach of anti-EGFR monoclonal antibodies for brain tumours.

When administered to the brain, monoclonal antibodies have limited survival advantages (BBB) because of their inability to penetrate the blood-brain barrier. Their inability to overcome substantial obstacles is one of their drawbacks. Personal changes to antigens may potentially influence the efficacy of antibody binding [39].

STRATEGIES FOR ENHANCED DRUG DELIVERY TO BRAIN TUMORS

BBB Disruption

Due to the ineffectiveness, risks, uneven drug distribution, slow action, and adverse effects associated with blood-brain barrier (BBB) disruption and irritative chemotherapy, clinicians generally avoid intrathecal injections [40]. A more effective strategy involves iatrogenic disruption of the BBB (BBBD) prior to administering systemic chemotherapy. This approach enhances drug concentrations within the brain parenchyma by combining BBBD with chemotherapy [41]. Recent studies have explored MRI-guided, ultrasound-induced BBBD as a promising technique [42]. The conventional method for achieving temporary BBB disruption involves the controlled infusion of heated mannitol into either the vertebral artery or the internal carotid artery. Vasoactive agents, such as leukotriene C4 (LTC4), bradykinin (BK), and various potassium channel agonists, have also been investigated. However, research indicates that these molecules do not selectively increase permeability in normal brain capillaries.

Osmotic Blood Brain Barrier Disruption

Osmotic disruption of the blood-brain barrier (BBB) has also been proposed as a method to deliver gene-carrying recombinant adenoviral vectors to intracranial malignancies and to facilitate the detection of brain metastases using magnetic resonance imaging (MRI) agents (Fig. **4**) [43]. This disruption occurs when elevated sugar concentrations in brain capillaries cause endothelial cells to contract and absorb water, temporarily opening the tight junctions of the BBB. The first 20 to 30 minutes following disruption are critical for the diffusion of therapeutic agents that would otherwise struggle to penetrate the BBB. This technique has shown efficacy in treating cerebral lymphomas, malignant gliomas, primary non-AIDS CNS lymphomas, and germ cell tumors disseminated throughout the central nervous system. However, despite its therapeutic potential, this approach is not without risks.

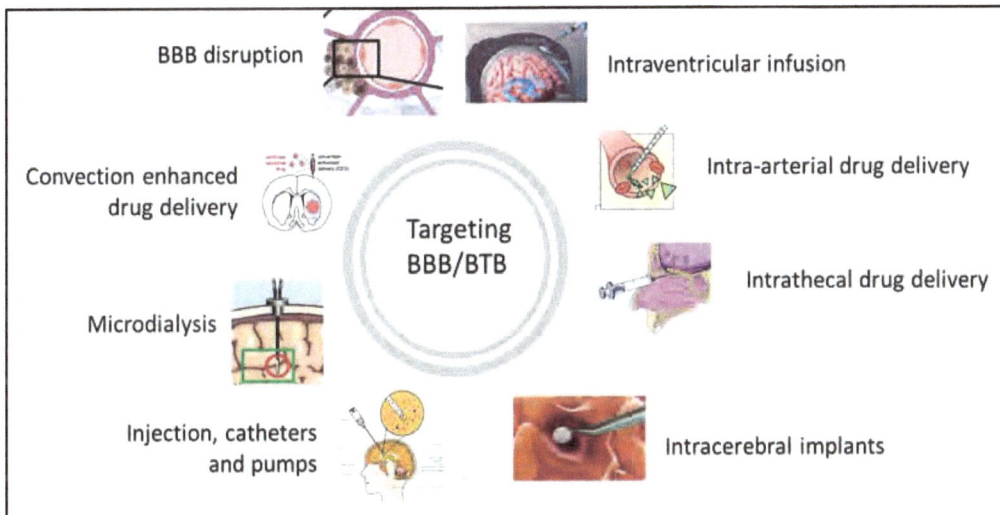

Fig. (4). Different strategies for enhanced drug delivery to brain tumors by targeting BBB/ BTB.

Biochemical BBB Disruption

Recent chemical agents used to disrupt BBB include derivatives of typical vasoactive substances, such as leukotriene C4, interleukin-2, and bradykinin. Clinical trials have begun for RMP-7, also known as bradykinin, which increases permeability at the intercellular junction [44]. The possibility that RMP-7 may enhance transport across the normal BBB has been suggested in another study [45]. The effectiveness of vasoactive leukotriene therapy for brain tumours or damaged brain capillaries seems to be dose-dependent [46]. The differential vulnerability of the pathological blood-brain barrier compared to the normal one is a contentious issue surrounding the use of this approach [47]. While some research found temporary increases in permeability, other investigations found statistically significant increases [48].

Intraventricular Infusion

Optimizing the distribution volume of medications is essential for effective intraventricular infusion, as it helps prevent backflow. The use of implanted reservoirs for intraventricular injection can enhance the dispersion of drugs throughout the cerebrospinal fluid (CSF). However, in cases where CSF flow abnormalities are present, current delivery methods may lead to reduced distribution volumes and increased toxicity [49]. A commonly used device for this purpose is the Ommaya reservoir, a plastic reservoir implanted subcutaneously into the scalp *via* a tube. It is connected to the brain's ventricles through an output catheter. Drug delivery is achieved by subcutaneous injection of therapeutic solutions into the reservoir, followed by physical compression through the scalp to facilitate infusion into the ventricles [50]. Zara *et al.* [51] proposed direct intracerebral injection as a strategy to bypass the blood-brain barrier (BBB). Notably, all the aforementioned approaches share a common feature, *i.e.*, the therapeutic targets are located on the ventricular surface. Evidence suggests that certain macromolecular drugs are capable of reaching these surface targets effectively [52, 53].

Convection Enhanced Drug Delivery

Convection-enhanced delivery (CED) has garnered significant attention as an alternative to conventional systemic drug administration for gliomas, which often fails due to the restrictive nature of the blood-brain barrier (BBB) [54]. CED involves the continuous infusion of therapeutic agents under positive pressure, allowing for improved local distribution within brain tissue. An alternative therapeutic strategy involves the local delivery of agents *via* implants placed at the disease site, using either biodegradable or non-biodegradable polymer-based delivery systems [55, 56]. CED, by utilizing positive pressure techniques, effectively prevents recurrence by enhancing drug penetration into the target tissue.

Studies have reported that fluid convection rather than passive diffusion maintains a pressure gradient during interstitial infusion, thereby improving the distribution of both large and small molecules within brain and tumor tissues. The combination of convection and diffusion significantly increases drug concentrations in the brain compared to systemic administration [57]. Moreover, the use of high-viscosity infusates has been found to reduce infusion time and minimize adverse effects. This is due to the efficient formation of convection currents, which enable broader distribution volumes in a shorter duration [58].

Intra-Arterial Drug Delivery

Intra-arterial (IA) drug delivery offers a route that bypasses the need to alter the blood-brain barrier (BBB). Unlike intravenous administration, IA delivery avoids first-pass metabolism, resulting in increased drug concentrations within the circulatory system. However, despite this pharmacokinetic advantage, IA therapy alone has not demonstrated significant improvements in prognosis for brain cancer patients [59, 60].

For effective cancer treatment, an ideal pharmaceutical agent for intravenous administration should possess the ability to rapidly cross the BBB or blood-tissue barrier (BTB), bind selectively to target tissue components, or undergo localized metabolism. The primary benefit of this approach lies in achieving higher plasma concentrations in perfused tissues during the initial circulation following drug injection. When combined with transient osmotic BBB disruption, intravenous chemotherapy can enhance drug delivery to the central nervous system (CNS), thereby preserving neurocognitive function and reducing systemic toxicity [61]. However, elevated tissue concentrations *via* intravenous routes may also increase the risk of local toxicity.

Clinical studies investigating IA chemotherapy have reported limited improvements in survival outcomes for patients with brain tumors [62 - 64]. This is likely attributable to the suboptimal physicochemical properties of most drugs, which fail to meet the essential criteria for effective IA administration. In a notable study, Joshi *et al.* [65] successfully measured real-time concentrations of mitoxantrone delivered intra-arterially in rabbit brain tissue. Intrathecal infusions of monoclonal antibodies have also been employed in the treatment of carcinomatous meningitis, building on a long history of intrathecal drug delivery [66]. Clinical examples include intraventricular injection (IVI) of aminoglycoside and glycopeptide antibiotics for meningitis, IVI of baclofen for meningeal metastases, and intrathecal injection (ITI) of baclofen for spasticity management [67]. A common feature across these cases is the proximity of therapeutic targets to the ventricular surface. Furthermore, certain macromolecular drugs have demonstrated the ability to reach these surface-level targets effectively [68, 69].

Injection, Catheters, and Pumps

Injecting straight into the brain allows us to circumvent the BBB. Intracerebroventricular injection of nerve growth factor into cerebrospinal fluid

(CSF) is one such example for patients with Alzheimer's disease [70]. A serious health risk may be posed by injecting high doses of anticancer medication. Solid lipid nanoparticles (SLNs) have the potential to be used as transport carriers for delivering pharmaceuticals to the brain, as demonstrated by increased drug levels in the brain following intravenous administration [71 - 74]. When used as a device for intracerebral medication administration *via* catheter with a pump, the Ommaya reservoir can deliver bolus injections of anticancer agents directly into the cerebrospinal fluid (CSF) [75]. Intravenous injection of 3.0 mM oligodeoxynucleotide polyplexes into mice increased circulation time, decreased clearance rate, and boosted brain uptake by a factor of ten [76]. The polyplexes undergo modification using monoclonal antibody 8D3, which is selective for the transferring receptor, and are encapsulated in pegylated liposomes. A study by Miller Landon *et al.* [77] demonstrated the use of a direct catheter for drug delivery to the central nervous system. Its numerous uses include aseptically delivering drugs and therapeutic agents into CSF, regulating CSF temperature, and accessing the spinal canal or ventricular region to sample CSF and monitor intracranial pressure [69]. Chen *et al.* demonstrated that leak-back from the larger cannulas (28 and 22 gauge) was significantly greater than that from the smallest cannula (32 gauge) [78]. This may be attributed to the formation of low-resistance pathways. According to research by Bauman *et al.* [79], backflow distance varies by a factor of four to five times the catheter's outer diameter, even when the flow rate remains constant. Catheters have also been applied in clinical and preclinical settings for targeted drug delivery, including recent use in patients with brain cancer to evaluate pharmacokinetics within tumours [80 - 82].

Microdialysis

Microdialysis enables access to target regions of the brain for both local and systemic drug delivery [83]. By sampling extracellular fluid through tiny, flexible catheters with semipermeable membranes, it is possible to measure drug concentrations *in situ*. This method has recently been applied to patients with brain cancer and is widely used in preclinical studies to assess pharmacokinetics at the tumour site. Microdialysis allows a medication to passively diffuse over a semipermeable membrane [84 - 86]. For example, it has been used to sample extracellular medication concentrations in experimental brain tumors [87, 88] and to distribute pharmaceuticals to the surrounding tissue. Yang *et al.* [76] used microdialysis to study the pharmacokinetics of boronophenylalanine-fructose (BPA-f) in gliomas with ultrasound-induced BBBD. This study's results provide credence to the idea that intracerebral microdialysis could be used to evaluate drug transport characteristics across the blood-brain barrier (BBB) at a sonicated area. After FUS-induced BBB breakdown, microdialysis can be used to evaluate metabolism and pharmacokinetics at a specific brain location, allowing for the acquisition of selective information [89].

Intracerebral Implants

Intracerebral implants can be customized in terms of shape, size, biodegradability, and drug release rate. These implants function as drug reservoirs or polymer matrices, offering several therapeutic advantages. Polymers play a critical role in protecting active pharmaceutical ingredients from degradation, enabling controlled release, and minimizing systemic adverse effects. This regulated drug delivery system holds promise for treating neurodegenerative disorders, such as Parkinson's and Huntington's disease, as well as other central nervous system (CNS) conditions [90 - 92].

One notable example is a disc-shaped wafer composed of the biodegradable polymer poly[bis(p-carboxyphenoxy)]propane-sebacic acid, embedded with the chemotherapeutic agent BCNU (1,3-bis(2-chloroethyl)-1-nitrosourea; carmustine). Clinical studies have shown that patients with newly diagnosed or recurrent malignant gliomas experienced a two-month improvement in survival following treatment with this implant [93 - 95]. Further trials have investigated the combination of BCNU with other chemotherapeutic agents, such as paclitaxel and cisplatin. The co-administration of Gliadel (BCNU wafer) and Temozolomide has demonstrated favorable safety and tolerability in patients with recurrent high-grade gliomas [96, 97].

Boer *et al.* [98] developed a miniature controlled-delivery device for CNS implantation, enabling zero-order vasopressin release into the cerebrospinal fluid. Fortin *et al.* [99] introduced a polymeric drug delivery system capable of transporting various therapeutic agents to the CNS. Dave *et al.* [100] conducted pharmacokinetic studies using Sprague-Dawley rats with orthotopically implanted C6 gliomas and normal brain tissue to evaluate letrozole, a third-generation aromatase inhibitor. Letrozole exhibited linear pharmacokinetics up to a dose of 8 mg/kg, but became non-linear at 12 mg/kg. The $AUC_{0-8}h$, plasma Cmax, and extracellular fluid (ECF) concentrations in normal brain tissue showed inverse trends. The relative brain distribution coefficients (AUCECF/AUCplasma, unbound) ranged from 0.03 to 0.98. Notably, letrozole absorption in tumor tissue was 1.5 to 2 times higher than in tumor-free regions.

BIOLOGICAL SYSTEMS

RNA Interference

Medications, such as procarbazine, vincristine, temozolomide, and dexamethasone, target brain tumours by interfering with rapid cytoplasmic translation and RNA interference (RNAi) processes (Fig. **5**). Recent research works have paved the way for the developing of new approaches for modifying protein synthesis by degrading messenger RNAs with the help of small interfering RNAs (siRNAs) [101]. MicroRNAs (miRNAs) have been discovered by researchers to play a crucial role in various neurological pathways and stem cell biology. When used in conjunction with small interfering RNAs (siRNAs), miRNAs stop translation [102]. To selectively halt gene translation, RNA may be delivered in a variety of shapes and forms. Some examples of these structures include arrays of carbon nanotubes, dendrimers, copolymers, nanocomposite spheres, multifunctional QDs adorned with nucleic acids, and auNPs decorated with nucleic acids.

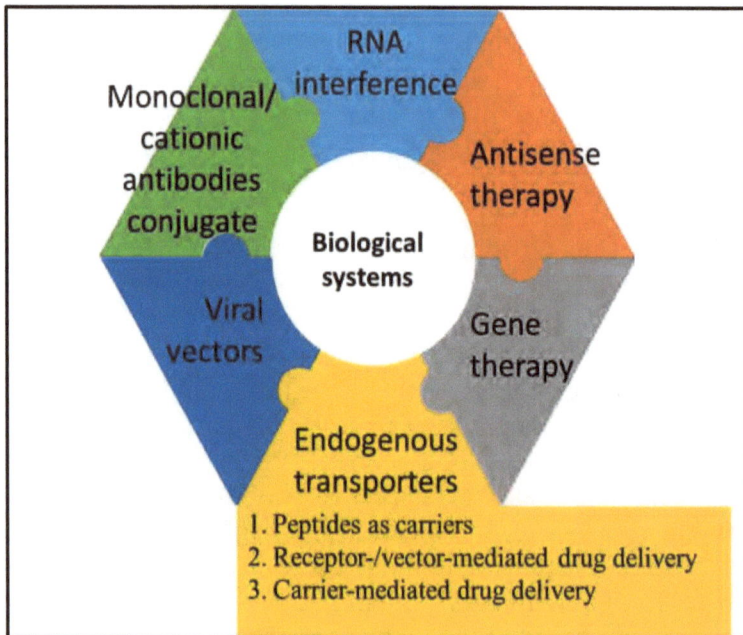

Fig. (5). Strategies for enhanced drug delivery to brain tumors using biological systems.

Antisense Therapy

Brain tumor-bearing mice were administered 8D3/83-14 MAb immunoliposomes *via* intravenous injection, together with plasmid DNA encoding the epidermal growth factor receptor (EGFR) antisense messenger RNA. A 79% reduction in immune-reactive EGFR protein was achieved by encapsulated mRNA [103]. The permeability of the BBB *via* NgRHl can be altered by various agents, including liposomes, inorganic compounds, small interfering RNAs, peptides, aptamers, peptide mimetics, proteins, antibodies, antisense, and external guide sequences [104, 105]. The injection of 15-20 nm superparamagnetic iron oxide nanoparticles, which are also fluorescent, has enabled magnetic resonance imaging of gene transcription in mouse brains [106, 107]. The use of CED to transport agents, such as antisense oligonucleotides, viral vectors, conjugates, or monoclonal antibodies, has shown great promise [108, 109]. Recent research has demonstrated that antisense oligonucleotides delivered using polycefin, targeting a tumor-specific angiogenic marker, significantly reduce tumor angiogenesis and improve survival rates in animal models [110, 111]. The delivery of antisense oligonucleotides and plasmids to brain tumors has been shown to be advantageous using poly(butyl cyanoacrylate) nanoparticles, in addition to small-molecule medicines [112].

Gene Therapy

Nanopharmaceuticals represent a relatively new and promising subfield within pharmaceutical sciences, involving the design and engineering of nanoparticles for diverse applications, such as cell therapy, vaccination, and gene therapy [113, 114]. Among these, gene therapy has emerged as a major focus, particularly in clinical trials targeting cancer patients [115]. Angiogenesis is a well-established driver of cancer progression, and gene therapy offers a strategic method for delivering antiangiogenic agents with the added benefit of sustained expression. Dickson *et al.* [116] conducted a comprehensive analysis of antiangiogenic gene therapy approaches, highlighting their therapeutic potential. Robson *et al.* [117] emphasized the synergistic use of gene therapy with radiotherapy. While radiotherapy alone often fails to completely eradicate tumors, gene therapy may enhance its efficacy, potentially allowing for reduced radiation doses.

Oncolytic gene therapy enables the insertion of novel genes or the replacement of defective or inactive tumor suppressor genes [118]. Although clinical trials have explored the transduction of p53, a key tumor suppressor, a significant proportion of cancer cells failed to internalize the viral vector and express p53 at the target site [119]. Additionally, antiviral agents, such as acyclovir, target the viral thymidine kinase enzyme expressed in the DNA of cancer cells. This therapeutic strategy induces DNA chain termination, ultimately leading to cancer cell death [120, 121].

Monoclonal/Cationic Antibodies Conjugate

Gene editing technologies have enabled the production of fusion proteins and monoclonal antibody (mAb)/avidin or mAb/streptavidin (SA) fusion genes for targeted drug delivery. For instance, the OX26 mouse monoclonal antibody can be used to target the transferrin receptor (TfR) in rats, facilitating drug transport across the blood-brain barrier (BBB). Similarly, genetically engineered chimeric human insulin receptor monoclonal antibodies (HIRmAbs) have demonstrated potential for drug delivery to the human brain. Both the original mouse HIRmAb and its chimeric counterpart function similarly, with the latter showing efficient absorption in primate brains [122].

Immunotherapy includes molecular antibody biology, phage display technology, and hybridoma techniques for producing monoclonal antibodies. mAbs are currently the preferred molecules for generating immunoliposomes due to their stability and enhanced binding avidity attributed to the presence of dual antigen-binding sites. Neuwelt [123] described a conjugated monoclonal antibody capable of crossing the BBB to deliver therapeutic agents. U.S. Patent 6287792 [124] outlines the use of avidin-biotin technology to couple drugs with antibody-based transport vectors. Fusion proteins, created by genetically linking drugs to mAb transport vectors, are also under investigation for targeted delivery.

Gold (Au) nanoparticles have shown superior signal contrast compared to antibody-fluorescent dye conjugates, enhancing imaging and therapeutic precision [125]. Multiple studies have confirmed the uptake of monoclonal antibodies following intravenous (IV) administration in malignant gliomas. For example, nimotuzumab, a radiolabeled, EGFR-specific humanized mAb, demonstrated tumor uptake in phase I/II clinical trials [126].

Ongoing research is exploring the efficacy of EGFR-targeting antibodies in glioma treatment. Therapeutic strategies include radiation therapy (RT) combined with cetuximab and temozolomide, or nimotuzumab administered alone or alongside RT, particularly for diffuse intrinsic pontine gliomas and recurrent cases [127]. Bevacizumab, a recombinant humanized mAb targeting VEGF-A, is also being evaluated for both newly diagnosed and recurrent malignant gliomas, either as monotherapy or in combination with other treatments.

ENDOGENOUS TRANSPORTERS

Peptides as Carriers

Parenteral formulations of peptide and protein medicines are commercially available (Fig. **5**). Oral administration, however, faces challenges due to enzymatic degradation in the gastrointestinal tract. Modifying the surface characteristics of carriers allows for targeted medication delivery [128, 129]. For example, nanoparticles containing proteins, nucleic acids, small molecules, and peptides may efficiently evade the immune system and enter targeted organs [130].

Modifying a polymer-based system allows for the formulation of copolymerized peptide nanoparticles. In this innovative method for therapeutic peptide administration *via* drug-polymer conjugates, covalent bonds are formed between the drug moiety and the carrier, regardless of whether physical entrapment is present or not.

Receptor-/Vector-Mediated Drug Delivery

Ligands, such as insulin, lactoferrin, and transferrin, can facilitate receptor-mediated transport of therapeutic agents across the blood-brain barrier (BBB) [131 - 134]. This process typically follows a three-stage pathway of receptor-mediated endocytosis, beginning on the luminal (blood-facing) side of the endothelial cell and progressing to the abluminal (brain-facing) side through intracellular migration and exocytosis [135].

The cellular machinery involved in this transport is capable of four primary functions [136]. After internalization, receptors and their bound ligands may undergo retroendocytosis, allowing them to be recycled back to the cell surface. Additionally, receptor-ligand complexes can translocate within the cell to different regions of the plasma membrane, facilitating targeted delivery. These mechanisms are essential for efficient receptor-mediated transport. Following dissociation of the receptor-ligand complex, lysosomal degradation may occur, with components either being recycled to the cell membrane or processed for cellular use. It has also been hypothesized that Angioprep-2 could benefit from enhanced endocytosis in the presence of low-density lipoprotein receptor-related protein 1 (LRP1), a key mediator of transcytosis across the BBB [137].

Carrier-Mediated Drug Delivery

Through a process known as carrier-mediated transport (CMT), drugs can cross from the circulation into the brain *via* passive or active mechanisms. Active efflux transport is the term used to describe the process of drug removal from the brain [138]. According to Banks *et al.* [139], PTS-1 actively effluxes endogenous peptides, such as Tyr-Pro-Trp-Gly-NH2, into the bloodstream. The CMT systems play a mediating role in the transport of nutrients with low molecular weights across the BBB. Examples of transporters include GLUT1, which transports D-glucose; MCT1, which transports monocarboxylic acids; LAT1, responsible for large neutral amino acids; EAAT, which transports excitatory amino acids; and various transporters for organic cations [140, 141]. Many drugs are believed to reach the brain *via* passive diffusion and CMT, allowing them to pass over the blood-brain barrier (BBB). Choline and an endogenous hydrophilic amine have been found to be absorbed *via* a carrier-mediated transport (CMT) pathway.

Viral Vectors

Oncolytic viruses, whether synthetic or naturally occurring, are engineered to selectively infect and destroy cancer cells. Unlike gene therapy, which uses viral vectors as carriers for therapeutic genes, oncolytic virotherapy employs the virus itself as an active therapeutic agent [142]. This selectivity arises from the fact that most cancer cells possess impaired antiviral defense mechanisms, allowing viruses to replicate more efficiently in malignant cells than in healthy ones [143].

One notable advancement in this field is the third-generation oncolytic herpes simplex virus type 1 (HSV-1), known as G47Δ, which was developed from its predecessor, G207, to enhance anticancer efficacy [144]. In G47Δ, the insertion of the LacZ gene from *Escherichia coli* inactivates a critical component of ribonucleotide reductase (RR), further attenuating viral replication in normal cells while preserving its oncolytic potential [145, 146]. Another promising vector is DNX-2401 (also known as tasadenoturev or Delta-24-RGD), an adenovirus-based therapy engineered for glioma selectivity. This is achieved through two key modifications: (i) alteration of the E1A protein to prevent binding and inactivation of the retinoblastoma (Rb) protein, and (ii) incorporation of an RGD motif to enhance tumor cell targeting. Given that nearly all gliomas exhibit dysregulated Rb pathways, DNX-2401 demonstrates strong tumor specificity [147 - 149].

PVSRIPO, a polio-rhinovirus chimera, is a live attenuated poliovirus type 1 in which the internal ribosome entry site (IRES) has been replaced with that of human rhinovirus type 2. This modification eliminates the virus's ability to translate its genome using host ribosomes and prevents neuronal infiltration, thereby enhancing safety [150, 151]. Reovirus (REOLYSIN), a double-stranded RNA virus, preferentially infects and lyses tumor cells by exploiting hyperactive Ras signaling pathways, a hallmark of many cancers. This mechanism facilitates selective viral replication and tumor cell disintegration [152, 153].

CHEMICAL SYSTEMS

Lipophilic Analogues

There is a substantial correlation between lipophilic analogues and cerebrovascular permeability [154]. Liposomes coated with PEG could impede antibody-target interaction due to steric hindrances. Therefore, connecting the ligand to the distal end of a few lipid-conjugated PEG molecules has been proposed as an alternative to attaching it to a lipid head groups on the outermost layer of a PEG-conjugated liposome (Fig. **6**). Immunoliposomes coupled with thiolated monoclonal antibodies were synthesised using bifunctional 2000-Da polyethylene glycol (PEG 2000), containing both lipids and maleimide [155].

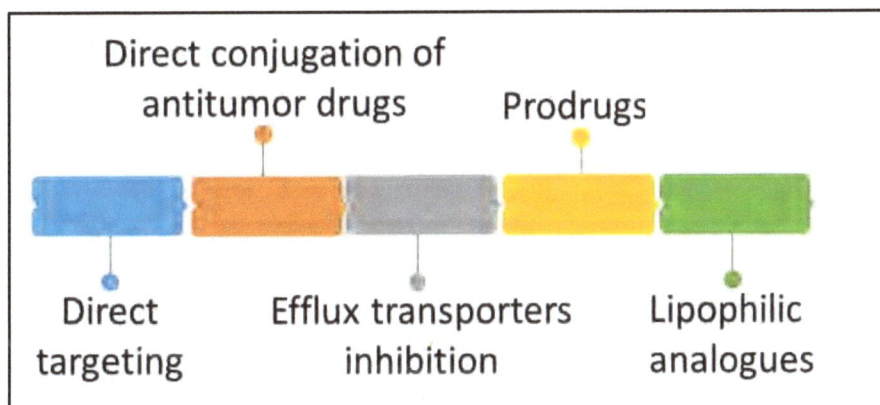

Fig. (6). Strategies for enhanced drug delivery to brain tumors using chemical systems.

Prodrugs

Prodrug design can enhance pharmacological absorption in the brain. Prodrugs are pharmacologically inactive molecules created by transient chemical modification of physiologically active compounds [156]. A variety of carboxylic acid as functional group containing drugs, including valproate, vigabatrin, levodopa, niflumic acid, GABA, and others, were investigated using such prodrug techniques [157]. Lipoamino acids (LAA) with a polar amino acid and alkyl chain as a linker moiety were attached with flurbiprofen, a neuroprotective treatment for Alzheimer's disease, to produce amphiphilic derivatives that exhibited membrane-like properties. These compounds could traverse the blood-brain barrier (BBB) and interact with other biological barriers [158].

Efflux Transporters Inhibition

Proteins like P-glycoprotein (Pgp), a multispecific organic ion, and a protein linked to breast cancer are transported by efflux transporters [159]. Anticancer medications cannot pass the blood-brain barrier without specific P-gp and other efflux transporter inhibitors. Pre-clinical investigations have shown increased brain penetration for paclitaxel, docetaxel, and imatinib [160]. The adverse impact profile of selective P-gp inhibitors from the current generation, namely tariquidar 13, zosuquidar 12, and elacridar 11, has improved [161].

Direct Conjugation of Antitumor Drugs

Antitumor drugs can be rendered more lipophilic through direct conjugation with efficient vectors capable of crossing the blood-brain barrier (BBB) (Fig. **6**). These vectors may include antibodies, peptides, protein carriers, or viral components. One innovative brain drug delivery system involves the combination of paclitaxel with the peptide Angiopep-2, which serves as a targeting vector. This approach has been investigated as a potential strategy to overcome the active efflux of paclitaxel mediated by P-glycoprotein (P-gp), a challenge attributed to the drug's high lipophilicity [162]. In comparative studies, polymeric prodrugs modified with targeting ligands, specifically transferrin-modified PXPEG (TFPXPEG), demonstrated superior antitumor efficacy compared to unmodified PXPEG, as evidenced by MTT assay results [163].

Direct Targeting

Direct targeting of the blood-brain barrier (BBB) has been shown to prolong the half-life of drugs in the cerebrospinal fluid (CSF) by reducing exposure to plasma enzymatic activity. For example, controlled-release polymer implants successfully delivered nerve growth factor (NGF) to the brain interstitium in rat models [164]. However, implantation procedures pose risks. The insertion of a catheter during surgery may cause mechanical damage to brain tissue, and the implant itself can contribute to localized injury. Additionally, injecting large volumes of fluid directly into the brain parenchyma at high pressure and velocity can result in tissue disruption and damage [165 - 167]. To overcome these limitations, convection-enhanced delivery (CED) has emerged as a promising technique. CED utilizes positive hydrostatic pressure to infuse therapeutic agents into brain tissue, enabling broader and more uniform distribution while minimizing mechanical trauma associated with direct injection [168].

Colloidal Systems

Enzyme inactivation for site-specific targeting of active compounds is the fundamental concept of chemical drug delivery systems [169]. Consequently, the medication is delivered site-specifically or with an improved delivery system by means of the newly attached moieties, which undergo multi-step enzymatic and/or chemical changes [170 - 173].

Liposomes

Lipid cakes or thin films are formed when stacks of liquid crystalline bilayers expand and become fluid. Repetition of this process leads to the formation of liposomes [174]. To prevent water from contacting the hydrocarbon core, hydrated lipid sheets self-assemble and divide at the bilayer boundaries, forming vesicular structures [175]. Liposomes are widely used as drug carriers due to their ability to protect pharmaceuticals from degradation [176, 177], minimize side effects, and deliver therapeutic agents to targeted sites with enhanced efficacy.

Brain-targeted PEGylated liposomes (Fig. **7**) can be further customized by conjugating monoclonal antibodies specific to human insulin receptors, transferrin receptors (*e.g.*, OX-26), or glial fibrillary acidic proteins [178]. To facilitate receptor-mediated transport of various therapeutics, including small molecules and nucleic acids, Pardridge and colleagues developed a series of immunoliposomes, which are PEGylated liposomes coupled with targeting antibodies [179 - 186].

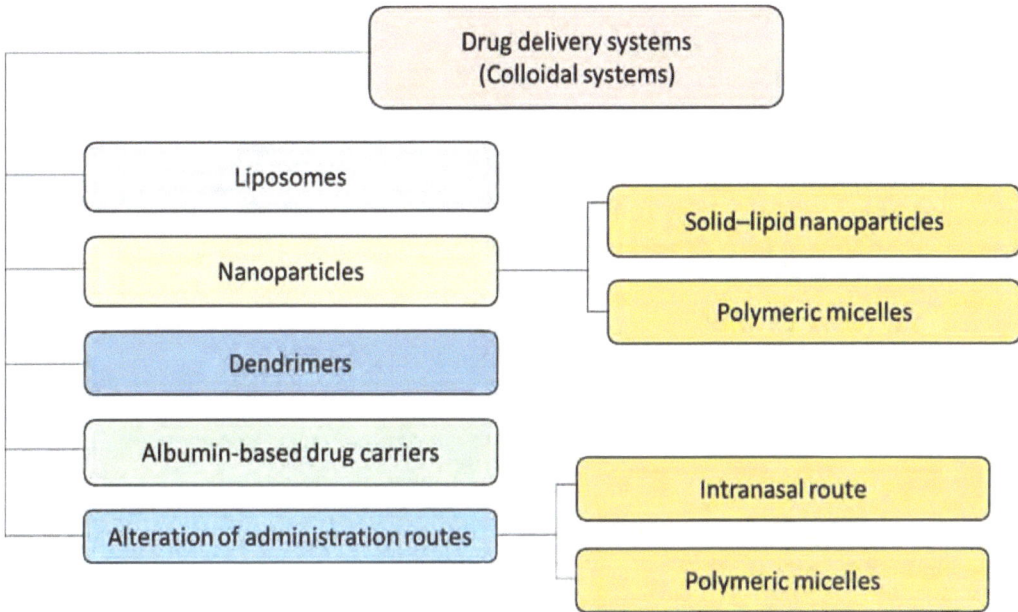

```
                        Drug delivery systems
                         (Colloidal systems)


        Liposomes
                                          Solid–lipid nanoparticles

        Nanoparticles
                                          Polymeric micelles

        Dendrimers


        Albumin-based drug carriers
                                          Intranasal route

        Alteration of administration routes
                                          Polymeric micelles
```

Fig. (7). Strategies for enhanced drug delivery to brain tumors using colloidal systems.

Nanoparticles

The size, composition, and structural characteristics of nanoparticles critically influence their transport mechanisms into the brain [187 - 189]. For instance, the Gli35 cancer and glioma cell lines express the mutant epidermal growth factor receptor variant III (EGFRvIII), but not the wild-type EGFR [190, 191]. Notably, brain-targeted nanocarriers (BNCs) were successfully and selectively delivered to rats bearing these tumor cells.

Intranasal administration of the analgesic peptide neurotoxin-I *via* poly(lactic acid) (PLA) nanoparticles demonstrated superior brain delivery compared to intravenous routes [192]. To address the limited transport of encapsulated protease inhibitors across the blood-brain barrier (BBB), PLA nanoparticles have been conjugated with the TAT cell-penetrating peptide, which inhibits P-glycoprotein efflux activity and enhances CNS penetration [193].

Cationic proteins, such as albumin and chitosan, have been explored for brain-targeted drug delivery (Fig. **7**). Albumin nanoparticles covalently linked to apolipoprotein E can cross the blood-brain barrier (BBB) *via* adsorptive-mediated transcytosis and bind to apolipoprotein receptors on brain endothelial cells [194, 212]. *In vivo* imaging studies using ligand-coupled nanoparticles have demonstrated optimal tumor distribution and favorable tumor-to-normal brain ratios [195]. Despite the widespread use of cell-penetrating peptides, they often lack selectivity between cancerous and noncancerous cells. To overcome this limitation, Gao *et al.* [213] employed a dual-targeting strategy using the AS1411 aptamer and a phage-displayed TGN peptide to specifically target both cancer cells and the BBB, resulting in improved tumor localization. A promising target for glioma therapy is the tumor-restricted receptor IL13Ra2, which can be selectively bound by the interleukin-13-derived peptide (IL-13p). Core-shell nanoparticles (LTNPs) loaded with lapatinib, a dual tyrosine kinase inhibitor, have demonstrated effective uptake by U87 glioma cells, leading to cell growth inhibition and G2 phase arrest. *In vivo* studies confirmed that the enhanced permeability and retention (EPR) effect facilitated the accumulation and distribution of LTNPs within glioma tissues [196].

Solid Lipid Nanoparticles

A new class of carriers, called solid-lipid nanoparticles (SLN), was developed to replace more conventional ones, such as emulsions, liposomes, and polymeric nanoparticles [197, 198]. A range of biocompatible and biodegradable lipids may be used to make SLN, such as lipid acids like stearic acid and palmitic acid, hard fats like glyceryl behenate, cetylpalmitate, and tristearin, and triglycerides like tristearin, tripalmitin, and trilaurin [199]. Furthermore, the majority of emulsifiers that have been authorised by drug regulatory bodies, such as poloxamer 188, sodium glycocholate, lecithin, and polysorbate 80, are compatible with SLN formulations [200]. Several studies have reported that solid-lipid nanoparticles (SLNs) and pegylated SLNs may increase the brain concentration of anticancer medicines, including doxorubicin (DOX), paclitaxel (PCL), and camptothecin (CA) [201 - 205].

Polymeric Micelles

Micellar nanoparticles are formed from amphiphilic polymers that possess a hydrophobic core surrounded by a hydrophilic shell in an aqueous environment

[206]. These structures self-assemble at a defined critical micelle concentration (CMC), typically resulting in polymeric micelles with diameters ranging from 10 to 100 nm [207]. Pluronic block copolymers, comprising two polyethylene glycol (PEG) blocks and one polypropylene glycol (PPG) block, serve as the primary hydrophobic components forming the micelle core. PEG-PPG-PEG copolymers have demonstrated the ability to penetrate endothelial cell membranes in cultured brain microvessels, leading to downregulation of P-glycoprotein (P-gp) expression [208]. Nanoparticles coated with polysorbate 80 can cross the blood-brain barrier (BBB) *via* receptor-mediated endocytosis. These particles selectively adsorb plasma proteins, such as apolipoproteins E and B, facilitating their transport into the brain [209]. Polysorbate 80-coated poly(butyl cyanoacrylate) (PBCA) nanoparticles have successfully delivered drugs like tacrine and loperamide to the CNS [210, 211].

Dendrimers

Dendrimers are polymeric molecules consisting of multiple branches radiating from a central core [212, 213]. The outer surface formed by the branch termini, along with the surrounding shell, defines the structure's interface with its environment [214 - 216]. For the treatment of brain disorders, polyamidoamine dendrimers have been extensively studied [217 - 219]. Polyanionic PAMAM dendrimers are considered an appealing option due to their low tissue deposition and rapid serosal transfer rates across the adult rat colon *in vitro* [220]. PAMAM and surface-modified PAMAM are transported across cell monolayers using cellular internalisation *via* endocytosis [221]. Tumor-targeting selectivity of dendrimers is enhanced when glucose is conjugated to their structure [222].

Albumin-Based Drug Carriers

Albumin possesses active tumor-targeting capabilities and serves as a versatile drug carrier in anticancer drug delivery systems [223]. Its natural tendency to accumulate in solid tumors has prompted extensive research into albumin-based delivery platforms for targeted cancer therapy [224]. Albumin facilitates endothelial transcytosis *via* interaction with the albumin receptor (gp60) and naturally binds hydrophobic molecules through reversible, non-covalent interactions, making it an ideal carrier for such compounds [225]. By physically or covalently binding to therapeutic peptides or proteins, albumin enhances drug stability and prolongs systemic half-life [226 - 233]. One notable example is nab-paclitaxel, a nanoparticle formulation in which paclitaxel is encapsulated within albumin nanoparticles. These particles, typically measuring 100–200 nm in diameter, are stable, negatively charged, and capable of penetrating the leaky vasculature of tumor beds, thereby enhancing drug accumulation at the tumor site

[234]. Given the significant therapeutic activity observed in metastatic breast cancer with the combination of paclitaxel and bevacizumab [235 - 237], albumin-bound paclitaxel has also been co-administered with bevacizumab to enhance treatment efficacy.

DRUG DELIVERY METHODS TO OVERCOME THE BBB

Nanoparticles

Nano drug delivery systems (NDDS) offer distinct advantages in therapeutic applications, particularly for targeting brain tumors. Optimizing physicochemical properties, such as solubility, particle size, surface charge, and shape, is essential for improving pharmacokinetics and tissue distribution. NDDS also facilitates combination therapies, enabling synergistic effects that enhance treatment efficacy [238].

Common design strategies for NDDS include optimizing physicochemical properties, bypassing the blood-brain barrier (BBB) and blood–tumor barrier (BTB), incorporating stimulus-responsive functional groups, and targeting specific organelles. Due to poor lymphatic drainage and leaky vasculature in tumors, nanocarriers can passively accumulate at tumor sites *via* the enhanced permeability and retention (EPR) effect [239]. However, particle size plays a critical role in biodistribution; nanostructures smaller than 10 nm are rapidly cleared by the kidneys, while those larger than 200 nm are more readily sequestered by the liver and eliminated by the reticuloendothelial system (RES), reducing their circulation time [240, 241].

Brain tumors present unique challenges due to their smaller vascular apertures, typically ranging from 7 to 100 nm, significantly narrower than those found in peripheral tumors. Consequently, the EPR effect is less pronounced in brain tumors. Nanoparticle shape also influences cellular uptake; rod-shaped nanoparticles larger than 100 nm exhibit superior internalization compared to spherical, cylindrical, or cubical forms [242], while spherical nanoparticles under 100 nm demonstrate optimal absorption [243].

Nanoparticles can cross the BBB through mechanisms, such as the EPR effect, receptor-mediated transcytosis, and endocytosis [244 - 246]. Despite these promising mechanisms, clinical studies have shown that nanoparticles often fail to achieve therapeutic concentrations within brain tumors [247, 248]. For example, glioblastoma multiforme (GBM) may exhibit either intact or disrupted BBB niches, and features, such as hyperoxia and elevated interstitial pressure, further hinder nanoparticle penetration.

Although current nanoparticle-based therapies have not yet yielded consistent clinical success in treating brain cancers, their potential remains significant. Nanoparticles with optimized properties, particularly those enabling sustained drug release, may serve as valuable adjuncts to existing treatment modalities [249, 250].

Focused Ultrasound

Sonoporation, also known as microbubble-mediated focused ultrasound (FUS), is a non-invasive technique that enables precise drug delivery to brain tumors [251 - 253]. When combined with conventional chemotherapeutics, monoclonal antibodies, nanoparticles, and gene-based therapies, FUS significantly expands the therapeutic possibilities for central nervous system malignancies [254].

The therapeutic window for microbubble-mediated FUS is governed by the kinetics of blood-brain barrier (BBB) closure following disruption. Smaller molecules benefit from a longer therapeutic window, whereas the BBB tends to reseal within a few hours for nanoparticles and other larger agents [255]. Importantly, FUS can be applied across the entire tumor region, mitigating the challenges posed by the heterogeneous BBB phenotype commonly observed in brain tumors.

ATP-binding cassette (ABC) transporters are frequently expressed in brain tumors, such as glioblastoma multiforme (GBM), medulloblastoma, and diffuse intrinsic pontine glioma (DIPG), contributing to drug resistance. FUS has demonstrated the ability to enhance drug accumulation in these tumors by transiently disrupting the BBB. Its non-invasive nature and capacity for targeted delivery make FUS particularly promising for treating diffuse infiltrative tumors like GBM and DIPG.

The first clinical trial using MRI-guided FUS in GBM patients was recently completed [256]. In two patients whose tumor tissue was analyzed, temozolomide concentrations were found to be 1.5 to 7 times higher in sonicated regions compared to unsonicated tissue. Notably, no adverse effects were reported among treated patients.

Several clinical investigations have explored the use of implanted ultrasound devices. For example, the SonoCloud device (CarThera) has been evaluated in phase I trials in combination with systemic carboplatin administration [257, 258]. However, due to anatomical limitations, these devices are not suitable for targeting the pontine region, which limits their utility in treating DIPG. Ongoing *in vivo* studies are investigating the potential of FUS in immunotherapy applications. Given its non-invasive nature and compatibility with a wide range of therapeutic agents, FUS holds considerable promise as a platform for delivering immunological and pharmacological treatments to various brain tumors.

Convection Enhanced Delivery

Convection-enhanced delivery (CED) presents a promising strategy for direct drug administration to the brain. This technique involves the placement of one or more intracranial catheters connected to an external infusion pump, enabling the controlled infusion of therapeutic agents into target tissues *via* a defined pressure gradient [259, 260]. By facilitating local drug delivery, CED allows for higher therapeutic concentrations within the brain parenchyma while minimizing systemic toxicity.

CED utilizes pressure-driven bulk flow to achieve uniform distribution of pharmaceutical solutions across larger tissue volumes. Importantly, the volume of infusion is independent of the molecular size and weight of the therapeutic agent, making this approach suitable for a wide range of compounds [261 - 263]. The ability of drugs to traverse several centimeters of tissue enhances treatment coverage, particularly in tumors with low vascular density.

Clinical trials investigating CED have primarily focused on glioblastoma multiforme (GBM) and diffuse intrinsic pontine glioma (DIPG) [264, 265]. In a multicenter phase III trial involving 276 patients with recurrent GBM, no significant difference in median survival was observed between those treated with GLIADEL wafers (carmustine implants) and those receiving CED with cintredekin besudotox (IL13-PE38QQR) [266]. For DIPG, only phase I trials have been conducted to date, with no notable improvements in survival outcomes.

Despite its potential, CED faces several challenges. First, the pressure gradient generated during infusion may lead to uneven drug distribution and concentration across the treatment area. Second, catheter placement can cause tissue injury, reflux, and intrinsic backflow of solutes and air bubbles. Third, the heterogeneous tissue environment may reduce drug retention due to rapid efflux of certain agents [267]. Additionally, elevated and variable interstitial fluid pressure within tumors can further hinder effective drug delivery. If these limitations can be addressed through improved catheter design, infusion protocols, and drug formulations, CED could become a viable method for bypassing the blood-brain barrier and enhancing therapeutic outcomes in the treatment of brain tumors.

Intranasal Delivery

Intranasal delivery is another option for crossing the blood-brain barrier. By passing through the nasal cavity, substances can bypass the blood-brain barrier

(BBB). From the neuroepithelium of the nasal cavity to the central nervous system, the medications pass *via* paracellular, transcellular, and neuronal routes. Modifying medication formulations using tools, such as nanoparticles, cyclodextrins, and liposomes, may enhance drug bioavailability. Furthermore, medications delivered intranasally bypass first-pass metabolism, which is an additional advantage. However, a drawback of intranasal delivery is that it can only administer a limited amount of the drug [268 - 270]. The intranasal administration of perillyl alcohol has been explored in the management of malignant gliomas [271, 272]. When combined, these techniques enhance drug absorption at the targeted tumor site and improve specificity [273, 274]. Given the paucity of research on intranasal drug administration for primary brain tumor therapy, drawing firm conclusions about its suitability as a method to bypass the blood-brain barrier remains challenging.

Intra-Arterial Drug Delivery

Intra-arterial (IA) drug delivery involves administering medications directly into an artery proximal to the tumor site, following cannulation of the targeted blood vessel [275, 276]. This approach allows for localized drug infusion, potentially increasing therapeutic concentrations at the tumor site while minimizing systemic exposure. To facilitate drug penetration across the blood-brain barrier (BBB), hyperosmolar agents, such as mannitol, may be co-administered to induce temporary BBB disruption [277].

In a small cohort of patients with ependymoma, IA administration of carmustine, bevacizumab, and cetuximab yielded therapeutic responses [278, 279]. However, significant toxicity was observed with agents, such as cisplatin, carmustine, and epipodophyllotoxin, raising concerns about the safety profile of this delivery method.

The efficacy of IA drug delivery in glioblastoma multiforme (GBM) has been evaluated in several clinical studies (Fig. 7). Patients treated with nimustine, bevacizumab, or carboplatin, often in combination with conventional chemotherapeutics, reported survival durations ranging from approximately 20 weeks to 10 months [280 - 283]. Despite these efforts, IA drug delivery has shown limited effectiveness and a notable risk of toxicity in the treatment of primary brain tumors [284].

Immunotherapy

Cancer immunotherapy aims not at the direct destruction of tumor cells but at enhancing the patient's antitumor immune response. While the immune system typically protects normal cells from autoimmune reactions, this regulatory mechanism is often compromised within tumor microenvironments, allowing tumors to evade immune surveillance. Immune checkpoint inhibitors, such as nivolumab, which block inhibitory pathways, have shown promise in restoring immune activity against tumors.

Over the past decade, forty clinical trials have investigated nivolumab in brain tumors, resulting in 34 unique combination regimens. Nivolumab has been used in combination 48 times and as monotherapy in nine instances. However, in a phase II trial (NCT02550249), nivolumab alone did not demonstrate significant therapeutic benefit for patients with glioblastoma multiforme (GBM). In combination therapies, nivolumab has been administered alongside agents, such as temozolomide, epacadostat, bevacizumab, ipilimumab, modified immune effector cells, and personalized anticancer vaccines. Radiation therapy and temozolomide are the most commonly combined agents with pembrolizumab-based regimens. Other combinations include vaccines, bevacizumab, and the cyclin-dependent kinase inhibitor abemaciclib. For instance, in a phase II trial (NCT02287428), pembrolizumab was combined with radiation therapy and a personalized neoantigen vaccine (NeoVax) for the treatment of GBM [285].

Ipilimumab has been evaluated in fourteen clinical studies, yielding ten different combination strategies. The most common combinations involved ipilimumab with nivolumab, temozolomide, radiation therapy, or bevacizumab. Currently, seven ipilimumab studies are ongoing, with three additional trials actively recruiting participants. Tremelimumab, a potential alternative to ipilimumab, is under investigation in a phase II trial (NCT02794883), either as monotherapy or in combination with durvalumab, an anti-PD-L1 checkpoint inhibitor.

Avelumab, a monoclonal antibody targeting PD-L1 (CD274), is another agent under study. PD-L1 expression is induced by cytokines, particularly interferons, and is found on various cell types. The ability of interferons to upregulate PD-L1 may contribute to the phenomenon of "adaptive resistance" in tumors, wherein infiltrating T-cells release proinflammatory cytokines that paradoxically increase PD-L1 expression and suppress immune responses [286]. Several clinical trials are exploring combinations with interleukins (ILs), which serve immunomodulatory roles. For example, natural killer (NK) cells have been studied both alone and in combination with dendritic cell (DC) vaccination to assess the effects of IL-2 and IL-15 (NCT01235845 and NCT01875601). Additionally, urelumab, a monoclonal antibody targeting CD137, acts as an agonist to stimulate immune cell activation and enhance antitumor responses.

Vaccines

When targeting multiple tumor antigens simultaneously, vaccines offer a practical approach. Rather than designing several distinct CAR constructs, it is more efficient to include multiple peptides within a single vaccine formulation. With careful design, this strategy has the potential to target tumors in a personalized manner, thereby reducing the likelihood of side effects. The versatility of this class is demonstrated by the identification of 67 unique vaccine formulations across 94 clinical trials. Survivin (baculoviral inhibitor of apoptosis repeat-containing 5), NY-ESO-1, WT1, IDH1, and cytomegalovirus antigens are the top five targets of vaccine development. Each vaccine formulation, typically comprising three to five peptides, aims at distinct targets. An intriguing approach to vaccine development included delivering attenuated bacterial cells to dendritic cells to simultaneously target EGFRviii and NY-ESO-1 (NCT01967758). Radiation therapy and alkylating chemicals are the most common combinations with vaccines. A phase three trial (NCT01759810) is now investigating a combinational strategy that involves stem cells, T-cell transplantation, and DC immunisation. Another experimental design that progressed to phase three included immunising DCs with glioblastoma stem-cell RNA, survivin, and hTERT in combination with temozolomide therapy (NCT03548571).

Radiopharmaceuticals: A Targeted Radiation Therapy

Radiopharmaceuticals can be further subdivided based on chemical makeup into those carrying radiolabeled antibodies and those containing radiolabeled small molecules or peptides. Among these, two medications are macromolecular complexes (liposomal rhenium Re-186), four are radioantibody conjugates, two are peptides, and one is a small molecule. The short radiolabeled peptide yttrium Y 90-DOTA-tyr3-octreotide is now being studied in several tumor types, including medulloblastoma (NCT02441088 and NCT03273712), and has progressed to Phase 2 of the clinical studies. Additionally, two phase 2 studies on inoperable/progressive meningiomas are currently evaluating the radiolabeled peptide lutetium Lu 177 dotatate (NCT03971461 and NCT04082520).

Histone Deacetylase Inhibitors

Histone deacetylases (HDACs) are a family of enzymes that regulate gene expression by modifying chromatin structure and influencing various cellular processes. HDACs catalyze the removal of acetyl groups from N-acetyl lysine residues in histones, a key mechanism in epigenetic regulation. This deacetylation promotes chromatin condensation and suppresses transcription by enhancing histone-DNA binding. The human genome encodes 18 distinct HDACs, each exhibiting unique expression patterns and functional roles across different tissues and cell types [287 - 289].

HDAC inhibitors (HDACis) have been investigated both as monotherapies and in combination with other treatments, including radiation therapy, temozolomide, and, less frequently, bevacizumab. Notably, pretreatment of breast cancer cells with HDACis, such as vorinostat and valproic acid, enhanced trastuzumab-mediated antibody-dependent cellular phagocytosis (ADCP) and trastuzumab-independent cytotoxicity in various brain tumor models. Mechanistically, both agents induced immunogenic cell death and downregulated the anti-apoptotic protein MCL1. Additionally, valproic acid upregulated the activating Fc-gamma receptor IIA (CD32A) on monocytes, further enhancing immune-mediated tumor clearance [290].

Several clinical trials have explored HDACis in brain tumor therapy, including NCT03243461, NCT01236560, and NCT02265770. The first has progressed to phase II/III, while the latter two have reached phase III [291 - 293]. Valproic acid has primarily been studied in combination with temozolomide or other chemotherapeutic agents, whereas vorinostat has been evaluated alongside both bevacizumab and temozolomide.

An innovative HDACi-based compound, tinostamustine, combines the alkylating agents bendamustine and vorinostat into a single fusion molecule, offering dual mechanisms of action. Additionally, panobinostat, a potent pan-HDAC inhibitor, has been reformulated into a water-soluble nanoformulation (MTX110) for administration *via* convection-enhanced delivery (CED). This method overcomes the limitations of poorly soluble oral formulations, which fail to cross the blood-brain barrier (BBB) and pose risks of systemic toxicity. CED enables targeted delivery of MTX110 into brain tissue, enhancing therapeutic efficacy while minimizing off-target effects [294].

scRNA-seq for Cancer Heterogeneity

DNA sequencing techniques and single-cell RNA (scRNA) are the most used technologies for revealing cellular heterogeneity. Tumour subtype identification, cancer cell state characterization, cellular subpopulation phenotyping and lineage tracking, as well as differential expression analysis are all enabled by scRNA sequencing [295]. After determining the likelihood of recurrence, the new technology used to address cancer heterogeneity enables the identification of uncommon cell subpopulations within the tumour mass, thereby improving precision medicine [296].

Divergent survival probabilities are defined by single-cell data, which enhances the clinical prognostic assessment of every patient and informs treatment plans. Numerous cancer types and metastases, such as advanced non-small cell lung cancer, paired lymph nodes, and primary triple-negative breast cancer (TNBC) tumors, have been subjected to single-cell profiling to study the tumor microenvironment and intratumoral heterogeneity [297, 298].

FUTURE PERSPECTIVE

In recent years, cell-based therapies have experienced a rapid surge in popularity, ranking as the third most promising experimental approach in cancer treatment. Both immunotherapies and targeted therapies increasingly focus on enzymes, particularly kinases, and cell surface receptors. While a small subset of targets is shared among various ligands, transporters, and proteins, the number of clinical trials investigating these classes continues to grow annually. Our research network has identified several novel agents beyond the conventional therapeutic landscape, including anti-PD-1 checkpoint inhibitors, underscoring the innovative nature of these approaches in brain tumor treatment. These biologically driven therapies constitute the central focus of this discussion. However, delivering checkpoint blockade (CB) agents, typically monoclonal antibodies, into the brain remains challenging due to their limited ability to cross the blood-brain barrier (BBB).

Chimeric antigen receptor (CAR) cell therapies represent a particularly diverse and rapidly expanding field, with at least 17 distinct targeting strategies currently under investigation, excluding those that target mutant forms of antigens. The domains of transcriptional and metabolic modulation have also seen notable advancements. Interestingly, metformin, a widely used antidiabetic agent, is among several drugs being repurposed for oncological applications. Moreover, combination therapy has emerged as a promising strategy, drawing parallels with HIV treatment protocols where three-drug regimens are commonly employed to achieve viral suppression. This model suggests that rationally designed combination therapies may hold significant potential in enhancing the efficacy of brain tumor treatments [299, 300].

REFERENCES

[1]   Kreso A, Dick JE. Evolution of the cancer stem cell model. Cell Stem Cell 2014; 14(3): 275-91.
        [http://dx.doi.org/10.1016/j.stem.2014.02.006] [PMID: 24607403]

[2]   Andlin-Sobocki P, Jönsson B, Wittchen HU, Olesen J. Cost of disorders of the brain in Europe. Eur J Neurol 2005; 12(s1) (Suppl. 1): 1-27.
        [http://dx.doi.org/10.1111/j.1468-1331.2005.01202.x] [PMID: 15877774]

[3]   Rohilla R, Garg T, Jitender B, Goyal AK, Rath G. Development. Optimization and characterization of glycyrrhetinic acid–chitosan nanoparticles of atorvastatin for liver targeting 2014; 23(7): 2290-7.
        [http://dx.doi.org/10.3109/10717544.2014.977460]

[4]   Sung H, Ferlay J, Siegel RL, *et al.* Global cancer statistics 2020: Globocan estimates of incidence and mortality worldwide for 36 cancers in 185 countries. CA Cancer J Clin 2021; 71(3): 209-49.
        [http://dx.doi.org/10.3322/caac.21660] [PMID: 33538338]

[5]   Reifenberger G, Wirsching HG, Knobbe-Thomsen CB, Weller M. Advances in the molecular genetics of gliomas — implications for classification and therapy. Nat Rev Clin Oncol 2017; 14(7): 434-52.
        [http://dx.doi.org/10.1038/nrclinonc.2016.204] [PMID: 28031556]

[6]   Lin S, Xu H, Zhang A, *et al.* Prognosis analysis and validation of m^{6}A signature and tumor immune microenvironment in glioma. Front Oncol 2020; 10: 541401.
        [http://dx.doi.org/10.3389/fonc.2020.541401] [PMID: 33123464]

[7]   Zhao Y, Yue P, Peng Y, *et al.* Recent advances in drug delivery systems for targeting brain tumors. Drug Deliv 2023; 30(1): 1-18.
        [http://dx.doi.org/10.1080/10717544.2022.2154409] [PMID: 36597214]

[8]   Martins C, Pacheco C, Faria P, Sarmento B. Nanomedicine approaches for treating glioblastoma. Nanomedicine (Lond) 2023; 18(18): 1135-8.
        [http://dx.doi.org/10.2217/nnm-2023-0163] [PMID: 37593960]

[9]   Hampton L, Rogers LJ, Bauer MA, *et al.* 1211 defining the role of PCK2 in T cell metabolic plasticity in glioblastoma. Regular and young investigator award abstracts 2023.
        [http://dx.doi.org/10.1136/jitc-2023-SITC2023.1211]

[10]   Chelliah SS, Paul EAL, Kamarudin MNA, Parhar I. Challenges and perspectives of standard therapy and drug development in high-grade gliomas. Molecules 2021; 26(4): 1169.
        [http://dx.doi.org/10.3390/molecules26041169] [PMID: 33671796]

[11]   Conniot J, Talebian S, Simões S, Ferreira L, Conde J. Revisiting gene delivery to the brain: silencing and editing. Biomater Sci 2021; 9(4): 1065-87.
        [http://dx.doi.org/10.1039/D0BM01278E] [PMID: 33315025]

[12]   Sanai N, Berger MS. Glioma extent of resection and its impact on patient outcome. Neurosurgery 2008; 62(4): 753-66.
        [http://dx.doi.org/10.1227/01.neu.0000318159.21731.cf] [PMID: 18496181]

[13]   Ibarra LE. Cellular Trojan horses for delivery of nanomedicines to brain tumors: where do we stand and what is next? Nanomedicine (Lond) 2021; 16(7): 517-22.
        [http://dx.doi.org/10.2217/nnm-2021-0034] [PMID: 33634710]

[14]   Ashby LS, Smith KA, Stea B. Gliadel wafer implantation combined with standard radiotherapy and concurrent followed by adjuvant temozolomide for treatment of newly diagnosed high-grade glioma: a systematic literature review. World J Surg Oncol 2016; 14(1): 225.
        [http://dx.doi.org/10.1186/s12957-016-0975-5] [PMID: 27557526]

[15]   Behin A, Hoang-Xuan K, Carpentier AF, Delattre JY. Primary brain tumours in adults. Lancet 2003; 361(9354): 323-31.
        [http://dx.doi.org/10.1016/S0140-6736(03)12328-8] [PMID: 12559880]

[16]   Wolburg H, Wolburg-Buchholz K, Liebner S, Engelhardt B. Claudin-1, claudin-2 and claudin-11 are present in tight junctions of choroid plexus epithelium of the mouse. Neurosci Lett 2001; 307(2): 77-80.
        [http://dx.doi.org/10.1016/S0304-3940(01)01927-9] [PMID: 11427304]

[17]   Joó F. The blood-brain barrier *in vitro*: The second decade. Neurochem Int 1993; 23(6): 499-521.
        [http://dx.doi.org/10.1016/0197-0186(93)90098-P] [PMID: 8281119]

[18]   Donelli MG, Zucchetti M, D'Incalci M. Do anticancer agents reach the tumor target in the human brain? Cancer Chemother Pharmacol 1992; 30(4): 251-60.
        [http://dx.doi.org/10.1007/BF00686291] [PMID: 1643692]

[19]   Pardridge WM, Triguero D, Yang J, Cancilla PA. Comparison of *in vitro* and *in vivo* models of drug

transcytosis through the blood-brain barrier. J Pharmacol Exp Ther 1990; 253(2): 884-91.
[http://dx.doi.org/10.1016/S0022-3565(25)13060-7] [PMID: 2338660]

[20]   Azarmi M, Maleki H, Nikkam N, Malekinejad H. Transcellular brain drug delivery: A review on recent advancements. Int J Pharm 2020; 586: 119582.
[http://dx.doi.org/10.1016/j.ijpharm.2020.119582] [PMID: 32599130]

[21]   Alexander JJ. Blood-brain barrier (BBB) and the complement landscape. Mol Immunol 2018; 102: 26-31.
[http://dx.doi.org/10.1016/j.molimm.2018.06.267] [PMID: 30007547]

[22]   Saidijam M, Karimi Dermani F, Sohrabi S, Patching SG. Efflux proteins at the blood–brain barrier: review and bioinformatics analysis. Xenobiotica 2018; 48(5): 506-32.
[http://dx.doi.org/10.1080/00498254.2017.1328148] [PMID: 28481715]

[23]   Mojarad-Jabali S, Farshbaf M, Walker PR, *et al.* An update on actively targeted liposomes in advanced drug delivery to glioma. Int J Pharm 2021; 602: 120645.
[http://dx.doi.org/10.1016/j.ijpharm.2021.120645] [PMID: 33915182]

[24]   Caro C, Avasthi A, Paez-Muñoz JM, Pernia Leal M, García-Martín ML. Passive targeting of high-grade gliomas *via* the EPR effect: a closed path for metallic nanoparticles? Biomater Sci 2021; 9(23): 7984-95.
[http://dx.doi.org/10.1039/D1BM01398J] [PMID: 34710207]

[25]   Chandana SR, Movva S, Arora M, Singh T. Primary brain tumors in adults. Am Fam Physician 2008; 77(10): 1423-30.
[PMID: 18533376]

[26]   Sarkaria JN, Hu LS, Parney IF, *et al.* Is the blood–brain barrier really disrupted in all glioblastomas? A critical assessment of existing clinical data. Neuro-oncol 2018; 20(2): 184-91.
[http://dx.doi.org/10.1093/neuonc/nox175] [PMID: 29016900]

[27]   Alifieris C, Trafalis DT. Glioblastoma multiforme: Pathogenesis and treatment. Pharmacol Ther 2015; 152(152): 63-82.
[http://dx.doi.org/10.1016/j.pharmthera.2015.05.005] [PMID: 25944528]

[28]   Armstrong TS. Head's up on the treatment of malignant glioma patients. Oncol Nurs Forum 2009; 36(5): E232-40.
[http://dx.doi.org/10.1188/09.ONF.E232-E240] [PMID: 19726382]

[29]   Wen PY, Kesari S. Malignant gliomas in adults. N Engl J Med 2008; 359(5): 492-507.
[http://dx.doi.org/10.1056/NEJMra0708126] [PMID: 18669428]

[30]   Kleihues P, Ohgaki H. Primary and secondary glioblastomas: From concept to clinical diagnosis. Neuro-oncol 1999; 1(1): 44-51.
[http://dx.doi.org/10.1093/neuonc/1.1.44] [PMID: 11550301]

[31]   Dolecek TA, Propp JM, Stroup NE, Kruchko C. CBTRUS statistical report: primary brain and central nervous system tumors diagnosed in the United States in 2005-2009. Neuro-oncol 2012; 14(Suppl 5) (Suppl. 5): v1-v49.
[http://dx.doi.org/10.1093/neuonc/nos218] [PMID: 23095881]

[32]   Ostrom QT, Gittleman H, Farah P, *et al.* CBTRUS statistical report: Primary brain and central nervous system tumors diagnosed in the United States in 2006-2010. Neuro-oncol 2013; 15(Suppl 2) (Suppl. 2): ii1-ii56.
[http://dx.doi.org/10.1093/neuonc/not151] [PMID: 24137015]

[33]   MolnarP. Classification of Primary Brain Tumors: Molecular Aspects. www.intechopen.com. Available from: https://www.intechopen.com/chapters/20839 (accessed 2011-05-31).
[http://dx.doi.org/10.5772/22484]

[34]   Elazab A, Wang C, Gardezi SJS, *et al.* GP-GAN: Brain tumor growth prediction using stacked 3D generative adversarial networks from longitudinal MR Images. Neural Netw 2020; 132: 321-32.

[http://dx.doi.org/10.1016/j.neunet.2020.09.004] [PMID: 32977277]

[35]   Park SH, Won J, Kim SI, *et al*. Molecular testing of brain tumor. J Pathol Transl Med 2017; 51(3): 205-23.
       [http://dx.doi.org/10.4132/jptm.2017.03.08] [PMID: 28535583]

[36]   Han L, Liu C, Qi H, *et al*. Systemic delivery of monoclonal antibodies to the central nervous system for brain tumor therapy. Adv Mater 2019; 31(19): 1805697.
       [http://dx.doi.org/10.1002/adma.201805697] [PMID: 30773720]

[37]   Szabo E, Schneider H, Seystahl K, *et al*. Autocrine VEGFR1 and VEGFR2 signaling promotes survival in human glioblastoma models *in vitro* and *in vivo*. Neuro-oncol 2016; 18(9): 1242-52.
       [http://dx.doi.org/10.1093/neuonc/now043] [PMID: 27009237]

[38]   Ferrara N, Hillan KJ, Novotny W. Bevacizumab (Avastin), a humanized anti-VEGF monoclonal antibody for cancer therapy. Biochem Biophys Res Commun 2005; 333(2): 328-35.
       [http://dx.doi.org/10.1016/j.bbrc.2005.05.132] [PMID: 15961063]

[39]   Pichaivel M, Anbumani G, Theivendren P, Gopal, M. An Overview of Brain Tumor 2022. www.intechopen.com. Available from: https://www.intechopen.com/chapters/79979
       [http://dx.doi.org/10.5772/intechopen.100806]

[40]   Fortin D, Desjardins A, Benko A, Niyonsega T, Boudrias M. Enhanced chemotherapy delivery by intraarterial infusion and blood-brain barrier disruption in malignant brain tumors. Cancer 2005; 103(12): 2606-15.
       [http://dx.doi.org/10.1002/cncr.21112] [PMID: 15880378]

[41]   Muldoon LL, Neuwelt EA. BR96–DOX immunoconjugate targeting of chemotherapy in brain tumor models. J Neurooncol. 2003;65(1):49–62.
       [http://dx.doi.org/10.1023/A:1026234130830]

[42]   Neuwelt EA, Bauer B, Fahlke C, Fricker G, Iadecola C, Janigro D, *et al*. The blood-brain barrier and cancer: transporters, treatment, and Trojan horses. Clin Cancer Res. 2007;13(6):1663–74.
       [http://dx.doi.org/10.1158/1078-0432.CCR-06-2966]

[43]   de Boer AB, Gaillard P. *In vitro* models of the blood-brain barrier: When to use which? Curr Med Chem Cent Nerv Syst Agents 2002; 2(3): 203-9.
       [http://dx.doi.org/10.2174/1568015023358012]

[44]   Helga MCM, Blom-Roosemalen Marijke, van Oosten, *et al*. The influence of cytokines on the integrity of the blood-brain barrier *in vitro*. 1996; 64(1): 37-43.
       [http://dx.doi.org/10.1016/0165-5728(95)00148-4]

[45]   Emerich DF, Snodgrass P, Pink M, Bloom F, Bartus RT. Central analgesic actions of loperamide following transient permeation of the blood brain barrier with Cereport™ (RMP-7). Brain Res 1998; 801(1-2): 259-66.
       [http://dx.doi.org/10.1016/S0006-8993(98)00571-X] [PMID: 9729419]

[46]   Mertsch K, Maas J. Blood-Brain Barrier Penetration and Drug Development from an Industrial Point of View. Curr Med Chem Cent Nerv Syst Agents 2002; 2(3): 187-201.
       [http://dx.doi.org/10.2174/1568015023358067]

[47]   Nakagawa H, Groothuis D, Blasberg RG. The effect of graded hypertonic intracarotid infusions on drug delivery to experimental RG-2 gliomas. Neurology 1984; 34(12): 1571-81.
       [http://dx.doi.org/10.1212/WNL.34.12.1571] [PMID: 6504329]

[48]   Neuwelt EA, Goldman DL, Dahlborg SA, *et al*. Primary CNS lymphoma treated with osmotic blood-brain barrier disruption: prolonged survival and preservation of cognitive function. J Clin Oncol 1991; 9(9): 1580-90.
       [http://dx.doi.org/10.1200/JCO.1991.9.9.1580] [PMID: 1875220]

[49]   Grossman SA, Trump DL, Chen DCP, Thompson G, Camargo EE. Cerebrospinal fluid flow abnormalities in patients with neoplastic meningitis. Am J Med 1982; 73(5): 641-7.

[http://dx.doi.org/10.1016/0002-9343(82)90404-1] [PMID: 6814249]

[50]   Begley DJ, Squires LK, Zloković BV, *et al.* Permeability of the blood-brain barrier to the immunosuppressive cyclic peptide cyclosporin A. J Neurochem 1990; 55(4): 1222-30.
[http://dx.doi.org/10.1111/j.1471-4159.1990.tb03128.x] [PMID: 2398356]

[51]   Kumar PB, Kadiri SK, Khobragade DS, *et al.* Synthesis, characterization and biological investigations of some new Oxadiazoles: *In-vitro* and *In-silico* approach. Results in Chemistry. 2024;7:101241.
[http://dx.doi.org/10.1016/j.rechem.2023.101241]

[52]   Erben M, Decker S, Franke H, Galla HJ. Electrical resistance measurements on cerebral capillary endothelial cells — a new technique to study small surface areas. 1995, 30 (4), 227–238.

[53]   Matthias H, Wegener J, Decker S, Engelbertz C, Galla H. Porcinechoroid plexus epithelial cells in culture: regulation of barrier properties and transport processes. Microsc Res Tech 2000; 52(1): 137-52.
[http://dx.doi.org/10.1002/1097-0029(20010101)52:1%3C137:aid-jemt15%3E3.0.co;2-j]

[54]   Sawyer AJ, Piepmeier JM, Saltzman WM. New methods for direct delivery of chemotherapy for treating brain tumors. Yale J Biol Med 2006; 79(3-4): 141-52.
[PMID: 17940624]

[55]   Guerin C, Olivi A, Weingart JD, Lawson HC, Brem H. Recent advances in brain tumor therapy: local intracerebral drug delivery by polymers. Invest New Drugs 2004; 22(1): 27-37.
[http://dx.doi.org/10.1023/b:drug.0000006172.65135.3e] [PMID: 14707492]

[56]   Raza SM, Pradilla G, Legnani FG, *et al.* Local delivery of antineoplastic agents by controlled-release polymers for the treatment of malignant brain tumours. Expert Opin Biol Ther 2005; 5(4): 477-94.
[http://dx.doi.org/10.1517/14712598.5.4.477] [PMID: 15934827]

[57]   Bobo RH, Laske DW, Akbasak A, Morrison PF, Dedrick RL, Oldfield EH. Convection-enhanced delivery of macromolecules in the brain. Proc Natl Acad Sci USA 1994; 91(6): 2076-80.
[http://dx.doi.org/10.1073/pnas.91.6.2076] [PMID: 8134351]

[58]   Vogelbaum MA. Convection enhanced delivery for treating brain tumors and selected neurological disorders: symposium review. J Neurooncol 2007; 83(1): 97-109.
[http://dx.doi.org/10.1007/s11060-006-9308-9] [PMID: 17203397]

[59]   Ashby S. L.; Shapiro, W. R. J Neurooncol 2001; 51(1): 67-86.
[http://dx.doi.org/10.1023/a:1006441104260] [PMID: 11349883]

[60]   Remsen LG, Trail PA, Hellström I, Hellström KE, Neuwelt EA. Enhanced delivery improves the efficacy of a tumor-specific doxorubicin immunoconjugate in a human brain tumor xenograft model. Neurosurgery 2000; 46(3): 704-9.
[http://dx.doi.org/10.1097/00006123-200003000-00034] [PMID: 10719867]

[61]   Neuwelt EA. Mechanisms of disease: the blood-brain barrier. Neurosurgery 2004; 54(1): 131-40.
[http://dx.doi.org/10.1227/01.NEU.0000097715.11966.8E] [PMID: 14683550]

[62]   Dropcho EJ, Rosenfeld SS, Vitek J, Guthrie BL, Morawetz RB. Phase II study of intracarotid or selective intracerebral infusion of cisplatin for treatment of recurrent anaplastic gliomas. J Neurooncol 1998; 36(2): 191-8.
[http://dx.doi.org/10.1023/a:1005871721697] [PMID: 9525819]

[63]   Gundersen S, Lote K, Watne K. A retrospective study of the value of chemotherapy as adjuvant therapy to surgery and radiotherapy in grade 3 and 4 gliomas. Eur J Cancer 1998; 34(10): 1565-9.
[http://dx.doi.org/10.1016/s0959-8049(98)00146-4] [PMID: 9893629]

[64]   Hirano Y, Mineura K, Mizoi K, Tomura N. Therapeutic results of intra-arterial chemotherapy in patients with malignant glioma. Int J Oncol 1998; 13(3): 537-42.
[http://dx.doi.org/10.3892/ijo.13.3.537] [PMID: 9683790]

[65]   Joshi S, Reif R, Wang M, *et al.* Intra-arterial mitoxantrone delivery in rabbits: an optical

pharmacokinetic study. Neurosurgery 2011; 69(3): 706-12.
[http://dx.doi.org/10.1227/neu.0b013e3182181b67] [PMID: 21430588]

[66]   Brown MT, Coleman RE, Friedman AH, *et al.* Intrathecal 131I-labeled antitenascin monoclonal antibody 81C6 treatment of patients with leptomeningeal neoplasms or primary brain tumor resection cavities with subarachnoid communication: phase I trial results. Clin Cancer Res 1996; 2(6): 963-72.
[PMID: 9816257]

[67]   Menacherry SD, Hubert W, Justice JB. *In vivo* calibration of microdialysis probes for exogenous compounds. 1992, 64 (6), 577–583.

[68]   Erben M, Decker S, Franke H, Galla HJ. Electrical resistance measurements on cerebral capillary endothelial cells — a new technique to study small surface areas. 1995, 30 (4), 227–238.

[69]   Gath U, Hakvoort A, Wegener J, Decker S, Galla HJ. Porcine choroid plexus cells in culture: expression of polarized phenotype, maintenance of barrier properties and apical secretion of CSF-components. Eur J Cell Biol 1997; 74(1): 68-78.
[PMID: 9309392]

[70]   Olson L, Nordberg A, Von Holst H, *et al.* Nerve growth factor affects 11C-nicotine binding, blood flow, EEG, and verbal episodic memory in an Alzheimer patient (case report). J Neural Transm Park Dis Dement Sect 1992; 4(1): 79-95.
[http://dx.doi.org/10.1007/bf02257624] [PMID: 1540306]

[71]   Bargoni A, Cavalli R, Zara GP, Fundarò A, Caputo O, Gasco MR. Transmucosal transport of tobramycin incorporated in solid lipid nanoparticles (SLN) after duodenal administration to rats. Part II--tissue distribution. Pharmacol Res 2001; 43(5): 497-502.
[http://dx.doi.org/10.1006/phrs.2001.0813] [PMID: 11394943]

[72]   Cavalli R, Zara GP, Caputo O, Bargoni A, Fundarò A, Gasco MR. Transmucosal transport of tobramycin incorporated in SLN after duodenal administration to rats. Part I-a pharmacokinetic study. Pharmacol Res 2000; 42(6): 541-5.
[http://dx.doi.org/10.1006/phrs.2000.0737] [PMID: 11058406]

[73]   Fundarò A, Cavalli R, Bargoni A, Vighetto D, Zara GP, Gasco MR. Non-stealth and stealth solid lipid nanoparticles (SLN) carrying doxorubicin: pharmacokinetics and tissue distribution after i.v. administration to rats. Pharmacol Res 2000; 42(4): 337-43.
[http://dx.doi.org/10.1006/phrs.2000.0695] [PMID: 10987994]

[74]   Zara GP, Cavalli R, Bargoni A, Fundarò A, Vighetto D, Gasco MR. Intravenous administration to rabbits of non-stealth and stealth doxorubicin-loaded solid lipid nanoparticles at increasing concentrations of stealth agent: pharmacokinetics and distribution of doxorubicin in brain and other tissues. J Drug Target 2002; 10(4): 327-35.
[http://dx.doi.org/10.1080/10611860290031868] [PMID: 12164381]

[75]   Sundheimer JK, Benzel J, Longuespée R, *et al.* Experimental insights and recommendations for successfully performing cerebral microdialysis with hydrophobic drug candidates. Clin Transl Sci. 2025;18(4):e70226.
[http://dx.doi.org/10.1111/cts.70226]

[76]   Ko YT, Bhattacharya R, Bickel U. Liposome encapsulated polyethylenimine/ODN polyplexes for brain targeting. J Control Release 2009; 133(3): 230-7.
[http://dx.doi.org/10.1016/j.jconrel.2008.10.013] [PMID: 19013203]

[77]   Miller L, Meythaler J, Peduzzi J. Direct central nervous system catheter and temperature control system. EP Patent 1,600,186; 2005.

[78]   Chen MY, Lonser RR, Morrison PF, Governale LS, Oldfield EH. Variables affecting convection-enhanced delivery to the striatum: a systematic examination of rate of infusion, cannula size, infusate concentration, and tissue-cannula sealing time. J Neurosurg 1999; 90(2): 315-20.
[http://dx.doi.org/10.3171/jns.1999.90.2.0315] [PMID: 9950503]

[79]   Bauman MA, Gillies GT, Raghavan R, Brady ML, Pedain C. Physical Characterization of Neurocatheter Performance in a Brain Phantom Gelatin with Nanoscale Porosity: Steady-State and Oscillatory Flows. Nanotechnology 2004; 15: 92-7.
[http://dx.doi.org/10.1088/0957-4484/15/1/018]

[80]   de Lange ECM, de Vries JD, Zurcher C, Danhof M, de Boer AG, Breimer DD. The use of intracerebral microdialysis for the determination of pharmacokinetic profiles of anticancer drugs in tumor-bearing rat brain. Pharm Res 1995; 12(12): 1924-31.
[http://dx.doi.org/10.1023/a:1016239822287] [PMID: 8786967]

[81]   Olson JJ, Blakeley JO, Grossman SA, Weingart J, Rashid A, Supko J. Differences in the distribution of methotrexate into high grade gliomas following intravenous administration, as monitored by microdialysis, are associated with blood brain barrier integrity. J Clin Oncol 2006; 24(18) (Suppl.): 1548-8.
[http://dx.doi.org/10.1200/jco.2006.24.18_suppl.1548]

[82]   Portnow J, Badie B, Chen M, Liu A, Blanchard S, Synold TW. The neuropharmacokinetics of temozolomide in patients with resectable brain tumors: potential implications for the current approach to chemoradiation. Clin Cancer Res 2009; 15(22): 7092-8.
[http://dx.doi.org/10.1158/1078-0432.CCR-09-1349] [PMID: 19861433]

[83]   Jayachandra Babu R, Dayal PP, Pawar K, Singh M. Nose-to-brain transport of melatonin from polymer gel suspensions: a microdialysis study in rats. J Drug Target 2011; 19(9): 731-40.
[http://dx.doi.org/10.3109/1061186x.2011.558090] [PMID: 21428693]

[84]   de Lange EC, Danhof M, de Boer AG, Breimer DD. Methodological considerations of intracerebral microdialysis in pharmacokinetic studies on drug transport across the blood-brain barrier. Brain Res Brain Res Rev 1997; 25(1): 27-49.
[http://dx.doi.org/10.1016/s0165-0173(97)00014-3] [PMID: 9370049]

[85]   Breimer DD, de Boer BAG, Breimer DD. Microdialysis for pharmacokinetic analysis of drug transport to the brain. Adv Drug Deliv Rev 1999; 36(2-3): 211-27.
[http://dx.doi.org/10.1016/s0169-409x(98)00089-1] [PMID: 10837717]

[86]   Parsons LH, Justice JB Jr. Quantitative approaches to *in vivo* brain microdialysis. Crit Rev Neurobiol 1994; 8(3): 189-220.
[PMID: 7923396]

[87]   Devineni D, Klein-Szanto A, Gallo JM. *In vivo* microdialysis to characterize drug transport in brain tumors: analysis of methotrexate uptake in rat glioma-2 (RG-2)-bearing rats. Cancer Chemother Pharmacol 1996; 38(6): 499-507.
[http://dx.doi.org/10.1007/s002800050518] [PMID: 8823490]

[88]   Nakashima M, Shibata S, Tokunaga Y, *et al. In vivo* microdialysis study of the distribution of cisplatin into brain tumour tissue after intracarotid infusion in rats with 9L malignant glioma. J Pharm Pharmacol 1997; 49(8): 777-80.
[http://dx.doi.org/10.1111/j.2042-7158.1997.tb06111.x] [PMID: 9379355]

[89]   Yang F-Y, Lin Y-L, Chou F-I, *et al.* Pharmacokinetics of BPA in gliomas with ultrasound induced blood-brain barrier disruption as measured by microdialysis. PLoS One 2014; 9(6): e100104.
[http://dx.doi.org/10.1371/journal.pone.0100104] [PMID: 24936788]

[90]   Benoit J-P, Faisant N, Venier-Julienne M-C, Menei P. Development of microspheres for neurological disorders: from basics to clinical applications. J Control Release 2000; 65(1-2): 285-96.
[http://dx.doi.org/10.1016/s0168-3659(99)00250-3] [PMID: 10699288]

[91]   Menei P, Benoit J-P, Boisdron-Celle M, Fournier D, Mercier P, Guy G. Drug targeting into the central nervous system by stereotactic implantation of biodegradable microspheres. Neurosurgery 1994; 34(6): 1058-64.
[http://dx.doi.org/10.1227/00006123-199406000-00016] [PMID: 8084391]

[92]   Mittal S, Cohen A, Maysinger D. *In vitro* effects of brain derived neurotrophic factor released from microspheres. Neuroreport 1994; 5(18): 2577-82.
[http://dx.doi.org/10.1097/00001756-199412000-00043] [PMID: 7696608]

[93]   Brem H, Ewend MG, Piantadosi S, Greenhoot J, Burger PC, Sisti M. The safety of interstitial chemotherapy with BCNU-loaded polymer followed by radiation therapy in the treatment of newly diagnosed malignant gliomas: phase I trial. J Neurooncol 1995; 26(2): 111-23.
[http://dx.doi.org/10.1007/bf01060217] [PMID: 8787853]

[94]   Ewend MG, Williams JA, Tabassi K, *et al.* Local delivery of chemotherapy and concurrent external beam radiotherapy prolongs survival in metastatic brain tumor models. Cancer Res 1996; 56(22): 5217-23.
[PMID: 8912860]

[95]   Valtonen S, Timonen U, Toivanen P, *et al.* Interstitial chemotherapy with carmustine-loaded polymers for high-grade gliomas: a randomized double-blind study. Neurosurgery 1997; 41(1): 44-8.
[http://dx.doi.org/10.1097/00006123-199707000-00011] [PMID: 9218294]

[96]   Sheleg SV, Korotkevich EA, Zhavrid EA, *et al.* Local chemotherapy with cisplatin-depot for glioblastoma multiforme. J Neurooncol 2002; 60(1): 53-9.
[http://dx.doi.org/10.1023/a:1020288015457] [PMID: 12416546]

[97]   Vukelja SJ, Anthony SP, Arseneau JC, *et al.* Phase 1 study of escalating-dose OncoGel (ReGel/paclitaxel) depot injection, a controlled-release formulation of paclitaxel, for local management of superficial solid tumor lesions. Anticancer Drugs 2007; 18(3): 283-9.
[http://dx.doi.org/10.1097/cad.0b013e328011a51d] [PMID: 17264760]

[98]   Boer GJ, van der Woude TP, Kruisbrink J, van Heerikhuize J. Successful ventricular application of the miniaturized controlled-delivery Accurel technique for sustained enhancement of cerebrospinal fluid peptide levels in the rat. J Neurosci Methods 1984; 11(4): 281-9.
[http://dx.doi.org/10.1016/0165-0270(84)90090-6] [PMID: 6513586]

[99]   Fortin D, Desjardins A, Benko A, Niyonsega T, Boudrias M. Enhanced chemotherapy delivery by intraarterial infusion and blood-brain barrier disruption in malignant brain tumors: the Sherbrooke experience. Cancer 2005; 103(12): 2606-15.
[http://dx.doi.org/10.1002/cncr.21112] [PMID: 15880378]

[100] Dave N, Gudelsky GA, Desai PB. The pharmacokinetics of letrozole in brain and brain tumor in rats with orthotopically implanted C6 glioma, assessed using intracerebral microdialysis. Cancer Chemother Pharmacol 2013; 72(2): 349-57.
[http://dx.doi.org/10.1007/s00280-013-2205-y] [PMID: 23748921]

[101] Hannon GJ. RNA interference. Nature 2002; 418(6894): 244-51.
[http://dx.doi.org/10.1038/418244a] [PMID: 12110901]

[102] Hébert SS, De Strooper B. miRNAs in Neurodegeneration. Science 2007; 317(5842): 1179-80.
[http://dx.doi.org/10.1126/science.1148530] [PMID: 17761871]

[103] Zhang Y, Hwa JL, Boado RJ, Pardridge WM. Receptor-Mediated Delivery of an Antisense Gene to Human Brain Cancer Cells. The journal of gene medicine 2002, 4 (2), 183–194.
[http://dx.doi.org/10.1002/jgm.255]

[104] Daneman R, Barres B. Permeability of blood-brain barrier. Available from: https://patents.google.com/patent/WO2007137303A8/en

[105] Patel N, Addo RT. Ruhi U, Mohammed NU, D'Souza M, Jobe L. The effect of antisense to nf-kb in an albumin microsphere formulation on the progression of left-ventricular remodeling associated with chronic volume overload in rats. J Drug Target 2014; 22(9): 796-804.
[http://dx.doi.org/10.3109/1061186X.2014.921927] [PMID: 24892743]

[106] Liu CH, Kim YR, Ren JQ, Eichler F, Rosen BR, Liu PK. Imaging cerebral gene transcripts in live animals. J Neurosci 2007; 27(3): 713-22.

[http://dx.doi.org/10.1523/JNEUROSCI.4660-06.2007] [PMID: 17234603]

[107] Masotti A, Vicennati P, Boschi F, Calderan L, Sbarbati A, Ortaggi G. A novel near-infrared indocyanine dye-polyethylenimine conjugate allows DNA delivery imaging *in vivo*. Bioconjug Chem 2008; 19(5): 983-7.
[http://dx.doi.org/10.1021/bc700356f] [PMID: 18429627]

[108] Béduneau A, Saulnier P, Benoit JP. Active targeting of brain tumors using nanocarriers. Biomaterials 2007; 28(33): 4947-67.
[http://dx.doi.org/10.1016/j.biomaterials.2007.06.011] [PMID: 17716726]

[109] Schneider T, Becker A, Ringe K, Reinhold A, Firsching R, Sabel BA. Brain tumor therapy by combined vaccination and antisense oligonucleotide delivery with nanoparticles. J Neuroimmunol 2008; 195(1-2): 21-7.
[http://dx.doi.org/10.1016/j.jneuroim.2007.12.005] [PMID: 18304655]

[110] Ljubimova JY, Fujita M, Khazenzon NM, *et al*. Nanoconjugate based on polymalic acid for tumor targeting. Chem Biol Interact 2008; 171(2): 195-203.
[http://dx.doi.org/10.1016/j.cbi.2007.01.015] [PMID: 17376417]

[111] Ljubimova JY, Fujita M, Ljubimov AV, Torchilin VP, Black KL, Holler E. Poly(malic acid) nanoconjugates containing various antibodies and oligonucleotides for multitargeting drug delivery. Nanomedicine (Lond) 2008; 3(2): 247-65.
[http://dx.doi.org/10.2217/17435889.3.2.247] [PMID: 18373429]

[112] Bawarski WE, Chidlowsky E, Bharali DJ, Mousa SA. Emerging nanopharmaceuticals. Nanomedicine 2008; 4(4): 273-82.
[http://dx.doi.org/10.1016/j.nano.2008.06.002] [PMID: 18640076]

[113] Jain KK. Nanomedicine: application of nanobiotechnology in medical practice. Med Princ Pract 2008; 17(2): 89-101.
[http://dx.doi.org/10.1159/000112961] [PMID: 18287791]

[114] Kim YK, Xing L, Chen BA, Xu F, Jiang HL, Zhang C. Aerosol delivery of programmed cell death protein 4 using polysorbitol-based gene delivery system for lung cancer therapy. J Drug Target 2014; 22(9): 829-38.
[http://dx.doi.org/10.3109/1061186X.2014.932796] [PMID: 24983766]

[115] Cusack JC Jr, Tanabe KK. Cancer gene therapy. Surg Oncol Clin N Am 1998; 7(3): 421-69.
[http://dx.doi.org/10.1016/S1055-3207(18)30255-2] [PMID: 9624212]

[116] Dickson PV, Nathwani AC, Davidoff AM. Delivery of antiangiogenic agents for cancer gene therapy. Technol Cancer Res Treat 2005; 4(4): 331-41.
[http://dx.doi.org/10.1177/153303460500400403] [PMID: 16029054]

[117] Robson T, Worthington J, McKeown SR, Hirst DG. Radiogenic therapy: novel approaches for enhancing tumor radiosensitivity. Technol Cancer Res Treat 2005; 4(4): 343-61.
[http://dx.doi.org/10.1177/153303460500400404] [PMID: 16029055]

[118] Dent P, Yacoub A, Park M, *et al*. Searching for a cure: gene therapy for glioblastoma. Cancer Biol Ther 2008.
[http://dx.doi.org/10.4161/cbt.7.9.6408]

[119] Lang FF, Bruner JM, Fuller GN, *et al*. Phase I trial of adenovirus-mediated p53 gene therapy for recurrent glioma: biological and clinical results. J Clin Oncol 2003; 21(13): 2508-18.
[http://dx.doi.org/10.1200/JCO.2003.21.13.2508] [PMID: 12839017]

[120] Klatzmann D, Valéry CA, Bensimon G, *et al*. A phase I/II study of herpes simplex virus type 1 thymidine kinase "suicide" gene therapy for recurrent glioblastoma. Hum Gene Ther 1998; 9(17): 2595-604.
[http://dx.doi.org/10.1089/hum.1998.9.17-2595] [PMID: 9853526]

[121] Rainov NG. A phase III clinical evaluation of herpes simplex virus type 1 thymidine kinase and

ganciclovir gene therapy as an adjuvant to surgical resection and radiation in adults with previously untreated glioblastoma multiforme. Hum Gene Ther 2000; 11(17): 2389-401.
[http://dx.doi.org/10.1089/104303400750038499] [PMID: 11096443]

[122]   Green NM. Avidin. Adv Protein Chem 1975; 29: 85-133.
[http://dx.doi.org/10.1016/S0065-3233(08)60411-8] [PMID: 237414]

[123]   Neuwelt, E. A. Method for diagnostically imaging lesions in the brain inside a blood-brain barrier. Available from: https://patents.google.com/patent/US5059415

[124]   Pardridge, W. M.; Boado, R. J. Drug delivery of antisense oligonucleotides and peptides to tissues *in vivo* and to cells using avidin-biotin technology. Available from: https://patents.google.com/patent/US6287792B1/en

[125]   Aaron J, Nitin N, Travis K, Kumar S, *et al.* Plasmon resonance coupling of metal nanoparticles for molecular imaging of carcinogenesis *in vivo*. 2007; 12(3): 034007-7.
[http://dx.doi.org/10.1117/1.2737351] [PMID: 17614715]

[126]   Crombet Ramos T, Figueredo J, Catala M, *et al.* Treatment of high-grade glioma patients with the humanized anti-epidermal growth factor receptor (EGFR) antibody h-R3: Report from a phase I/II trial. Cancer Biol Ther 2006; 5(4): 375-9.
[http://dx.doi.org/10.4161/cbt.5.4.2522] [PMID: 16575203]

[127]   Combs SE, Heeger S, Haselmann R, Edler L, Debus J, Schulz-Ertner D. Treatment of primary glioblastoma multiforme with cetuximab, radiotherapy and temozolomide (GERT) – phase I/II trial: study protocol. BMC Cancer 2006; 6(1): 133.
[http://dx.doi.org/10.1186/1471-2407-6-133] [PMID: 16709245]

[128]   Kreuter J. Nanoparticulate systems for brain delivery of drugs. Adv Drug Deliv Rev 2001; 47(1): 65-81.
[http://dx.doi.org/10.1016/S0169-409X(00)00122-8] [PMID: 11251246]

[129]   Yamamoto H, Kuno Y, Sugimoto S, Takeuchi H, Kawashima Y. Surface-modified PLGA nanosphere with chitosan improved pulmonary delivery of calcitonin by mucoadhesion and opening of the intercellular tight junctions. J Control Release 2005; 102(2): 373-81.
[http://dx.doi.org/10.1016/j.jconrel.2004.10.010] [PMID: 15653158]

[130]   Cohen H, Levy RJ, Gao J, *et al.* Sustained delivery and expression of DNA encapsulated in polymeric nanoparticles. Gene Ther 2000; 7(22): 1896-905.
[http://dx.doi.org/10.1038/sj.gt.3301318] [PMID: 11127577]

[131]   King GL, Johnson SM. Receptor-mediated transport of insulin across endothelial cells. Science 1985; 227(4694): 1583-6.
[http://dx.doi.org/10.1126/science.3883490] [PMID: 3883490]

[132]   Maratos-Flier E, Kao CY, Verdin EM, King GL. Receptor-mediated vectorial transcytosis of epidermal growth factor by Madin-Darby canine kidney cells. J Cell Biol 1987; 105(4): 1595-601.
[http://dx.doi.org/10.1083/jcb.105.4.1595] [PMID: 3312235]

[133]   Roberts RL, Fine RE, Sandra A. Receptor-mediated endocytosis of transferrin at the blood-brain barrier. J Cell Sci 1993; 104(2): 521-32.
[http://dx.doi.org/10.1242/jcs.104.2.521] [PMID: 8505377]

[134]   Bickel U, Yoshikawa T, Pardridge WM. Delivery of peptides and proteins through the blood–brain barrier. Adv Drug Deliv Rev 2001; 46(1-3): 247-79.
[http://dx.doi.org/10.1016/S0169-409X(00)00139-3] [PMID: 11259843]

[135]   Pardridge WM. Blood-brain barrier biology and methodology. J Neurovirol 1999; 5(6): 556-69.
[http://dx.doi.org/10.3109/13550289909021285] [PMID: 10602397]

[136]   Vyas SP, Sihorkar V. Endogenous carriers and ligands in non-immunogenic site-specific drug delivery. Adv Drug Deliv Rev 2000; 43(2-3): 101-64.
[http://dx.doi.org/10.1016/S0169-409X(00)00067-3] [PMID: 10967224]

[137] Demeule M, Currie JC, Bertrand Y, *et al.* Involvement of the low-density lipoprotein receptor-related protein in the transcytosis of the brain delivery vector Angiopep-2. J Neurochem 2008; 106(4): 1534-44.
[http://dx.doi.org/10.1111/j.1471-4159.2008.05492.x] [PMID: 18489712]

[138] Engelhardt B. Molecular mechanisms involved in T cell migration across the blood–brain barrier. J Neural Transm (Vienna) 2006; 113(4): 477-85.
[http://dx.doi.org/10.1007/s00702-005-0409-y] [PMID: 16550326]

[139] Banks WA, Kastin AJ, Ehrensing CA. Endogenous peptide Tyr-Pro-Trp-Gly-NH$_2$ (Tyr-W-MIF-1) is transported from the brain to the blood by peptide transport system-1. J Neurosci Res 1993; 35(6): 690-5.
[http://dx.doi.org/10.1002/jnr.490350611] [PMID: 8105102]

[140] Allen DD, Geldenhuys WJ. Molecular modeling of blood–brain barrier nutrient transporters: In silico basis for evaluation of potential drug delivery to the central nervous system. Life Sci 2006; 78(10): 1029-33.
[http://dx.doi.org/10.1016/j.lfs.2005.06.004] [PMID: 16126231]

[141] Tsuji A. Small molecular drug transfer across the blood-brain barrier *via* carrier-mediated transport systems. NeuroRx 2005; 2(1): 54-62.
[http://dx.doi.org/10.1602/neurorx.2.1.54] [PMID: 15717057]

[142] Fukuhara H, Ino Y, Todo T. Oncolytic virus therapy: A new era of cancer treatment at dawn. Cancer Sci 2016; 107(10): 1373-9.
[http://dx.doi.org/10.1111/cas.13027] [PMID: 27486853]

[143] Platanias LC. Mechanisms of type-I- and type-II-interferon-mediated signalling. Nat Rev Immunol 2005; 5(5): 375-86.
[http://dx.doi.org/10.1038/nri1604] [PMID: 15864272]

[144] Todo T, Martuza RL, Rabkin SD, Johnson PA. Oncolytic herpes simplex virus vector with enhanced MHC class I presentation and tumor cell killing. Proc Natl Acad Sci USA 2001; 98(11): 6396-401.
[http://dx.doi.org/10.1073/pnas.101136398] [PMID: 11353831]

[145] Aghi M, Visted T, DePinho RA, Chiocca EA. Oncolytic herpes virus with defective ICP6 specifically replicates in quiescent cells with homozygous genetic mutations in p16. Oncogene 2008; 27(30): 4249-54.
[http://dx.doi.org/10.1038/onc.2008.53] [PMID: 18345032]

[146] Kaur B, Chiocca EA, Cripe TP, Oncolytic T. Oncolytic HSV-1 virotherapy: clinical experience and opportunities for progress. Curr Pharm Biotechnol 2012; 13(9): 1842-51.
[http://dx.doi.org/10.2174/138920112800958814] [PMID: 21740359]

[147] Fueyo J, Alemany R, Gomez-Manzano C, *et al.* Preclinical characterization of the antiglioma activity of a tropism-enhanced adenovirus targeted to the retinoblastoma pathway. J Natl Cancer Inst 2003; 95(9): 652-60.
[http://dx.doi.org/10.1093/jnci/95.9.652] [PMID: 12734316]

[148] Maher EA, Furnari FB, Bachoo RM, *et al.* Malignant glioma: genetics and biology of a grave matter. Genes Dev 2001; 15(11): 1311-33.
[http://dx.doi.org/10.1101/gad.891601] [PMID: 11390353]

[149] Philbrick B, Adamson DC. DNX-2401: an investigational drug for the treatment of recurrent glioblastoma. Expert Opin Investig Drugs 2019; 28(12): 1041-9.
[http://dx.doi.org/10.1080/13543784.2019.1694000] [PMID: 31726894]

[150] Gromeier M, Alexander L, Wimmer E. Internal ribosomal entry site substitution eliminates neurovirulence in intergeneric poliovirus recombinants. Proc Natl Acad Sci USA 1996; 93(6): 2370-5.
[http://dx.doi.org/10.1073/pnas.93.6.2370] [PMID: 8637880]

[151] Dobrikova EY, Goetz C, Walters RW, *et al.* Attenuation of neurovirulence, biodistribution, and

shedding of a poliovirus:rhinovirus chimera after intrathalamic inoculation in Macaca fascicularis. J Virol 2012; 86(5): 2750-9.
[http://dx.doi.org/10.1128/JVI.06427-11] [PMID: 22171271]

[152]   Strong JE, Coffey MC, Tang D, Sabinin P, Lee PW. The molecular basis of viral oncolysis: usurpation of the Ras signaling pathway by reovirus. EMBO J 1998; 17(12): 3351-62.
[http://dx.doi.org/10.1093/emboj/17.12.3351] [PMID: 9628872]

[153]   Coffey MC, Strong JE, Forsyth PA, Lee PWK. Reovirus therapy of tumors with activated Ras pathway. Science 1998; 282(5392): 1332-4.
[http://dx.doi.org/10.1126/science.282.5392.1332] [PMID: 9812900]

[154]   Siegal T, Zylber-Katz E. Strategies for increasing drug delivery to the brain: focus on brain lymphoma. Clin Pharmacokinet 2002; 41(3): 171-86.
[http://dx.doi.org/10.2165/00003088-200241030-00002] [PMID: 11929318]

[155]   Cole SPC, Bhardwaj G, Gerlach JH, *et al.* Overexpression of a transporter gene in a multidrug-resistant human lung cancer cell line. Science 1992; 258(5088): 1650-4.
[http://dx.doi.org/10.1126/science.1360704] [PMID: 1360704]

[156]   Namba H, Iwadate Y, Iyo M, *et al.* Glucose and methionine uptake by rat brain tumor treated with prodrug-activated gene therapy. Nucl Med Biol 1998; 25(3): 247-50.
[http://dx.doi.org/10.1016/S0969-8051(97)00171-6] [PMID: 9620630]

[157]   Savolainen J, Edwards JE, Morgan ME, McNamara PJ, Anderson BD. Effects of a P-glycoprotein inhibitor on brain and plasma concentrations of anti-human immunodeficiency virus drugs administered in combination in rats. Drug Metab Dispos 2002; 30(5): 479-82.
[http://dx.doi.org/10.1124/dmd.30.5.479] [PMID: 11950774]

[158]   Pignatello R, Pantò V, Salmaso S, *et al.* Flurbiprofen derivatives in Alzheimer's disease: synthesis, pharmacokinetic and biological assessment of lipoamino acid prodrugs. Bioconjug Chem 2008; 19(1): 349-57.
[http://dx.doi.org/10.1021/bc700312y] [PMID: 18072715]

[159]   Begley, D. J. Efflux mechanisms in the CNS: A powerful influnec on drug distribution in the brain. King's College London. 2004; 83-97. Available from: https://kclpure.kcl.ac.uk/portal/en/publications/efflux-mechanisms-in-the-cns-a-powerful-influnec-on-drug-distribu

[160]   Blakeley J. Drug delivery to brain tumors. Curr Neurol Neurosci Rep 2008; 8(3): 235-41.
[http://dx.doi.org/10.1007/s11910-008-0036-8] [PMID: 18541119]

[161]   Breedveld P, Beijnen JH, Schellens JHM. Use of P-glycoprotein and BCRP inhibitors to improve oral bioavailability and CNS penetration of anticancer drugs. Trends Pharmacol Sci 2006; 27(1): 17-24.
[http://dx.doi.org/10.1016/j.tips.2005.11.009] [PMID: 16337012]

[162]   Régina A, Demeule M, Ché C, *et al.* Antitumour activity of ANG1005, a conjugate between paclitaxel and the new brain delivery vector Angiopep-2. Br J Pharmacol 2008; 155(2): 185-97.
[http://dx.doi.org/10.1038/bjp.2008.260] [PMID: 18574456]

[163]   Nam JP, Park JK, Son DH, *et al.* Evaluation of polyethylene glycol-conjugated novel polymeric anti-tumor drug for cancer therapy. Colloids Surf B Biointerfaces 2014; 120: 168-75.
[http://dx.doi.org/10.1016/j.colsurfb.2014.04.013] [PMID: 24918700]

[164]   Krewson CE, Klarman ML, Saltzman WM. Distribution of nerve growth factor following direct delivery to brain interstitium. Brain Res 1995; 680(1-2): 196-206.
[http://dx.doi.org/10.1016/0006-8993(95)00261-N] [PMID: 7663977]

[165]   Kabanov AV, Batrakova EV, Miller DW. Pluronic® block copolymers as modulators of drug efflux transporter activity in the blood–brain barrier. Adv Drug Deliv Rev 2003; 55(1): 151-64.
[http://dx.doi.org/10.1016/S0169-409X(02)00176-X] [PMID: 12535579]

[166]   Litman T, Skovsgaard T, Stein WD. Pumping of drugs by P-glycoprotein: a two-step process? J

Pharmacol Exp Ther 2003; 307(3): 846-53.
[http://dx.doi.org/10.1124/jpet.103.056960] [PMID: 14534356]

[167]   Cserr HF, Patlak CS. Secretion and Bulk Flow of Interstitial Fluid. Handb Exp Pharmacol 1992; 103: 245-61.
[http://dx.doi.org/10.1007/978-3-642-76894-1_9]

[168]   Laquintana V, Trapani A, Denora N, Wang F, Gallo JM, Trapani G. New strategies to deliver anticancer drugs to brain tumors. Expert Opin Drug Deliv 2009; 6(10): 1017-32.
[http://dx.doi.org/10.1517/17425240903167942] [PMID: 19732031]

[169]   Modgill V, Garg T, Goyal A, Rath G. Transmucosal delivery of linagliptin for the treatment of type- 2 diabetes mellitus by ultra-thin nanofibers. Curr Drug Deliv 2015; 12(3): 323-32.
[http://dx.doi.org/10.2174/1567201811666141117144332] [PMID: 25410375]

[170]   Fellner S, Bauer B, Miller DS, *et al.* Transport of paclitaxel (Taxol) across the blood-brain barrier *in vitro* and *in vivo*. J Clin Invest 2002; 110(9): 1309-18.
[http://dx.doi.org/10.1172/JCI0215451] [PMID: 12417570]

[171]   Gupta SP. QSAR studies on drugs acting at the central nervous system. Chem Rev 1989; 89(8): 1765-800.
[http://dx.doi.org/10.1021/cr00098a007]

[172]   Martin YC. Exploring QSAR: Hydrophobic, Electronic, and Steric Constants C. Hansch, A. Leo, and D. Hoekman. American Chemical Society, Washington, DC. 1995. Xix + 348 pp. 22 × 28.5 cm. Exploring QSAR: Fundamentals and Applications in Chemistry and Biology. C. Hansch and A. Leo. American Chemical Society, Washington, DC. 1995. Xvii + 557 pp. 18.5 × 26 cm. ISBN 0-841--2993-7 (set). $99.95 (set). J Med Chem 1996; 39(5): 1189-90.
[http://dx.doi.org/10.1021/jm950902o]

[173]   Van de Waterbeemd H, Smith DA, Beaumont K, Walker DK. Property-based design: optimization of drug absorption and pharmacokinetics. J Med Chem 2001; 44(9): 1313-33.
[http://dx.doi.org/10.1021/jm000407e] [PMID: 11311053]

[174]   Garg T. Current nanotechnological approaches for an effective delivery of bio-active drug molecules in the treatment of acne. Artif Cells Nanomed Biotechnol 2016; 44(1): 98-105.
[http://dx.doi.org/10.3109/21691401.2014.916715] [PMID: 24844191]

[175]   Chrai SS, Murari R, Ahmad I. Liposomes (a review), part one: manufacturing issues. Biopharm Int 2001; 14(11).

[176]   Garg T, Kumar Goyal A. Iontophoresis: drug delivery system by applying an electrical potential across the skin. Drug Deliv Lett 2012; 2(4): 270-80.
[http://dx.doi.org/10.2174/2210304x11202040005]

[177]   Garg T, Goyal K, Liposomes A. Targeted and controlled delivery system. Drug Deliv Lett 2014; 4(1): 62-71.
[http://dx.doi.org/10.2174/221030311113036660015]

[178]   Pardridge WM. Vector-mediated drug delivery to the brain. Adv Drug Deliv Rev 1999; 36(2-3): 299-321.
[http://dx.doi.org/10.1016/S0169-409X(98)00087-8] [PMID: 10837722]

[179]   Boado RJ, Zhang Y, Zhang Y, Pardridge WM. Humanization of anti-human insulin receptor antibody for drug targeting across the human blood–brain barrier. Biotechnol Bioeng 2007; 96(2): 381-91.
[http://dx.doi.org/10.1002/bit.21120] [PMID: 16937408]

[180]   Budai L, Hajdú M, Budai M, *et al.* Gels and liposomes in optimized ocular drug delivery: Studies on ciprofloxacin formulations. Int J Pharm 2007; 343(1-2): 34-40.
[http://dx.doi.org/10.1016/j.ijpharm.2007.04.013] [PMID: 17537601]

[181]   Dubey S, Suraj MR, Goni T, *et al.* 3D QSAR studies of 3, 16 and 17 position modifications in steroidal derivatives for CNS anticancer activity. Curr Res Chem. 2023;15(1):1–4.

[http://dx.doi.org/10.3923/crc.2023.1.4]

[182] Huwyler J, Wu D, Pardridge WM. Brain drug delivery of small molecules using immunoliposomes. Proc Natl Acad Sci USA 1996; 93(24): 14164-9.
[http://dx.doi.org/10.1073/pnas.93.24.14164] [PMID: 8943078]

[183] Lee HJ, Engelhardt B, Lesley J, Bickel U, Pardridge WM. Targeting rat anti-mouse transferrin receptor monoclonal antibodies through blood-brain barrier in mouse. J Pharmacol Exp Ther 2000; 292(3): 1048-52.
[http://dx.doi.org/10.1016/S0022-3565(24)35388-1] [PMID: 10688622]

[184] Olivier JC, Huertas R, Lee HJ, Calon F, Pardridge WM. Synthesis of pegylated immunonanoparticles. Pharm Res 2002; 19(8): 1137-43.
[http://dx.doi.org/10.1023/A:1019842024814] [PMID: 12240939]

[185] Pardridge WM, Boado RJ, Kang YS. Vector-mediated delivery of a polyamide ("peptide") nucleic acid analogue through the blood-brain barrier *in vivo*. Proc Natl Acad Sci USA 1995; 92(12): 5592-6.
[http://dx.doi.org/10.1073/pnas.92.12.5592] [PMID: 7777554]

[186] Zhang Y, Pardridge WM. Conjugation of brain-derived neurotrophic factor to a blood–brain barrier drug targeting system enables neuroprotection in regional brain ischemia following intravenous injection of the neurotrophin. Brain Res 2001; 889(1-2): 49-56.
[http://dx.doi.org/10.1016/S0006-8993(00)03108-5] [PMID: 11166685]

[187] Garg T, Murthy RSR, Kumar Goyal A, Arora S, Malik B. Development, optimization and evaluation of porous chitosan scaffold formulation of gliclazide for the treatment of type-2 diabetes mellitus. Drug Deliv Lett 2012; 2(4): 251-61.
[http://dx.doi.org/10.2174/2210304x11202040003]

[188] Kaur R, Garg T, K Goyal A, Rath G. Development, optimization and evaluation of electrospun nanofibers: tool for targeted vaginal delivery of antimicrobials against urinary tract infections. Curr Drug Deliv 2016; 13(5): 754-63.
[http://dx.doi.org/10.2174/1567201812666150212123348] [PMID: 25675338]

[189] Kumar A, Garg T, Sarma GS, Rath G, Goyal AK. Optimization of combinational intranasal drug delivery system for the management of migraine by using statistical design. Eur J Pharm Sci 2015; 70: 140-51.
[http://dx.doi.org/10.1016/j.ejps.2015.01.012] [PMID: 25676136]

[190] Chertok B, David AE, Huang Y, Yang VC. Glioma selectivity of magnetically targeted nanoparticles: A role of abnormal tumor hydrodynamics. J Control Release 2007; 122(3): 315-23.
[http://dx.doi.org/10.1016/j.jconrel.2007.05.030] [PMID: 17628157]

[191] Tsutsui Y, Tomizawa K, Nagita M, *et al.* Development of bionanocapsules targeting brain tumors. J Control Release 2007; 122(2): 159-64.
[http://dx.doi.org/10.1016/j.jconrel.2007.06.019] [PMID: 17692421]

[192] Cheng Q-Y, Feng J, Li F-Z. [Brain delivery of neurotoxin-I-loaded nanoparticles through intranasal administration]. Yao Xue Xue Bao 2008; 43(4): 431-4.
[PMID: 18664209]

[193] Rao KS, Reddy MK, Horning JL, Labhasetwar V. TAT-conjugated nanoparticles for the CNS delivery of anti-HIV drugs. Biomaterials 2008; 29(33): 4429-38.
[http://dx.doi.org/10.1016/j.biomaterials.2008.08.004] [PMID: 18760470]

[194] Zensi A, Begley D, Pontikis C, *et al.* Albumin nanoparticles targeted with Apo E enter the CNS by transcytosis and are delivered to neurones. J Control Release 2009; 137(1): 78-86.
[http://dx.doi.org/10.1016/j.jconrel.2009.03.002] [PMID: 19285109]

[195] Yin X, He Z, Ge W, Zhao Z. Application of aptamer-functionalized nanomaterials in targeting therapeutics of typical tumors. Front Bioeng Biotechnol. 2023;11:1092901
[http://dx.doi.org/10.3389/fbioe.2023.1092901]

[196] Gao H, Wang Y, Chen C, *et al.* Incorporation of lapatinib into core–shell nanoparticles improves both the solubility and anti-glioma effects of the drug. Int J Pharm 2014; 461(1-2): 478-88.
[http://dx.doi.org/10.1016/j.ijpharm.2013.12.016] [PMID: 24368101]

[197] Mu¨ller R, Lucks J. Arzneistofftra¨ger aus festen lipidteilchen, feste lipidnanospha¨ren (sln). European patent, 605497; 1996. Available from: https://patents.google.com/patent/EP0605497A1/un

[198] Zhu X, Huang S, Huang H, *et al. In vitro and in vivo* anti-cancer effects of targeting and photothermal sensitive solid lipid nanoparticles. J Drug Target 2014; 22(9): 822-8.
[http://dx.doi.org/10.3109/1061186X.2014.931405] [PMID: 24964053]

[199] Müller RH. Solid lipid nanoparticles R. (SLN) for controlled drug delivery â€" a review of the state of the art. Eur J Pharm Biopharm 2000; 50(1): 161-77.
[http://dx.doi.org/10.1016/S0939-6411(00)00087-4] [PMID: 10840199]

[200] Smith A, Hunneyball M. Evaluation of poly(lactic acid) as a biodegradable drug delivery system for parenteral administration. Int J Pharm 1986; 30(2-3): 215-20.
[http://dx.doi.org/10.1016/0378-5173(86)90081-5]

[201] Gupta B, Levchenko TS, Torchilin VP. TAT peptide-modified liposomes provide enhanced gene delivery to intracranial human brain tumor xenografts in nude mice. Oncol Res 2006; 16(8): 351-9.
[http://dx.doi.org/10.3727/000000006783980946] [PMID: 17913043]

[202] Feng B, Tomizawa K, Michiue H, *et al.* Delivery of sodium borocaptate to glioma cells using immunoliposome conjugated with anti-EGFR antibodies by ZZ-His. Biomaterials 2009; 30(9): 1746-55.
[http://dx.doi.org/10.1016/j.biomaterials.2008.12.010] [PMID: 19121537]

[203] Chertok B, Moffat BA, David AE, *et al.* Iron oxide nanoparticles as a drug delivery vehicle for MRI monitored magnetic targeting of brain tumors. Biomaterials 2008; 29(4): 487-96.
[http://dx.doi.org/10.1016/j.biomaterials.2007.08.050] [PMID: 17964647]

[204] Du J, Lu WL, Ying X, *et al.* Dual-targeting topotecan liposomes modified with tamoxifen and wheat germ agglutinin significantly improve drug transport across the blood-brain barrier and survival of brain tumor-bearing animals. Mol Pharm 2009; 6(3): 905-17.
[http://dx.doi.org/10.1021/mp800218q] [PMID: 19344115]

[205] Yanagië H, Ogata A, Sugiyama H, Eriguchi M, Takamoto S, Takahashi H. Application of drug delivery system to boron neutron capture therapy for cancer. Expert Opin Drug Deliv 2008; 5(4): 427-43.
[http://dx.doi.org/10.1517/17425247.5.4.427] [PMID: 18426384]

[206] Gill KK, Kaddoumi A, Nazzal S. PEG–lipid micelles as drug carriers: physiochemical attributes, formulation principles and biological implication. J Drug Target 2015; 23(3): 222-31.
[http://dx.doi.org/10.3109/1061186X.2014.997735] [PMID: 25547369]

[207] Sharma AK, Garg T, Goyal AK, Rath G. Role of microemuslions in advanced drug delivery. Artif Cells Nanomed Biotechnol 2015; 1-9.
[http://dx.doi.org/10.3109/21691401.2015.1012261] [PMID: 25711493]

[208] Batrakova EV, Kabanov AV. Pluronic block copolymers: Evolution of drug delivery concept from inert nanocarriers to biological response modifiers. J Control Release 2008; 130(2): 98-106.
[http://dx.doi.org/10.1016/j.jconrel.2008.04.013] [PMID: 18534704]

[209] Gelperina S, Maksimenko O, Khalansky A, *et al.* Drug delivery to the brain using surfactant-coated poly(lactide-co-glycolide) nanoparticles: Influence of the formulation parameters. Eur J Pharm Biopharm 2010; 74(2): 157-63.
[http://dx.doi.org/10.1016/j.ejpb.2009.09.003] [PMID: 19755158]

[210] Alyautdin RN, Petrov VE, Langer K, Berthold A, Kharkevich DA, Kreuter J. Delivery of loperamide across the blood-brain barrier with polysorbate 80-coated polybutylcyanoacrylate nanoparticles. Pharm Res 1997; 14(3): 325-8.

[http://dx.doi.org/10.1023/A:1012098005098] [PMID: 9098875]

[211]   Wilson B, Samanta M, Santhi K, Kumar K, Paramakrishnan N, Suresh B. Targeted delivery of tacrine into the brain with polysorbate 80-coated poly(n-butylcyanoacrylate) nanoparticles. Eur J Pharm Biopharm 2008; 70(1): 75-84.
[http://dx.doi.org/10.1016/j.ejpb.2008.03.009] [PMID: 18472255]

[212]   Arya R, Jain S, Paliwal S, *et al.* BACE1 inhibitors: A promising therapeutic approach for the management of Alzheimer's disease. Asian Pacific Journal of Tropical Biomedicine. 2024; 14(9):p 369-381.
[http://dx.doi.org/10.4103/apjtb.apjtb_192_24]

[213]   Gao H, Qian J, Cao S, *et al.* Precise glioma targeting of and penetration by aptamer and peptide dual-functioned nanoparticles. Biomaterials 2012; 33(20): 5115-23.
[http://dx.doi.org/10.1016/j.biomaterials.2012.03.058] [PMID: 22484043]

[214]   Garg T, Singh O, Arora S, Murthy RSR. Scaffold: A novel carrier for cell and drug delivery. critical reviews™ in therapeutic drug carrier systems 2012, 29 (1), 1–63.
[http://dx.doi.org/10.1615/CritRevTherDrugCarrierSyst.v29.i1.10]

[215]   Kim JH, Yoon HJ, Sim J, Ju SY, Jang WD. The effects of dendrimer size and central metal ions on photosensitizing properties of dendrimer porphyrins. J Drug Target 2014; 22(7): 610-8.
[http://dx.doi.org/10.3109/1061186X.2014.928717] [PMID: 24955617]

[216]   Esfand R, Tomalia DA. Poly(amidoamine) (PAMAM) dendrimers: from biomimicry to drug delivery and biomedical applications. Drug Discov Today 2001; 6(8): 427-36.
[http://dx.doi.org/10.1016/S1359-6446(01)01757-3] [PMID: 11301287]

[217]   Kailasan A, Yuan Q, Yang H. Synthesis and characterization of thermoresponsive polyamidoamine–polyethylene glycol–poly(d,l-lactide) core–shell nanoparticles. Acta Biomater 2010; 6(3): 1131-9.
[http://dx.doi.org/10.1016/j.actbio.2009.08.036] [PMID: 19716444]

[218]   Sarkar K, Yang H. Encapsulation and extended release of anti-cancer anastrozole by stealth nanoparticles. Drug Deliv 2008; 15(5): 343-6.
[http://dx.doi.org/10.1080/10717540802035343] [PMID: 18763165]

[219]   Yang H, Lopina ST. Stealth dendrimers for antiarrhythmic quinidine delivery. J Mater Sci Mater Med 2007; 18(10): 2061-5.
[http://dx.doi.org/10.1007/s10856-007-3144-0] [PMID: 17558476]

[220]   Wiwattanapatapee R, Carreño-Gómez B, Malik N, Duncan R. Anionic PAMAM dendrimers rapidly cross adult rat intestine *in vitro*: a potential oral delivery system? Pharm Res 2000; 17(8): 991-8.
[http://dx.doi.org/10.1023/A:1007587523543] [PMID: 11028947]

[221]   Jevprasesphant R, Penny J, Attwood D, D'Emanuele A. Transport of dendrimer nanocarriers through epithelial cells *via* the transcellular route. J Control Release 2004; 97(2): 259-67.
[http://dx.doi.org/10.1016/j.jconrel.2004.03.022] [PMID: 15196753]

[222]   Zhang XQ, Intra J, Salem AK. Conjugation of polyamidoamine dendrimers on biodegradable microparticles for nonviral gene delivery. Bioconjug Chem 2007; 18(6): 2068-76.
[http://dx.doi.org/10.1021/bc070116l] [PMID: 17848077]

[223]   Garg T, Kumar A, Rath G, Goyal AK. Gastroretentive drug delivery systems for therapeutic management of peptic ulcer. Crit Rev Ther Drug Carrier Syst 2014; 31(6): 531-57.
[http://dx.doi.org/10.1615/CritRevTherDrugCarrierSyst.2014011104] [PMID: 25271775]

[224]   Garg T, Rath G, Goyal AK. Comprehensive review on additives of topical dosage forms for drug delivery. Drug Deliv 2015; 22(8): 969-87.
[http://dx.doi.org/10.3109/10717544.2013.879355] [PMID: 24456019]

[225]   Garg T, Rath G, Goyal AK. Ancient and advanced approaches for the treatment of an inflammatory autoimmune disease-psoriasis. Crit Rev Ther Drug Carrier Syst 2014; 31(4): 331-64.

[http://dx.doi.org/10.1615/CritRevTherDrugCarrierSyst.2014010122] [PMID: 25072198]

[226] Desai N, Trieu V, Yao Z, *et al.* Increased antitumor activity, intratumor paclitaxel concentrations, and endothelial cell transport of cremophor-free, albumin-bound paclitaxel, ABI-007, compared with cremophor-based paclitaxel. Clin Cancer Res 2006; 12(4): 1317-24.
[http://dx.doi.org/10.1158/1078-0432.CCR-05-1634] [PMID: 16489089]

[227] Foote M. Using nanotechnology to improve the characteristics of antineoplastic drugs: Improved characteristics of nab-paclitaxel compared with solvent-based paclitaxel. Biotechnol Annu Rev (Amst) 2007; 13: 345-57.
[http://dx.doi.org/10.1016/S1387-2656(07)13012-X] [PMID: 17875482]

[228] Hawkins MJ, Soon-Shiong P, Desai N. Protein nanoparticles as drug carriers in clinical medicine. Adv Drug Deliv Rev 2008; 60(8): 876-85.
[http://dx.doi.org/10.1016/j.addr.2007.08.044] [PMID: 18423779]

[229] Henderson IC, Bhatia V. Nab-paclitaxel for breast cancer: a new formulation with an improved safety profile and greater efficacy. Expert Rev Anticancer Ther 2007; 7(7): 919-43.
[http://dx.doi.org/10.1586/14737140.7.7.919] [PMID: 17627452]

[230] Kratz F. Albumin as a drug carrier: Design of prodrugs, drug conjugates and nanoparticles. J Control Release 2008; 132(3): 171-83.
[http://dx.doi.org/10.1016/j.jconrel.2008.05.010] [PMID: 18582981]

[231] Tomao S. Albumin-bound formulation of paclitaxel (Abraxane® ABI-007) in the treatment of breast cancer. Int J Nanomedicine 2009; 99: 99.
[http://dx.doi.org/10.2147/IJN.S3061]

[232] Paál K, Müller J, Hegedûs L. High affinity binding of paclitaxel to human serum albumin. Eur J Biochem 2001; 268(7): 2187-91.
[http://dx.doi.org/10.1046/j.1432-1327.2001.02107.x] [PMID: 11277943]

[233] Purcell M, Neault JF, Tajmir-Riahi HA. Interaction of taxol with human serum albumin. Biochim Biophys Acta Protein Struct Mol Enzymol 2000; 1478(1): 61-8.
[http://dx.doi.org/10.1016/S0167-4838(99)00251-4] [PMID: 10719175]

[234] Garg T, Rath G, Goyal AK. Biomaterials-based nanofiber scaffold: targeted and controlled carrier for cell and drug delivery. J Drug Target 2015; 23(3): 202-21.
[http://dx.doi.org/10.3109/1061186X.2014.992899] [PMID: 25539071]

[235] Desai N, Trieu V, Damascelli B, Soon-Shiong P. SPARC Expression Correlates with Tumor Response to Albumin-Bound Paclitaxel in Head and Neck Cancer Patients. Transl Oncol 2009; 2(2): 59-64.
[http://dx.doi.org/10.1593/tlo.09109] [PMID: 19412420]

[236] Miller K, Wang M, Gralow J, *et al.* Paclitaxel plus bevacizumab *versus* paclitaxel alone for metastatic breast cancer. N Engl J Med 2007; 357(26): 2666-76.
[http://dx.doi.org/10.1056/NEJMoa072113] [PMID: 18160686]

[237] Wilson B, Lavanya Y, Priyadarshini SRB, Ramasamy M, Jenita JL. Albumin nanoparticles for the delivery of gabapentin: Preparation, characterization and pharmacodynamic studies. Int J Pharm 2014; 473(1-2): 73-9.
[http://dx.doi.org/10.1016/j.ijpharm.2014.05.056] [PMID: 24999053]

[238] Yeini E, Ofek P, Albeck N, *et al.* Targeting glioblastoma: advances in drug delivery and novel therapeutic approaches. Adv Ther (Weinh) 2021; 4(1): 2000124.
[http://dx.doi.org/10.1002/adtp.202000124]

[239] Subhan MA, Yalamarty SSK, Filipczak N, Parveen F, Torchilin VP. Recent advances in tumor targeting *via* EPR effect for cancer treatment. J Pers Med 2021; 11(6): 571.
[http://dx.doi.org/10.3390/jpm11060571] [PMID: 34207137]

[240] Kibria G, Hatakeyama H, Ohga N, Hida K, Harashima H. The effect of liposomal size on the targeted delivery of doxorubicin to Integrin $\alpha v \beta 3$-expressing tumor endothelial cells. Biomaterials 2013;

34(22): 5617-27.
[http://dx.doi.org/10.1016/j.biomaterials.2013.03.094] [PMID: 23623323]

[241]   Golombek SK, May JN, Theek B, *et al.* Tumor targeting *via* EPR: Strategies to enhance patient responses. Adv Drug Deliv Rev 2018; 130: 17-38.
[http://dx.doi.org/10.1016/j.addr.2018.07.007] [PMID: 30009886]

[242]   Jia W, Wang Y, Liu R, Yu X, Gao H. Shape transformable strategies for drug delivery. Adv Funct Mater 2021; 31(18): 2009765.
[http://dx.doi.org/10.1002/adfm.202009765]

[243]   Qiu Y, Liu Y, Wang L, *et al.* Surface chemistry and aspect ratio mediated cellular uptake of Au nanorods. Biomaterials 2010; 31(30): 7606-19.
[http://dx.doi.org/10.1016/j.biomaterials.2010.06.051] [PMID: 20656344]

[244]   Vieira D, Gamarra L. Getting into the brain: liposome-based strategies for effective drug delivery across the blood–brain barrier. Int J Nanomedicine 2016; 11: 5381-414.
[http://dx.doi.org/10.2147/IJN.S117210] [PMID: 27799765]

[245]   Allen TM, Cullis PR. Drug delivery systems: entering the mainstream. Science 2004; 303(5665): 1818-22.
[http://dx.doi.org/10.1126/science.1095833] [PMID: 15031496]

[246]   Zhao M, van Straten D, Broekman MLD, Préat V, Schiffelers RM. Nanocarrier-based drug combination therapy for glioblastoma. Theranostics 2020; 10(3): 1355-72.
[http://dx.doi.org/10.7150/thno.38147] [PMID: 31938069]

[247]   Bobo D, Robinson KJ, Islam J, Thurecht KJ, Corrie SR. Nanoparticle-Based Medicines: A Review of FDA-Approved Materials and Clinical Trials to Date. Pharm Res 2016; 33(10): 2373-87.
[http://dx.doi.org/10.1007/s11095-016-1958-5] [PMID: 27299311]

[248]   Arvanitis CD, Ferraro GB, Jain RK. The blood–brain barrier and blood–tumour barrier in brain tumours and metastases. Nat Rev Cancer 2020; 20(1): 26-41.
[http://dx.doi.org/10.1038/s41568-019-0205-x] [PMID: 31601988]

[249]   Etame AB, Diaz RJ, O'Reilly MA, *et al.* Enhanced delivery of gold nanoparticles with therapeutic potential into the brain using MRI-guided focused ultrasound. Nanomedicine 2012; 8(7): 1133-42.
[http://dx.doi.org/10.1016/j.nano.2012.02.003] [PMID: 22349099]

[250]   Sawyer AJ, Saucier-Sawyer JK, Booth CJ, *et al.* Convection-enhanced delivery of camptothecin-loaded polymer nanoparticles for treatment of intracranial tumors. Drug Deliv Transl Res 2011; 1(1): 34-42.
[http://dx.doi.org/10.1007/s13346-010-0001-3] [PMID: 21691426]

[251]   Hynynen K, McDannold N, Vykhodtseva N, Jolesz FA, Noninvasive MR. Noninvasive MR imaging-guided focal opening of the blood-brain barrier in rabbits. Radiology 2001; 220(3): 640-6.
[http://dx.doi.org/10.1148/radiol.2202001804] [PMID: 11526261]

[252]   McDannold N, Arvanitis CD, Vykhodtseva N, Livingstone MS. Temporary disruption of the blood-brain barrier by use of ultrasound and microbubbles: safety and efficacy evaluation in rhesus macaques. Cancer Res 2012; 72(14): 3652-63.
[http://dx.doi.org/10.1158/0008-5472.CAN-12-0128] [PMID: 22552291]

[253]   Burgess A, Hynynen K. Noninvasive and targeted drug delivery to the brain using focused ultrasound. ACS Chem Neurosci 2013; 4(4): 519-26.
[http://dx.doi.org/10.1021/cn300191b] [PMID: 23379618]

[254]   Etame AB, Diaz RJ, Smith CA, Mainprize TG, Hynynen K, Rutka JT. Focused ultrasound disruption of the blood-brain barrier: a new frontier for therapeutic delivery in molecular neurooncology. Neurosurg Focus 2012; 32(1): E3.
[http://dx.doi.org/10.3171/2011.10.FOCUS11252] [PMID: 22208896]

[255]   O'Reilly MA, Hough O, Hynynen K. Blood-brain barrier closure time after controlled ultrasound-

induced opening is independent of opening volume. J Ultrasound Med 2017; 36(3): 475-83.
[http://dx.doi.org/10.7863/ultra.16.02005] [PMID: 28108988]

[256]   Mainprize T, Lipsman N, Huang Y, *et al.* Blood-brain barrier opening in primary brain tumors with non-invasive mr-guided focused ultrasound: a clinical safety and feasibility study. Sci Rep 2019; 9(1): 321.
[http://dx.doi.org/10.1038/s41598-018-36340-0] [PMID: 30674905]

[257]   Carpentier A, Canney M, Vignot A, *et al.* Clinical trial of blood-brain barrier disruption by pulsed ultrasound. Sci Transl Med 2016; 8(343): 343re2.
[http://dx.doi.org/10.1126/scitranslmed.aaf6086] [PMID: 27306666]

[258]   Idbaih A, Canney M, Belin L, *et al.* Safety and feasibility of repeated and transient blood–brain barrier disruption by pulsed ultrasound in patients with recurrent glioblastoma. Clin Cancer Res 2019; 25(13): 3793-801.
[http://dx.doi.org/10.1158/1078-0432.CCR-18-3643] [PMID: 30890548]

[259]   de Vries NA, Beijnen JH, Boogerd W, van Tellingen O. Blood–brain barrier and chemotherapeutic treatment of brain tumors. Expert Rev Neurother 2006; 6(8): 1199-209.
[http://dx.doi.org/10.1586/14737175.6.8.1199] [PMID: 16893347]

[260]   Lonser RR, Sarntinoranont M, Morrison PF, Oldfield EH. Convection-enhanced delivery to the central nervous system. J Neurosurg 2015; 122(3): 697-706.
[http://dx.doi.org/10.3171/2014.10.JNS14229] [PMID: 25397365]

[261]   Bidros DS, Vogelbaum MA. Novel drug delivery strategies in neuro-oncology. Neurotherapeutics 2009; 6(3): 539-46.
[http://dx.doi.org/10.1016/j.nurt.2009.04.004] [PMID: 19560743]

[262]   Jahangiri A, Chin AT, Flanigan PM, Chen R, Bankiewicz K, Aghi MK. Convection-enhanced delivery in glioblastoma: a review of preclinical and clinical studies. J Neurosurg 2017; 126(1): 191-200.
[http://dx.doi.org/10.3171/2016.1.JNS151591] [PMID: 27035164]

[263]   Mehta AM, Sonabend AM, Bruce JN. Convection-Enhanced Delivery. Neurotherapeutics 2017; 14(2): 358-71.
[http://dx.doi.org/10.1007/s13311-017-0520-4] [PMID: 28299724]

[264]   Souweidane MM, Kramer K, Pandit-Taskar N, *et al.* Convection-enhanced delivery for diffuse intrinsic pontine glioma: a single-centre, dose-escalation, phase 1 trial. Lancet Oncol 2018; 19(8): 1040-50.
[http://dx.doi.org/10.1016/S1470-2045(18)30322-X] [PMID: 29914796]

[265]   CTG Labs - NCBI. www.clinicaltrials.gov. Available from: https://www.clinicaltrials.gov

[266]   Kunwar S, Chang S, Westphal M, *et al.* Phase III randomized trial of CED of IL13-PE38QQR *vs* Gliadel wafers for recurrent glioblastoma. Neuro-oncol 2010; 12(8): 871-81.
[http://dx.doi.org/10.1093/neuonc/nop054] [PMID: 20511192]

[267]   Muldoon LL, Soussain C, Jahnke K, *et al.* Chemotherapy delivery issues in central nervous system malignancy: a reality check. J Clin Oncol 2007; 25(16): 2295-305.
[http://dx.doi.org/10.1200/JCO.2006.09.9861] [PMID: 17538176]

[268]   Pires A, Fortuna A, Alves G, Falcão A. Intranasal drug delivery: how, why and what for? J Pharm Pharm Sci 2009; 12(3): 288-311.
[http://dx.doi.org/10.18433/J3NC79] [PMID: 20067706]

[269]   Van Woensel M, Wauthoz N, Rosière R, *et al.* Formulations for Intranasal Delivery of Pharmacological Agents to Combat Brain Disease: A New Opportunity to Tackle GBM? Cancers (Basel) 2013; 5(3): 1020-48.
[http://dx.doi.org/10.3390/cancers5031020] [PMID: 24202332]

[270]   League-Pascual JC, Lester-McCully CM, Shandilya S, *et al.* Plasma and cerebrospinal fluid pharmacokinetics of select chemotherapeutic agents following intranasal delivery in a non-human

primate model. J Neurooncol 2017; 132(3): 401-7.
[http://dx.doi.org/10.1007/s11060-017-2388-x] [PMID: 28290002]

[271]   Fonseca CO, Linden R, Futuro D, Gattass CR, Quirico-Santos T. Ras pathway activation in gliomas: a
        strategic target for intranasal administration of perillyl alcohol. Arch Immunol Ther Exp (Warsz)
        2008; 56(4): 267-76.
        [http://dx.doi.org/10.1007/s00005-008-0027-0] [PMID: 18726148]

[272]   da Fonseca CO, Schwartsmann G, Fischer J, *et al.* Preliminary results from a phase I/II study of
        perillyl alcohol intranasal administration in adults with recurrent malignant gliomas. Surg Neurol
        2008; 70(3): 259-66.
        [http://dx.doi.org/10.1016/j.surneu.2007.07.040] [PMID: 18295834]

[273]   Ye D, Zhang X, Yue Y, *et al.* Focused ultrasound combined with microbubble-mediated intranasal
        delivery of gold nanoclusters to the brain. J Control Release 2018; 286: 145-53.
        [http://dx.doi.org/10.1016/j.jconrel.2018.07.020] [PMID: 30009893]

[274]   Chen H, Chen CC, Acosta C, Wu SY, Sun T, Konofagou EE. A new brain drug delivery strategy:
        focused ultrasound-enhanced intranasal drug delivery. PLoS One 2014; 9(10): e108880.
        [http://dx.doi.org/10.1371/journal.pone.0108880] [PMID: 25279463]

[275]   Joshi S, Ellis JA, Ornstein E, Bruce JN. Intraarterial drug delivery for glioblastoma mutiforme. J
        Neurooncol 2015; 124(3): 333-43.
        [http://dx.doi.org/10.1007/s11060-015-1846-6] [PMID: 26108656]

[276]   Warren KE. Novel therapeutic delivery approaches in development for pediatric gliomas. CNS Oncol
        2013; 2(5): 427-35.
        [http://dx.doi.org/10.2217/cns.13.37] [PMID: 24511389]

[277]   Basso U, Lonardi S, Brandes AA. Is intra-arterial chemotherapy useful in high-grade gliomas? Expert
        Rev Anticancer Ther 2002; 2(5): 507-19.
        [http://dx.doi.org/10.1586/14737140.2.5.507] [PMID: 12382519]

[278]   Stewart DJ, Grahovac Z, Benoit B, *et al.* Intracarotid chemotherapy with a combination of 1,3-bis(2-
        chloroethyl)-1-nitrosourea (BCNU), cis-diaminedichloroplatinum (cisplatin), and 4'-O-demethyl-
        1-O-(4,6-O-2-thenylidene-beta-D-glucopyranosyl) epipodophyllotoxin (VM-26) in the treatment of
        primary and metastatic brain tumors. Neurosurgery 1984; 15(6): 828-33.
        [http://dx.doi.org/10.1227/00006123-198412000-00010] [PMID: 6392925]

[279]   Rajappa P, Krass J, Riina HA, Boockvar JA, Greenfield JP. Super-selective basilar artery infusion of
        bevacizumab and cetuximab for multiply recurrent pediatric ependymoma. Interv Neuroradiol 2011;
        17(4): 459-65.
        [http://dx.doi.org/10.1177/159101991101700410] [PMID: 22192550]

[280]   Happold C, Roth P, Wick W, *et al.* ACNU-based chemotherapy for recurrent glioma in the
        temozolomide era. J Neurooncol 2009; 92(1): 45-8.
        [http://dx.doi.org/10.1007/s11060-008-9728-9] [PMID: 18987781]

[281]   Vega F, Davila L, Chatellier G, *et al.* Treatment of malignant gliomas with surgery, intraarterial
        chemotherapy with ACNU and radiation therapy. J Neurooncol 1992; 13(2): 131-5.
        [http://dx.doi.org/10.1007/BF00172762] [PMID: 1331343]

[282]   Newton HB, Ann Slivka M, Stevens CL, *et al.* Intra-arterial carboplatin and intravenous etoposide for
        the treatment of recurrent and progressive non-GBM gliomas. J Neurooncol 2002; 56(1): 79-86.
        [http://dx.doi.org/10.1023/A:1014498225405] [PMID: 11949830]

[283]   Burkhardt JK, Riina H, Shin BJ, *et al.* Intra-arterial delivery of bevacizumab after blood-brain barrier
        disruption for the treatment of recurrent glioblastoma: progression-free survival and overall survival.
        World Neurosurg 2012; 77(1): 130-4.
        [http://dx.doi.org/10.1016/j.wneu.2011.05.056] [PMID: 22405392]

[284]   Muldoon LL, Pagel MA, Netto JP, Neuwelt EA. Intra-arterial administration improves temozolomide

delivery and efficacy in a model of intracerebral metastasis, but has unexpected brain toxicity. J Neurooncol 2016; 126(3): 447-54.
[http://dx.doi.org/10.1007/s11060-015-2000-1] [PMID: 26694547]

[285] Cloughesy TF, Mochizuki AY, Orpilla JR, *et al.* Neoadjuvant anti-PD-1 immunotherapy promotes a survival benefit with intratumoral and systemic immune responses in recurrent glioblastoma. Nat Med 2019; 25(3): 477-86.
[http://dx.doi.org/10.1038/s41591-018-0337-7] [PMID: 30742122]

[286] Sharpe AH, Pauken KE. The diverse functions of the PD1 inhibitory pathway. Nat Rev Immunol 2018; 18(3): 153-67.
[http://dx.doi.org/10.1038/nri.2017.108] [PMID: 28990585]

[287] Medema JP. Cancer stem cells: The challenges ahead. Nat Cell Biol 2013; 15(4): 338-44.
[http://dx.doi.org/10.1038/ncb2717] [PMID: 23548926]

[288] Ho TCS, Chan AHY, Ganesan A. Thirty Years of HDAC Inhibitors: 2020 Insight and Hindsight. J Med Chem 2020; 63(21): 12460-84.
[http://dx.doi.org/10.1021/acs.jmedchem.0c00830] [PMID: 32608981]

[289] Verza FA, Das U, Fachin AL, Dimmock JR, Marins M. Roles of Histone Deacetylases and Inhibitors in Anticancer Therapy. Cancers (Basel) 2020; 12(6): 1664.
[http://dx.doi.org/10.3390/cancers12061664] [PMID: 32585896]

[290] Laengle J, Kabiljo J, Hunter L, *et al.* Histone deacetylase inhibitors valproic acid and vorinostat enhance trastuzumab-mediated antibody-dependent cell-mediated phagocytosis. J Immunother Cancer 2020; 8(1): e000195.
[http://dx.doi.org/10.1136/jitc-2019-000195] [PMID: 31940587]

[291] Kramm, P. D. C. International Cooperative Phase III Trial of the HIT-HGG Study Group for the Treatment of High Grade Glioma, Diffuse Intrinsic Pontine Glioma, and Gliomatosis Cerebri in Children and Adolescents & it; 18 Years.(HIT-HGG-2013); Clinical trial registration NCT03243461; clinicaltrials.gov, 2022. Available from: https://clinicaltrials.gov/study/NCT03243461

[292] National Cancer Institute (NCI). A Randomized Phase II/III Study of Vorinostat and Local Irradiation OR Temozolomide and Local Irradiation OR Bevacizumab and Local Irradiation Followed by Maintenance Bevacizumab and Temozolomide in Children With Newly Diagnosed High-Grade Gliomas; Clinical trial registration NCT01236560; clinicaltrials.gov, 2024. Available from: https://clinicaltrials.gov/study/NCT01236560

[293] Clinical Trials Register. Available from: https://www.clinicaltrialsregister.eu/ctr-search/trial/201--002766-39/NL#summary

[294] Singleton WGB, Bienemann AS, Woolley M, *et al.* The distribution, clearance, and brainstem toxicity of panobinostat administered by convection-enhanced delivery. J Neurosurg Pediatr 2018; 22(3): 288-96.
[http://dx.doi.org/10.3171/2018.2.PEDS17663] [PMID: 29856296]

[295] Zhang Y, Song J, Zhao Z, *et al.* Single-cell transcriptome analysis reveals tumor immune microenvironment heterogenicity and granulocytes enrichment in colorectal cancer liver metastases. Cancer Lett 2020; 470: 84-94.
[http://dx.doi.org/10.1016/j.canlet.2019.10.016] [PMID: 31610266]

[296] Ho DWH, Tsui YM, Sze KMF, *et al.* Single-cell transcriptomics reveals the landscape of intra-tumoral heterogeneity and stemness-related subpopulations in liver cancer. Cancer Lett 2019; 459: 176-85.
[http://dx.doi.org/10.1016/j.canlet.2019.06.002] [PMID: 31195060]

[297] Yuan X, Ma S, Fa B, *et al.* A high-efficiency differential expression method for cancer heterogeneity using large-scale single-cell RNA-sequencing data. Front Genet (2022) 13:1063130.
[http://dx.doi.org/10.3389/fgene.2022.1063130]

[298] Liu T, Liu C, Yan M, *et al.* Single cell profiling of primary and paired metastatic lymph node tumors

in breast cancer patients. Nat Commun (2022) 13(1):6823.
[http://dx.doi.org/10.1038/s41467-022-34581-2]

[299]   Deeks SG, Lewin SR, Havlir DV. The end of AIDS: HIV infection as a chronic disease. Lancet 2013; 382(9903): 1525-33.
[http://dx.doi.org/10.1016/S0140-6736(13)61809-7] [PMID: 24152939]

[300]   Atta MG, De Seigneux S, Lucas GM. Clinical Pharmacology in HIV Therapy. Clin J Am Soc Nephrol 2019; 14(3): 435-44.
[http://dx.doi.org/10.2215/CJN.02240218] [PMID: 29844056]

CHAPTER 2

LDL Receptors and their Impact in Targeted Therapies for Brain Tumors

Aarti Tiwari[1] and **Pradeep Kumar Samal**[1,*]

Department of Pharmacy, Guru Ghasidas Vishwavidyalaya (Central University), Bilaspur, Chhattisgarh, India

Abstract: Around twenty months is the normal median survival time for those who have been diagnosed with a brain tumor by medical professionals. Brain tumors are responsible for around 1.6% of all known cases of tumors, and they are responsible for 2.5% of the total death rates. Brain tumors present a number of problems that need to be recognized and overcome before they can be properly treated. There are a number of barriers that are present in this scenario. These barriers include the blood-brain tumor barrier (BBTB), the blood-brain barrier (BBB), the presence of efflux pumps, the diversity of tumor cells, antibiotic resistance, the tumor microenvironment (TME), and cancer stem cells (CSCs), which cause immune evasion, as well as the infiltration and invasion of tumor cells. Treatment of brain tumors with receptor-mediated drug delivery systems that make use of targeted nanoparticles (NPs) is one of the most advantageous approaches. This is due to the fact that there is a strong desire to make use of the potential that is offered by these systems. Particularly in the field of medical administration, the emphasis is placed on the utilization of research in order to target particular receptors. A damaged blood-brain barrier is associated with increased levels of expression of low-density lipoprotein receptors, which are commonly referred to as LDLR. These receptors are found in both healthy and diseased brains. The influence of LDLR-mediated therapy in the treatment of brain tumors was the key topic of discussion that we focused on in this chapter.

Keyword: Brain tumor, Low-density lipoprotein receptor, Low-density lipoproteins, Nanoparticles.

INTRODUCTION

Brain Tumor

The phrase "intracranial tumor" is frequently employed when referring to a tumor that involves the brain. The abnormal buildup of tissue that is known as a brain

---

* **Corresponding author Pradeep Kumar Samal:** Department of Pharmacy, Guru Ghasidas Vishwavidyalaya (Central University), Bilaspur, Chhattisgarh, India; Tel: 8839536598, E-mail: samalpharmacology@rediffmail.com

Prashant Tiwari, Pankaj Kumar Singh & Sunil Kumar Kadiri (Eds.)
All rights reserved-© 2025 Bentham Science Publishers

tumor is characterised by the fast and uncontrolled proliferation of cells, which appears to be untouched by the regulatory systems that are normally considered to regulate normal cells. When it comes to brain tumors, there are two primary groups that may be distinguished: major and metastatic illnesses. The fact that there have been documented cases of more than 150 different forms of brain tumors is something that should be taken into consideration. Primary tumors are the most common type of tissue that may be found in the brain. Cancers that start in brain tissue or other structures in the body can metastasise to other parts of the brain, a condition known as secondary brain tumors. Primary brain tumors are also referred to as intracranial tumors, which is an alternate word. Primary tumors may be divided into two categories: two types of brain cancers: glial tumors composed of glial cells, and non-glial tumors that develop on or inside brain systems, including blood vessels, glands and nerves. Glial tumors are the more common kind of primary disease. The most common type of neoplasms is known as glial tumors. In addition, primary tumors can be categorised as either potentially useful or potentially harmful, depending on these characteristics. Malignancies that begin in other regions of the body, such as the breast or the lungs, and then metastasise to the brain, generally through the bloodstream, are referred to as metastatic brain tumors. The phrase "metastatic brain tumors" refers to three distinct types of malignancies that have spread to the brain. Malignant tumors are commonly believed to be cancerous and to have metastasised to other organs. It is estimated that each year, around 150,000 people are impacted by metastatic tumors that have spread to the brain. This is responsible for around twenty-five percent of all cancer patients who are impacted by the illness. People who have been diagnosed with lung cancer have a forty percent chance of getting brain tumors that spread to other parts of the body. Current treatment options include surgery, radiotherapy, chemotherapy, and targeted therapy, while future approaches are expected to focus on blood-brain barrier disruption, genomic and immune-genomic strategies, gene therapy, advanced immunotherapy, and the integration of AI to overcome existing challenges [1]. Individuals who were diagnosed with these tumors had an extremely poor probability of surviving over the course of history, with the average survival rate being only a few weeks [1 - 4]. Surgery is often used to treat brain tumors; however, the invasive nature of the treatment and the imprecise borders of the tumor pose certain problems that must be overcome in order to achieve complete removal of the tumor. On top of that, the rate of recurrence following surgical intervention is more than ninety percent [5]. Additionally, following surgical excision of the tumor, the current treatment regimen for brain tumors includes a sequential injection of radiation therapy and chemotherapy. The alkylating drug Temozolomide (TMZ) is among the most frequently used chemotherapy treatments for brain tumors. The process of methyl group transfer to the purine bases of DNA is what it uses to carry out its duties, which ultimately results in the death of cells. The different type of brain tumors is given in Fig. (**1**).

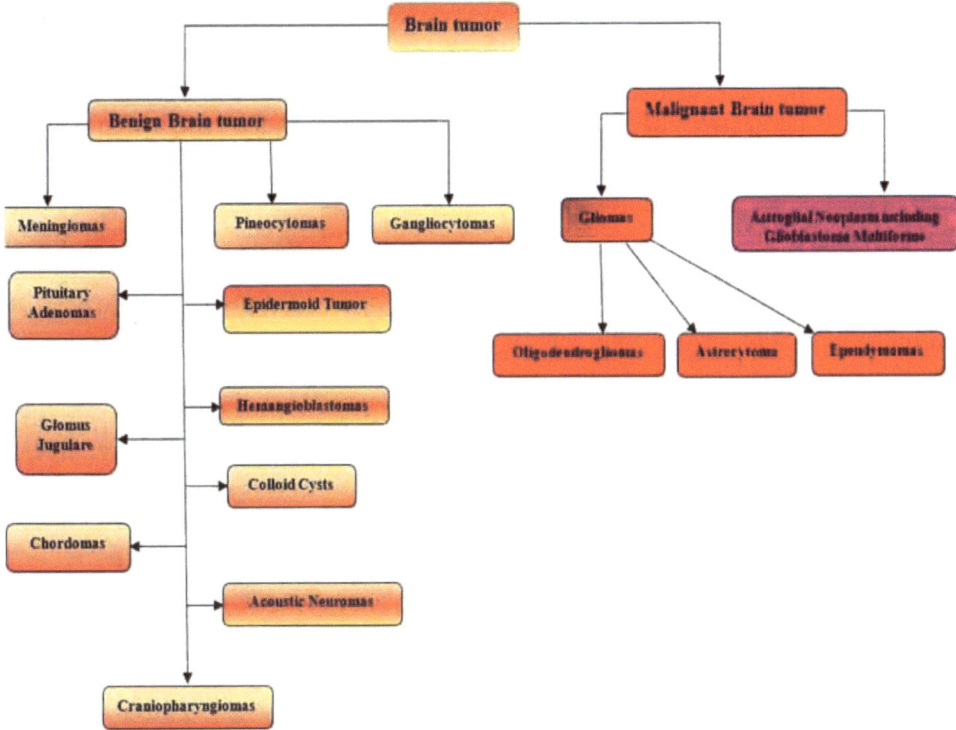

Fig. (1). Various types of brain tumors.

DIFFICULTIES IN THE DEVELOPMENT OF MEDICATIONS FOR THE BRAIN TUMOR

When compared to cancers located in other parts of the body, brain tumors pose a number of significant therapeutic obstacles. There are physiological obstacles that make it difficult for medications to enter the CNS and even more difficult for them to reach the tumor site. Two such barriers are the blood-brain tumor barrier (BBTB) and the blood-brain barrier (BBB), and over-expressed efflux pumps are also present. However, the inherent characteristics of brain tumors, such as their ability to invade and infiltrate surrounding tissues, their high degree of heterogeneity, their drug resistance, and their ability to elude the immune system due to the tumor microenvironment (TME) and cancer stem cells (CSC), further restrict the effectiveness of treatments. As a consequence, this ultimately leads to a high risk of treatment failure and recurrence of the targeted cancer. The survival rate for individuals with brain tumors who are having standard treatment is around twenty months, with survival rates of twenty-seven percent after two years and ten percent after five years, respectively [6].

Blood-Brain Tumor Barrier (BBTB) and Blood-Brain Barrier (BBB)

The blood-brain barrier (BBB) consists of neurones, pericytes, astrocytes, brain capillary endothelial cells (BCEC), and basement membranes. Facilitating the flow of both naturally occurring and artificially produced chemicals into the bloodstream and the brain is one of its crucial functions [7]. By using these systems, the BBB ensures that no macromolecular pharmaceuticals, and nearly all small-molecule drugs (98 percent), are able to penetrate the central nervous system. Paracellular barrier: The close proximity of BCEC cells prevents drugs from diffusing passively into the CNS. This restricts brain access to just those chemicals and tiny molecules that are hydrophilic or lipophilic. Second, the transcellular barrier: Compared to other brain cells, BCEC cells have decreased endocytosis activity, limiting the quantity of medicine that may cross the intercellular barrier. Enzyme barrier: The increased synthesis of enzymes by BCEC, including peptidase, phosphatase, nucleotidase, esterase, and cytochrome P450, increases the metabolic capacity of BBB cells and the rate at which drugs are degraded. Microglia, mastocytes, and macrophages are the components that make up the immunologic barrier, which is responsible for accelerating the clearance of drugs. The blood-brain barrier (BBB) causes an overexpression of several efflux proteins, including several ATP-binding cassette transporters (P-gp, BCRP, and MRPs) and solute carrier transporters. In addition to actively removing medicines, these proteins limit cellular permeability. It is also a crucial contributor to the development of drug resistance in brain tumors [8 - 10]. Angiogenesis causes the loss of BBB functions and integrity, which ultimately results in the development of BBTB when brain tumors are larger than $2mm^3$ [11]. One of the most well-established methods for the accumulation of nanoparticles is called enhanced permeability and retention (EPR), which involves passively targeting tumors. On the other hand, brain tumors contain vascular holes that are typically smaller (7-100 nm) and have a weaker EPR [12]. As a result of the EPR effect, medications continue to have difficulty reaching the sites of brain tumors. When treating brain tumors, BBTB is a major roadblock since it prevents medications from reaching the tumor's tissues [13, 14].

Stem Cells Originating from Brain Tumors

Subsets of brain tumor cells have characteristics similar to those of stem cells and exhibit markers such as CD133, A2B5, and EGFRvIII [15]. These characteristics occur in the stem cells. i. Aggressiveness: This includes resistance to chemoradiotherapy, invasive treatment, and migratory transmission. ii. Like normal stem cells or progenitor cells, cancer stem cells (CSCs) in some tumor tissues can divide and multiply into different types of cancer cells. iii. Multidrug resistance (MDR) is a phenomenon that occurs when cells repair DNA damage and excrete harmful substances, resulting in multidrug resistance [16 - 18]. Furthermore, clinical stem cells can promote angiogenesis, efflux transporter synthesis, and anti-apoptotic gene transcription. Nevertheless, due to their invasiveness, resistance, self-renewal, and differentiation, stem cells that penetrate the brain parenchyma will eventually lead to disease recurrence. Standard therapy is effective in killing the majority of tumor cells. Consequently, the elimination of tumor stem cells is essential for surmounting multidrug resistance and enhancing tumor treatment [19 - 23].

Microenvironment of the Brain Tumor

The tumor microenvironment consists of cells, stem cells, blood vessels, lymphatics, immune cells, fibroblasts, and extracellular matrix; it also supports tumor cell proliferation, division, angiogenesis, and metastasis. These are the main ways that the tumor microenvironment (TME) shields tumor cells from harm. Micro-vessel proliferation is inversely proportional to vascular endothelial growth factor activity. Tumor cells get nutrients from aberrant blood arteries by means of cytokines and growth factors. This leads to drug resistance and the proliferation of fibroblasts and macrophages. The cross-linking of the extracellular matrix—which comprises hyaluronic acid, proteoglycan, collagen, and stromal cell protein—leads to drug resistance. Additionally, this causes therapeutic resistance by preventing medications from reaching tumor cells. Extracellular matrix not only serves as a structural component, but it also facilitates the transportation of nutrients and oxygen, which in turn promotes the genesis and growth of tumors [24 - 26].

Invasion and Infiltration

The cells that make up a brain tumor behave in an aggressive manner against the tissues that surround them. During each of the following stages, a single brain tumor cell has the potential to infect normal tissues and develop into a tumor. To prevent endothelial cells from contacting the basement membrane, brain tumor cells go to the surrounding arteries and secrete glioma-derived proteins such as TGF-β2, reactive oxygen species, and proinflammatory peptides. In order to break down tight junctions, the factors activate matrix metalloproteinases (MMPs), which in turn downregulate claudin proteins. Overexpression of vascular endothelial growth factor (VEGF) causes endothelial cell migration and the formation of aberrant new blood vessels; these processes also harm the vascular basement membrane and the extracellular matrix. So, when blood vessels expand too quickly, they damage tight junctions, which allows the blood-brain barrier (BBB) to break down, allowing tumors with unclear boundaries and metastases to invade [27].

Immune Evasion

By preventing some chemicals from penetrating the brain, the blood-brain barrier (BBB) creates an environment free of danger and reduces the frequency of immune system attacks. Brain cells are particularly vulnerable to autoimmune attacks, which can have devastating effects. Quieting the immune system in the brain is common. Common lymphatic and dendritic cells, which normally present antigens, may not be present in the central nervous system. Because central nervous system immune surveillance is not regular, tumors cannot develop. There have been encouraging outcomes from immunotherapy when used for solid tumors such as melanoma and non-small cell lung cancer. Clinical trials including patients with brain tumors have not shown that the present treatment can increase survival rates. Simply put, the BBB prevents immune cells and antibodies from penetrating the central nervous system [28].

THE FUNCTION OF LDLRs IN THE MANAGEMENT OF BRAIN TUMORS

LDLR is among the several receptors that have been demonstrated to have an abnormally high level of expression in brain tumors. As a consequence of this, it has been effectively utilized as a target for the administration of some medications. It is possible that the utilization of some characteristics of LDLR might possibly increase its accumulation in the brain, which would result in an enhanced therapeutic response. For the purpose of constructing nano delivery systems that are successful, A thorough familiarity with the LDLR component's structure and content is essential [29].

LDLR Structure and Content

The transmembrane glycoprotein known as LDLR has 840 different amino acids. In contrast, the cytoplasmic side of the membrane contains just fifty amino acids per compartment. Apoprotein (apo) B binds LDL *via* its association with the cysteine-rich receptor binding domains of low-density lipoprotein receptors (LDLRs) [30, 31]. The process of endocytosis of LDL particles is the outcome of these interactions. There are seven different receptor subtypes that are part of the LDLR family. These include LDLRs, VLDLs, apoE receptor 2, MEGF7, LRP2, LRP1, LRP1B, LRP5, LRP6, and LRP11, which are distantly related. The acronyms LBD, YWTD, and epidermal growth factor precursor-like repeats, O-linked sugar, terminal membrane domain (TMD), and cytoplasmic are the six structural modules that make up the LDLR. Through the formation of a calcium ion crystal cage, the cysteine-rich complement-type repeats are responsible for stabilizing the LBD. In its cytoplasmic domain, ApoER2 possesses both a domain that is abundant in proline and a motif that is NPxY. The clathrin molecules that this domain interacts with are responsible for the formation of vesicles that endocytose receptors that are connected to ligands on certain NPs. The LDLR domain is filled with a significant number of serine and threonine residues [32 - 34].

Localisation and Biosynthesis of LDLR

While LDLR expression is typically modest in healthy neurones and brain tissues, it is highly expressed in glioma cells and cells lining the blood-brain barrier. Therefore, one important biomarker for glioblastoma is the overexpression of LDLR. Since the liver is in charge of a great deal of the body's metabolic functions, it is not surprising that the LDLR is also widely dispersed there. Rough endoplasmic reticulum (ER) membranes are found in all cells and are responsible for LDLR synthesis. The translation of LDLR mRNA is carried out by ribosomes that are connected to the rough endoplasmic reticulum. Next, the precursor integral membrane protein is transferred from the rough endoplasmic reticulum to the cell's Golgi apparatus after fusing together immature O-linked carbohydrate chains. In particular, the O-linked carbohydrate chains undergo expansion at this site. Following its release from the Golgi apparatus, the LDLR travels to the cell membrane *via* vesicles. Interaction between the cytosolic tail of the LDLR and the adaptin protein found in the clathrin-coated pits directs it to clathrin-rich regions of the cell membrane. The amino acid sequence [Asn-Pro-Val-Tyr] is located in this tail. Because it opens up channels for ions and extracellular chemicals, clathrin is a crucial component of the LDLR. This is something that the LDLR needs in order to endocytose ligands. It is in the form of vesicles that are enveloped with proteins which these substances make their way inside the cell [35 - 41].

The LDLR's Function

One mechanism that keeps cholesterol levels stable is the endocytosis of lipoproteins, which is carried out by the low-density lipoprotein receptor (LDLR). Beyond this, it is involved in extremely important processes like lipid metabolism, lipoprotein metabolism, cholesterol transport, and so on. In addition to its role in the binding and absorption of lipoproteins, such as apoB and E, this specific protein is responsible for a variety of other functions as well [42, 43]. The LDLR, which is in charge of removing circulating LDL and then distributing this information to the cells, is responsible for receiving this data. These LDLR actions are extremely important because they reduce the amount of cholesterol that is present in the circulation and make it possible for cells to employ cholesterol in the process of membrane synthesis. This is why it is essential that these activities take place. Mutations in the LDLR gene, which also lead to inappropriate cell wall production, are the source of a defective LDLR, which is the result of these mutations. In addition, these mutations induce the wrong cell wall formation. To accomplish its job, the LDLR collaborates with nearby enzymes and proteins [44 - 48]. To be more specific, adaptor proteins help ligand and signal trafficking by interacting with the cytosolic domains of LRP5, LRP6, and LDLR. The purpose of doing so is to make the situation more transparent. Cancer cells absorb a far larger amount of cholesterol than normal cells do because they can maintain their fast rate of expansion and replication. The reason behind this is that cancer cells have the remarkable ability to replicate themselves indefinitely. Conversely, LDLR overexpression provides an extra energy source, which enables the unchecked growth of gliomas and the quick production of lipid-dependent membranes. This is due to the fact that gliomas are tumors that are defined by genes. On the subject of cancer, the low-density lipoprotein receptor (LDLR) can do more than only induce tumors to form; it can also promote the growth and migration of cancer cells. Aside from that, it sets off other signalling pathways that play a role in cell proliferation, inflammation, and maturation. These pathways are necessary for the process. The process of cell proliferation is tied to these individual pathways [29, 49 - 56].

One of the downstream processes that the LDLR initiates is the Wnt/b-catenin signalling pathway. When LRP5, LRP6, and LRP8 are overexpressed, this pathway is activated [57]. A connection between the Wnt pathway and the process of cancer metastasis has been revealed. LRP5/6 is involved in this process. The communication channel that connects Wnt and β-catenin is also positively regulated by LRP8, which is another function of this protein [58 - 63]. Proteins linked to the Wnt pathway can affect cell proliferation and cell fate outcomes during oncogenesis by activating receptor-mediated signalling pathways. It is the formation of a cytoplasmic complex that inhibits receptor function that follows the binding of Wnt ligands to the cell-surface receptors frizzled (FZD) and LRP5/LRP6. Adenomatous polyposis coli (APC), axin 2 (AXIN2), and glycogen synthase kinase-3-b (GSK3b) team together to form this complex. Once Wnt ligands attach to FZD and LRP5, respectively, they initiate the Wnt/b-catenin signalling cascade, which in turn downregulates GSK3b [64 - 68]. Situated farther downstream, GSK3b controls the Wnt/b-catenin signal transduction pathway in addition to its serine/threonine protein kinase activity. Along with this, it helps regulate cell cycle progression, promote cell division, and start cell death by contributing to the creation of signal transduction cascades. The pathway recruits GSK3b to the plasma membrane, which in turn phosphorylates LRP6 and other pathway substrates. Among these substrates is B-catenin, an apparatus component that promotes cell-cell adhesion and which may represent a hitherto unrecognised link between inflammation and cancer. Changes in the Wnt pathway can cause the β–catenin–TCF transcription factor complex to activate oncogenes. This activation can occur in response to the changes. This may be a process that leads to cancer [69 - 73]. Additionally, VLDL is engaged in the Reelin pathway, which is responsible for providing extracellular signals to neurones that are migrating. This route is responsible for transmitting these signals. Among the big proteins that may be found in the extracellular matrix (ECM), reelin is one component. LRP8 and VLDL are two receptors that have a great affinity for the hormone Reelin. Throughout embryonic brain development, the Reelin pathway is responsible for controlling the neuronal layering of the forebrain. Additionally, macrophage surface expression of LRP1 is substantial, which is another interesting fact. The control of innate immunity and the immunological responses that follow is one of the numerous functions for which it is responsible. The synthesis of TNF-a and monocyte chemoattractant protein 1 (MCP1) is specifically under the control of LRP1, which is specifically responsible for this charge. In addition to this, it regulates the activation of the pathways that are associated with mitogen-activated protein kinases (MAPK) and c-Jun N-terminal kinase (JNK) [74].

LOW-DENSITY LIPOPROTEIN (LDL)

In order for the LDLR to bind, the LDL must first act as a ligand. A key component of adipose tissue, low-density lipoprotein (LDL) facilitates the synthesis of steroid hormones and plays a pivotal role in the transport of cholesterol throughout the body. This procedure, which results in a lipid protein, is responsible for transporting cholesterol esters all over the body. At an average wavelength of 3×106 Da, the mass of each spherical LDL molecule is about identical. At its centre, low-density lipoprotein (LDL) proteins contain around 1,500 cholesterol molecules bound to long-chain fatty acids. In LDL's central region, you may find these fatty acids. The monolayer of lipids that surrounds this arrangement is composed of eight hundred molecules of phospholipids and five hundred molecules of cholesterol that have not been esterified. Within the LDL molecule, there are a total of 4563 amino acid residues that function as constituents. By virtue of the fact that it comprises a core that is very hydrophobic and is surrounded by a shell that is hydrophilic, the molecule is categorized as an amphipathic one. The mechanism of receptor-mediated endocytosis allows LDLR to potentially take up LDL. The binding of LDL to LDLR and the buildup of LDL in coated pits activate this mechanism. Following this, clathrin-coated vesicles are utilized in order to transfer the ligand–receptor combination from the outside into the cells *via* transport. Endosomes get their contents during the process of vesicle loss of clathrin coating and transport to the endosomes [75 - 79]. Endosomes have a low pH (pH=5), which causes LDL to be released from LDLR. After this, LDL has completed its voyage, and it is returned to the plasma membrane by clathrin-coated pits that extend from the endosome's tubular area. The next step is for the endosomes to transfer the LDL to the lysosomes, where the built-in cholesteryl esters are hydrolysed to produce free cholesterol. After the LDL is eliminated, the free cholesterols are used to build new cell membranes. This process continues until the LDL is totally gone. This process happens rather often since every 10 minutes, one LDLR travels from the plasma membrane to the cell and back again. In order to improve their chances of survival, tumor cells are continuously changing and controlling the number of receptors positioned on the surface of their cell membrane. The goal is to increase their chances of surviving; thus, they do this. Tumor cells, in comparison to normal cells, tend to upregulate receptors linked to proliferation, adhesion, and nutrition and downregulate those linked to cell death and adherence. Both of these steps carry the risk of changing the cell's sensitivity to the host's regulatory systems. It is necessary to keep an eye on and control the number of distinct specialised receptors since certain of these receptors can induce cell cycle entry or cell death. For this reason, the quantity of these receptors is strictly regulated and closely watched [80]. Tumor cells overexpress certain receptors on their cell membranes so that ligands have a better chance of binding and inducing the desired cellular response. To encourage the tumor's development, this is done. A greater number of receptors on the surface of a cell makes it more sensitive to interactions with even a small number of ligands. Why? Because ligands have a better chance of activating receptors. Raising LDLR expression enhances the probability that the receptor will come into contact with LDL or other receptor-suitable ligands. This interaction might end either helping or hurting things [81 - 83]. The role of LDLR-targeted LDL-NPs in brain tumors is shown in Fig. (2).

USE OF TARGETED NPS IN LDLR FOR BRAIN TUMOR TREATMENT

After conducting an investigation, the Food and Drug Administration (FDA) came to the conclusion that TMZ is the most effective pharmacological therapy for the treatment of gliomas. Doxorubicin (DOX), paclitaxel (PTX), and daunorubicin are some of the active chemicals that are utilized in specific therapeutic approaches. According to the findings of a recent study, however, the effects of these medications are not nearly as powerful as was once supposed. Increasing the efficacy of these treatments might be accomplished through the implementation of a number of different methods. The first choice is the combination therapy, which is still being researched despite the fact that it needs further research, particularly in light of the toxicological profile that it carries. The encapsulating of pharmaceuticals in nanoparticles (NPs) in order to achieve targeted distribution is one of the various approaches that may be utilized to improve the efficacy of pharmacological treatments. Typically, the blood-brain barrier (BBB) expresses ligands that target protein receptors are integrated into nanoparticles (NPs), and then potential patients with glioma may benefit from a more successful course of therapy. Nanoparticles (NPs) have the potential to be more effective in treating gliomas when they are combined with a wide range of ligands that are often employed. In addition to LDL, lactoferrin, transferrin, and folate, these ligands are also present. LDLR-targeted nanoparticles are the focus of our attention in this respect because they ease the process of nanoparticles *via* the blood-brain barrier by means of receptor-mediated absorption [84]. Additionally, the utilization of ultrasmall nanoparticles makes it possible to achieve a more uniform administration and dispersion of the drug, which ultimately results in an increase in the treatment's effectiveness. The goal is to obtain better medication penetration across the permeable blood-brain barrier (BBB) and increase drug diffusion into intratumoral compartments. Injecting ultrasmall fluorescent core-shell silica nanoparticles functionalised with nontargeting (cRAD) peptides improved brain tumor transport and penetration. Additionally, there was a rise in accumulation, distribution, and retention of the nanoparticles [85, 86]. However, when compared to a control group that did not contain any nontargeting peptides, this was found to be significantly different. These findings add credibility to the hypothesis that ultrasmall nanoparticles, which are praised for their 'ultrasmall' size, have the potential to act as promising drug carriers for future therapeutic endeavors against gliomas. These investigations are expected to be carried out in the future. To treat the condition, LDL-targeted therapy is especially effective since LDLR is overexpressed in glioma cells, tumor-associated blood vessels, tumor tissue, and on the blood-brain barrier [87]. Several types of LDLR-targeted NPs used in brain tumors are listed in Table 1.

PROS AND CONS OF USING LDLR-TARGETING NANOPARTICLES

LDLR-targeting nanoparticles provide their recipients with a number of benefits. The high levels of cholesterol that are present in cancer cells are responsible for the increased rate of membrane production that is observed in these cells. As a consequence of this, they exhibit an inordinate amount of LDLR expression. When this is taken into consideration, the receptor becomes a target that is easily accessible. Nanoparticles designed to bind to LDLR can also bind to other receptors, such as folate receptors, when the binding of LDLR is disrupted by the alkylation of LDL lysine chains on the apoB-binding site. This is due to folate receptors not binding to LDLR [42, 94 - 97]. Due to the fact that they are biocompatible, nanoparticles that target LDLR and incorporate LDL ligands do not provoke immunological responses [98 - 100]. These molecules may be broken down in lysosomes, enabling the synthesis of fatty acids, cholesterol and amino acids. All of these compounds are recyclable. One of its distinguishing features is its inherent propensity to biodegrade. Nanoparticles that specifically bind to LDLR can effectively protect medicines by encapsulating them in a hydrophobic core, so preventing the degradation of the drugs. LDL NPs have a longer half-life compared to ordinary NPs since they are not degraded [101]. The nuclei of these particles are rather large, and they are capable of storing a significant quantity of medication, despite the fact that their size can range anywhere from 10 to 100 nanometers [102]. LDLR-targeting nanoparticles, on the other hand, restrict their use to the treatment of malignancies and pathological illnesses that include LDLRs. This is a significant limitation of the technology. Furthermore, the expression of these proteins in normal tissues is not consistent [103]. The existence of this receptor in normal tissues may generate apprehensions regarding the utilization of these nanoparticles because of the possibility of non-specific binding. This occurs because nanoparticles diminish the quantity of medication delivered to the targeted site, minimizing drug efficiency [92]. Altering the diameters of the particles and making use of PEG are two potential solutions to this problem. Nonetheless, there is still a probability that it will continue to have an impact on the situation.

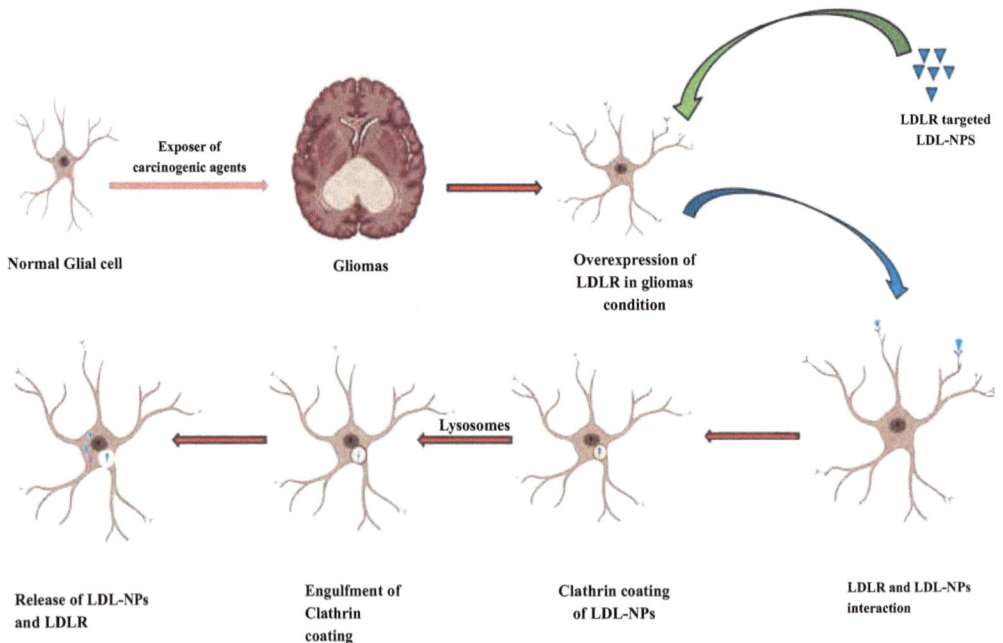

Fig. (2). Role of LDLR-Targeted LDL-NPs in brain tumor.

Table 1. Examples of LDLR-targeted LDLs in brain tumors.

S. No	Disease State/Study	Medication	Agent	Refs.
1	Human neuroblastoma	Donepezil	ApoE3-LDL	[88]
2	CNS tumor cell line *in vitro* study	Dox	LDL-receptor targeted liposomal drug (in combination with a statin)	[89]
3	Brain glioma/ *in vivo*	Paclitaxel	PNP-PTX	[90]
4	Glioblastoma multiforme *in vitro* study	Paclitaxel oleate	nLDL-PO	[91, 92]
5	Glioblastoma multiforme/*in vivo*	c-Met siRNA	c-Met siRNA-PEG/SLN	[93]

CONCLUSION

A brain tumor, regardless of its location, is considered one of the most challenging and possibly life-threatening forms of cancer. Brain tumors are notoriously difficult to treat and eliminate due to several factors, including the blood-brain barrier (BBB), immune evasion, tumor stem cells, tumor heterogeneity, and tumor invasiveness. The development of a multi-functional NDDS (Novel Drug Delivery System) for medication administration is crucial in order to effectively tackle the issues discussed in the previous paragraphs. LDL, also known as low-density lipoprotein, is distinguished by its tiny size, amphiphilic molecular surface, uptake by receptors, and prolonged lifespan. Combining therapeutic drugs with LDL might potentially target tumor cells that have an excessive expression of LDLR. Considering the presence of these activities, this is achievable. The utilization of nanoparticles to encapsulate low-density lipoprotein (LDL) would enhance the targeted administration of drugs and promote the wider application of LDL in cancer therapy. These nanoparticles serve as an effective transporter and ligand for targeted treatment. LDL-based nanoparticles are particularly advantageous for solid tumors and leukemia because of their reliance on LDLR-endocytosis, which allows them to effectively target cancer cells. LDL carriers possess the ability to specifically target certain groups of neurons and LDL-positive tumors that are present in the brain and can penetrate the blood-brain barrier. The distinguishing factor of these carriers is their capability. Given the substantial importance of potential applications in the field of biomedicine, it seemed that this issue had not been extensively explored. Neurological ailments are, in fact, the primary cause of disability on a global scale, with brain tumors being the most difficult conditions to treat. Nevertheless, both of these diseases can be efficiently treated by the LDLR targeting method commonly observed in LDL carriers. This sort of testing must be conducted promptly within the context of translational research.

AUTHORS' CONTRIBUTIONS

Aarti Tiwari contributed to the writing of the original draft, investigation, resource gathering, formal analysis, and data curation. Pradeep Kumar Samal was responsible for conceptualization, supervision, and editing of the manuscript.

REFERENCES

[1]   Liu Z, Tong L, Chen L, *et al.* Deep learning based brain tumor segmentation: a survey. Complex Intell Syst. 2023.
[http://dx.doi.org/10.1007/s40747-022-00815-5]

[2]   Wang P, Chung ACS. Relax and focus on brain tumor segmentation. Med Image Anal 2022; 75: 102259.
[http://dx.doi.org/10.1016/j.media.2021.102259] [PMID: 34800788]

[3]   Quail DF, Joyce JA. The microenvironmental landscape of brain tumors. Cancer Cell 2017; 31(3): 326-41.
[http://dx.doi.org/10.1016/j.ccell.2017.02.009] [PMID: 28292436]

[4]   Guzman G, Pellot K, Reed MR, Rodriguez A. CAR T-cells to treat brain tumors. Brain Res Bull 2023; 196: 76-98.
[http://dx.doi.org/10.1016/j.brainresbull.2023.02.014] [PMID: 36841424]

[5]   Kreatsoulas D, Damante M, Gruber M, Duru O, Elder JB. Supratotal surgical resection for low-grade glioma: a systematic review. Cancers (Basel) 2023; 15(9): 2493.
[http://dx.doi.org/10.3390/cancers15092493] [PMID: 37173957]

[6]   Ashby LS, Smith KA, Stea B. Gliadel wafer implantation combined with standard radiotherapy and concurrent followed by adjuvant temozolomide for treatment of newly diagnosed high-grade glioma: a systematic literature review. World J Surg Oncol 2016; 14(1): 225.
[http://dx.doi.org/10.1186/s12957-016-0975-5] [PMID: 27557526]

[7]   Zhao Y, Peng Y, Yang Z, *et al.* pH-redox responsive cascade-targeted liposomes to intelligently deliver doxorubicin prodrugs and lonidamine for glioma. Eur J Med Chem 2022; 235(4–6): 114281.
[http://dx.doi.org/10.1016/j.ejmech.2022.114281] [PMID: 35344903]

[8]   Azarmi M, Maleki H, Nikkam N, Malekinejad H. Transcellular brain drug delivery: A review on recent advancements. Int J Pharm 2020; 586: 119582.
[http://dx.doi.org/10.1016/j.ijpharm.2020.119582] [PMID: 32599130]

[9]   Pulgar VM. Transcytosis to cross the blood brain barrier, new advancements and challenges. Front Neurosci 2019; 12: 1019.
[http://dx.doi.org/10.3389/fnins.2018.01019] [PMID: 30686985]

[10]   Khalil A, Barras A, Boukherroub R, *et al.* Enhancing paracellular and transcellular permeability using nanotechnological approaches for the treatment of brain and retinal diseases. Nanoscale Horiz 2023; 9(1): 14-43.
[http://dx.doi.org/10.1039/D3NH00306J] [PMID: 37853828]

[11]   Mojarad-Jabali S, Farshbaf M, Walker PR, *et al.* An update on actively targeted liposomes in advanced drug delivery to glioma. Int J Pharm 2021; 602: 120645.
[http://dx.doi.org/10.1016/j.ijpharm.2021.120645] [PMID: 33915182]

[12]   Caro C, Avasthi A, Paez-Muñoz JM, Pernia Leal M, García-Martín ML. Passive targeting of high-grade gliomas *via* the EPR effect: a closed path for metallic nanoparticles? Biomater Sci 2021; 9(23): 7984-95.
[http://dx.doi.org/10.1039/D1BM01398J] [PMID: 34710207]

[13]   Saidijam M, Karimi Dermani F, Sohrabi S, Patching SG. Efflux proteins at the blood–brain barrier:

review and bioinformatics analysis. Xenobiotica 2018; 48(5): 506-32.
[http://dx.doi.org/10.1080/00498254.2017.1328148] [PMID: 28481715]

[14]   Correction to. "Efflux proteins at the blood–brain barrier: review and bioinformatics analysis" (Xenobiotica, 10.1080/00498254.2017.1328148). Xenobiotica 2021.

[15]   Smiley SB, Yun Y, Ayyagari P, *et al.* Development of CD133 targeting multi-drug polymer micellar nanoparticles for glioblastoma - *In vitro* evaluation in glioblastoma stem cells. Pharm Res 2021; 38(6): 1067-79.
[http://dx.doi.org/10.1007/s11095-021-03050-8] [PMID: 34100216]

[16]   Alcantara Llaguno S, Parada LF. Cancer stem cells in gliomas: evolving concepts and therapeutic implications. Curr Opin Neurol 2021; 34(6): 868-74.
[http://dx.doi.org/10.1097/WCO.0000000000000994] [PMID: 34581301]

[17]   M. A. Multipotent mesenchymal stromal cells (MSC) - Evolving role in cancer and leukemia therapy. Clin Lymphoma Myeloma Leuk 2011.

[18]   Andreeff M, Battula VL, Wang R, *et al.* 900 multipotent Mesenchymal Stromal Cells (MSC) - Evolving role in cancer and leukemia therapy. Clin Lymphoma Myeloma Leuk 2011; 11: S134-5.
[http://dx.doi.org/10.1016/j.clml.2011.05.033]

[19]   He Z, Zhang S. Tumor-associated macrophages and their functional transformation in the hypoxic tumor microenvironment. Front Immunol 2021; 12: 741305.
[http://dx.doi.org/10.3389/fimmu.2021.741305] [PMID: 34603327]

[20]   Petrova V, Annicchiarico-Petruzzelli M, Melino G, Amelio I. The hypoxic tumour microenvironment. Oncogenesis 2018; 7(1): 10.
[http://dx.doi.org/10.1038/s41389-017-0011-9] [PMID: 29362402]

[21]   Meng W, Hao Y, He C, Li L, Zhu G. Exosome-orchestrated hypoxic tumor microenvironment. Mol Cancer 2019; 18(1): 57.
[http://dx.doi.org/10.1186/s12943-019-0982-6] [PMID: 30925935]

[22]   Kim I, Choi S, Yoo S, Lee M, Kim IS. Cancer-associated fibroblasts in the hypoxic tumor microenvironment. Cancers (Basel) 2022; 14(14): 3321.
[http://dx.doi.org/10.3390/cancers14143321] [PMID: 35884382]

[23]   Tao J, Yang G, Zhou W, *et al.* Targeting hypoxic tumor microenvironment in pancreatic cancer. J Hematol Oncol 2021; 14(1): 14.
[http://dx.doi.org/10.1186/s13045-020-01030-w] [PMID: 33436044]

[24]   Wei X, Chen Y, Jiang X, *et al.* Mechanisms of vasculogenic mimicry in hypoxic tumor microenvironments. Mol Cancer 2021; 20(1): 7.
[http://dx.doi.org/10.1186/s12943-020-01288-1] [PMID: 33397409]

[25]   Li J, Yue Z, Tang M, *et al.* Strategies to reverse hypoxic tumor microenvironment for enhanced sonodynamic therapy. Adv Healthc Mater 2024; 13(1): 2302028.
[http://dx.doi.org/10.1002/adhm.202302028] [PMID: 37672732]

[26]   Butturini E, Carcereri de Prati A, Boriero D, Mariotto S. Tumor dormancy and interplay with hypoxic tumor microenvironment. Int J Mol Sci 2019; 20(17): 4305.
[http://dx.doi.org/10.3390/ijms20174305] [PMID: 31484342]

[27]   Zhou Y, Tan Z, Chen K, *et al.* Overexpression of SHCBP1 promotes migration and invasion in gliomas by activating the NF-κB signaling pathway. Mol Carcinog 2018; 57(9): 1181-90.
[http://dx.doi.org/10.1002/mc.22834]

[28]   He C, Ding H, Chen J, *et al.* Immunogenic cell death induced by chemoradiotherapy of novel pH-sensitive cargo-loaded polymersomes in glioblastoma. Int J Nanomedicine 2021; 16: 7123-35.
[http://dx.doi.org/10.2147/IJN.S333197] [PMID: 34712045]

[29]   Shi Y, Andhey PS, Ising C, *et al.* Overexpressing low-density lipoprotein receptor reduces tau-

associated neurodegeneration in relation to apoE-linked mechanisms. Neuron 2021; 109(15): 2413-2426.e7.
[http://dx.doi.org/10.1016/j.neuron.2021.05.034] [PMID: 34157306]

[30] Goldstein JL, Brown MS. Regulation of low-density lipoprotein receptors: implications for pathogenesis and therapy of hypercholesterolemia and atherosclerosis. Circulation 1987; 76(3): 504-7.
[http://dx.doi.org/10.1161/01.CIR.76.3.504] [PMID: 3621516]

[31] Malone JM. Regulation of low-density lipoprotein receptors: implications for pathogenesis and therapy of hypercholesterolemia and atherosclerosis. J Vasc Surg 1989; 9(1): 185.
[http://dx.doi.org/10.1016/0741-5214(89)90248-6]

[32] Lillis AP, Van Duyn LB, Murphy-Ullrich JE, Strickland DK. LDL receptor-related protein 1: unique tissue-specific functions revealed by selective gene knockout studies. Physiol Rev 2008; 88(3): 887-918.
[http://dx.doi.org/10.1152/physrev.00033.2007] [PMID: 18626063]

[33] Campion O, Thevenard-Devy J, Etique N, *et al.* Abstract PO020: The matricellular receptor LRP-1 acts as a pro-tumorigenic receptor by supporting tumor angiogenesis in a triple negative breast cancer model. Cancer Res 2021; 81(5_Supplement): PO020.
[http://dx.doi.org/10.1158/1538-7445.TME21-PO020]

[34] Lillis AP, Van Duyn LB, Murphy-ullrich JE, Dudley K, Strickland DK. The low density lipoprotein receptor-related protein 1: Unique tissue-specific functions revealed by selective gene knockout studies. NIH-PA Author Manuser 2009.

[35] Gravotta D, Perez Bay A, Jonker CTH, *et al.* Clathrin and clathrin adaptor AP-1 control apical trafficking of megalin in the biosynthetic and recycling routes. Mol Biol Cell 2019; 30(14): 1716-28.
[http://dx.doi.org/10.1091/mbc.E18-12-0811] [PMID: 31091172]

[36] Bucher D, Frey F, Sochacki KA, *et al.* Clathrin-adaptor ratio and membrane tension regulate the flat-to-curved transition of the clathrin coat during endocytosis. Nat Commun 2018; 9(1): 1109.
[http://dx.doi.org/10.1038/s41467-018-03533-0] [PMID: 29549258]

[37] López-Hernández T, Takenaka K, Mori Y, *et al.* Clathrin-independent endocytic retrieval of SV proteins mediated by the clathrin adaptor AP-2 at mammalian central synapses. eLife 2022; 11: e71198.
[http://dx.doi.org/10.7554/eLife.71198] [PMID: 35014951]

[38] Hirst J, Robinson MS. Clathrin and adaptors. Biochim Biophys Acta Mol Cell Res 1998; 1404(1-2): 173-93.
[http://dx.doi.org/10.1016/S0167-4889(98)00056-1] [PMID: 9714795]

[39] Lee D, Zhao X, Zhang F, Eisenberg E, Greene LE. Depletion of GAK/auxilin 2 inhibits receptor-mediated endocytosis and recruitment of both clathrin and clathrin adaptors. J Cell Sci 2005; 118(18): 4311-21.
[http://dx.doi.org/10.1242/jcs.02548] [PMID: 16155256]

[40] Ramírez-Santiago G, Robles-Valero J, Morlino G, *et al.* Clathrin regulates lymphocyte migration by driving actin accumulation at the cellular leading edge. Eur J Immunol 2016; 46(10): 2376-87.
[http://dx.doi.org/10.1002/eji.201646291] [PMID: 27405273]

[41] Yim YI, Scarselletta S, Zang F, *et al.* Exchange of clathrin, AP2 and epsin on clathrin-coated pits in permeabilized tissue culture cells. J Cell Sci 2005; 118(11): 2405-13.
[http://dx.doi.org/10.1242/jcs.02356] [PMID: 15923653]

[42] Go GW, Mani A. Low-density lipoprotein receptor (LDLR) family orchestrates cholesterol homeostasis. Yale J Biol Med 2012; 85(1): 19-28.
[PMID: 22461740]

[43] Banerjee P, Chan KC, Tarabocchia M, *et al.* Functional analysis of LDLR (low-density lipoprotein receptor) variants in patient lymphocytes to assess the effect of evinacumab in homozygous familial

hypercholesterolemia patients with a spectrum of LDLR activity. Arterioscler Thromb Vasc Biol 2019; 39(11): 2248-60.
[http://dx.doi.org/10.1161/ATVBAHA.119.313051] [PMID: 31578082]

[44]   Yang HX, Zhang M, Long SY, *et al.* Cholesterol in LDL receptor recycling and degradation. Clinica Chimica Acta. 2020.

[45]   Alcicek FC, Mohaissen T, Bulat K, *et al.* Sex-specific differences of adenosine triphosphate levels in red blood cells isolated from ApoE/LDLR double-deficient mice. Front Physiol 2022; 13: 839323.
[http://dx.doi.org/10.3389/fphys.2022.839323] [PMID: 35250640]

[46]   Aldana-Hernández P, Leonard KA, Zhao YY, Curtis JM, Field CJ, Jacobs RL. Dietary choline or trimethylamine N-oxide supplementation does not influence atherosclerosis development in Ldlr−/− and Apoe−/− male mice. J Nutr 2020; 150(2): 249-55.
[http://dx.doi.org/10.1093/jn/nxz214] [PMID: 31529091]

[47]   Abifadel M, Boileau C. Genetic and molecular architecture of familial hypercholesterolemia. J Intern Med 2023; 293(2): 144-65.
[http://dx.doi.org/10.1111/joim.13577] [PMID: 36196022]

[48]   Li M, Ma L, Chen Y, *et al.* Large-scale CRISPR screen of LDLR pathogenic variants. Research 2023.
[http://dx.doi.org/10.34133/research.0203]

[49]   Kim J, Castellano JM, Jiang H, *et al.* Overexpression of low-density lipoprotein receptor in the brain markedly inhibits amyloid deposition and increases extracellular A β clearance. Neuron 2009; 64(5): 632-44.
[http://dx.doi.org/10.1016/j.neuron.2009.11.013] [PMID: 20005821]

[50]   Castellano JM, Deane R, Gottesdiener AJ, *et al.* Low-density lipoprotein receptor overexpression enhances the rate of brain-to-blood Aβ clearance in a mouse model of β-amyloidosis. Proc Natl Acad Sci USA 2012; 109(38): 15502-7.
[http://dx.doi.org/10.1073/pnas.1206446109] [PMID: 22927427]

[51]   Wang C, Li A, Spellman R, *et al.* Effects of human LDLR overexpression on apoE-related tau pathology and brain dysfunction. J Immunol. 2020.

[52]   Song X, Zhang W, Yu N, Zhong X. PAQR3 facilitates the ferroptosis of diffuse large B-cell lymphoma *via* the regulation of LDLR-mediated PI3K/AKT pathway. Hematol Oncol 2024; 42(1): e3219.
[http://dx.doi.org/10.1002/hon.3219] [PMID: 37690092]

[53]   Mendiola AS, Tognatta R, Yan Z, Akassoglou K. ApoE and immunity in Alzheimer's disease and related tauopathies: Low-density lipoprotein receptor to the rescue. Neuron 2021; 109(15): 2363-5.
[http://dx.doi.org/10.1016/j.neuron.2021.07.013] [PMID: 34352209]

[54]   Maxwell KN, Fisher EA, Breslow JL. Overexpression of PCSK9 accelerates the degradation of the LDLR in a post-endoplasmic reticulum compartment. Proc Natl Acad Sci USA 2005; 102(6): 2069-74.
[http://dx.doi.org/10.1073/pnas.0409736102] [PMID: 15677715]

[55]   Sun Y, Feng Y, Zhang G, Xu Y. The endonuclease APE1 processes miR-92b formation, thereby regulating expression of the tumor suppressor LDLR in cervical cancer cells. Ther Adv Med Oncol 2019; 11: 1758835919855859.
[http://dx.doi.org/10.1177/1758835919855859] [PMID: 31320936]

[56]   Susan-Resiga D, Girard E, Essalmani R, *et al.* Asialoglycoprotein receptor 1 is a novel PCSK9-independent ligand of liver LDLR cleaved by furin. J Biol Chem 2021; 297(4): 101177.
[http://dx.doi.org/10.1016/j.jbc.2021.101177] [PMID: 34508778]

[57]   Roslan Z, Muhamad M, Selvaratnam L, Ab-Rahim S. The roles of low-density lipoprotein receptor-related proteins 5, 6, and 8 in cancer: a review. J Oncol 2019; 2019: 1-6.
[http://dx.doi.org/10.1155/2019/4536302] [PMID: 31031810]

[58]   Clevers H. Wnt/β-catenin signaling in development and disease. Cell 2006; 127(3): 469-80.

[http://dx.doi.org/10.1016/j.cell.2006.10.018] [PMID: 17081971]

[59] Kormish JD, Sinner D, Zorn AM. Interactions between SOX factors and Wnt/β-catenin signaling in development and disease. Dev Dyn 2010; 239(1): 56-68.
[http://dx.doi.org/10.1002/dvdy.22046] [PMID: 19655378]

[60] Perugorria MJ, Olaizola P, Labiano I, *et al.* Wnt–β-catenin signalling in liver development, health and disease. Nat Rev Gastroenterol Hepatol 2019; 16(2): 121-36.
[http://dx.doi.org/10.1038/s41575-018-0075-9] [PMID: 30451972]

[61] Liu F, Millar SE. Wnt/β-catenin signaling in oral tissue development and disease. J Dent Res 2010; 89(4): 318-30.
[http://dx.doi.org/10.1177/0022034510363373] [PMID: 20200414]

[62] Liu J, Xiao Q, Xiao J, *et al.* Wnt/β-catenin signalling: function, biological mechanisms, and therapeutic opportunities. Signal Transduct Target Ther 2022; 7(1): 3.
[http://dx.doi.org/10.1038/s41392-021-00762-6] [PMID: 34980884]

[63] Clevers H. Wnt/beta-catenin signaling in development and disease. Cell 2006; 127(3): 469-80.
[http://dx.doi.org/10.1016/j.cell.2006.10.018] [PMID: 17081971]

[64] Long HZ, Cheng Y, Zhou ZW, Luo HY, Wen DD, Gao LC. PI3K/AKT signal pathway: A target of natural products in the prevention and treatment of Alzheimer's disease and Parkinson's disease. Front Pharmacol 2021; 12: 648636.
[http://dx.doi.org/10.3389/fphar.2021.648636] [PMID: 33935751]

[65] McCubrey JA, Steelman LS, Bertrand FE, *et al.* GSK-3 as potential target for therapeutic intervention in cancer. Oncotarget 2014; 5(10): 2881-911.
[http://dx.doi.org/10.18632/oncotarget.2037] [PMID: 24931005]

[66] Jacobs KM, Bhave SR, Ferraro DJ, Jaboin JJ, Hallahan DE, Thotala D. GSK-3β: A bifunctional role in cell death pathways. Int J Cell Biol 2012; 2012: 1-11.
[http://dx.doi.org/10.1155/2012/930710] [PMID: 22675363]

[67] Li W, Wu M, Zhang Y, *et al.* Intermittent fasting promotes adult hippocampal neuronal differentiation by activating GSK-3β in 3xTg-AD mice. J Neurochem 2020; 155(6): 697-713.
[http://dx.doi.org/10.1111/jnc.15105]

[68] De Simone A, Tumiatti V, Andrisano V, Milelli A. Glycogen synthase kinase 3β: A new gold rush in anti-alzheimer's disease multitarget drug discovery? J Med Chem 2021; 64(1): 26-41.
[http://dx.doi.org/10.1021/acs.jmedchem.0c00931] [PMID: 33346659]

[69] Gao J, Huo Z, Song X, *et al.* EGFR mediates epithelial-mesenchymal transition through the Akt/GSK-3β/Snail signaling pathway to promote liver cancer proliferation and migration. Oncol Lett 2024.

[70] Chen JS, Huang JQ, Luo B, *et al.* PIK 3 CD induces cell growth and invasion by activating AKT / GSK - 3β/β-catenin signaling in colorectal cancer. Cancer Sci 2019; 110(3): 997-1011.
[http://dx.doi.org/10.1111/cas.13931]

[71] Carotenuto M, De Antonellis P, Liguori L, *et al.* H-Prune through GSK-3β interaction sustains canonical WNT/β-catenin signaling enhancing cancer progression in NSCLC. Oncotarget 2014; 5(14): 5736-49.
[http://dx.doi.org/10.18632/oncotarget.2169] [PMID: 25026278]

[72] Nagini S, Sophia J, Mishra R. Glycogen synthase kinases: Moonlighting proteins with theranostic potential in cancer. Semin Cancer Biol 2019; 56: 25-36.
[http://dx.doi.org/10.1016/j.semcancer.2017.12.010] [PMID: 29309927]

[73] Duda P, Akula SM, Abrams SL, *et al.* Targeting GSK3 and associated signaling pathways involved in cancer. Cells 2020; 9(5): 1110.
[http://dx.doi.org/10.3390/cells9051110] [PMID: 32365809]

[74] Gorovoy M, Gaultier A, Campana WM, Firestein GS, Gonias SL. Inflammatory mediators promote

production of shed LRP1/CD91, which regulates cell signaling and cytokine expression by macrophages. J Leukoc Biol 2010; 88(4): 769-78.
[http://dx.doi.org/10.1189/jlb.0410220] [PMID: 20610799]

[75] Wang S, Meng Y, Li C, Qian M, Huang R. Receptor-mediated drug delivery systems targeting to glioma. Nanomaterials (Basel) 2015; 6(1): 3.
[http://dx.doi.org/10.3390/nano6010003] [PMID: 28344260]

[76] Teixeira MI, Lopes CM, Amaral MH, Costa PC. Surface-modified lipid nanocarriers for crossing the blood-brain barrier (BBB): A current overview of active targeting in brain diseases. Colloids Surf B Biointerfaces 2023; 221: 112999.
[http://dx.doi.org/10.1016/j.colsurfb.2022.112999] [PMID: 36368148]

[77] Huang CW, Chuang CP, Chen YJ, *et al.* Integrin $\alpha_2\beta_1$-targeting ferritin nanocarrier traverses the blood–brain barrier for effective glioma chemotherapy. J Nanobiotechnology 2021; 19(1): 180.
[http://dx.doi.org/10.1186/s12951-021-00925-1] [PMID: 34120610]

[78] Choudhury H, Pandey M, Chin PX, *et al.* Transferrin receptors-targeting nanocarriers for efficient targeted delivery and transcytosis of drugs into the brain tumors: a review of recent advancements and emerging trends. Drug Deliv Transl Res 2018; 8(5): 1545-63.
[http://dx.doi.org/10.1007/s13346-018-0552-2] [PMID: 29916012]

[79] Zheng Z, Zhang J, Jiang J, *et al.* Remodeling tumor immune microenvironment (TIME) for glioma therapy using multi-targeting liposomal codelivery. J Immunother Cancer 2020; 8(2): e000207.
[http://dx.doi.org/10.1136/jitc-2019-000207] [PMID: 32817393]

[80] Correction: An LXR agonist promotes glioblastoma cell death through inhibition of an EGFR/AKT/SREBP-1/LDLR-dependent pathway. Cancer Discov 2012.

[81] A comparitive study to assess the level of distress among patients with cancer undergoing chemotherapy and radiation therapy in selected oncology centers Mangaluru. Indian J Public Heal Res Dev. 2020.

[82] Patil V, Kulkarni A, Rathod CV, Sanjivani CR. Comparitive study between 2% and 4% lignocaine nebulisation on pressor response to laryngoscopy and intubation. Indian J Public Heal Res Dev 2015.
[http://dx.doi.org/10.5958/0976-5506.2015.00077.7]

[83] Chung SH, Chae Y, Yu HS, Cho KS, Ham WS, Bang WJ, *et al.* Comparitive study of biochemical effect of salvage radiotherapy and salvage androgen deprivation therapy after radical prostatectomy. J Urol 1643; 2011.

[84] Lam FC, Morton SW, Wyckoff J, *et al.* Enhanced efficacy of combined temozolomide and bromodomain inhibitor therapy for gliomas using targeted nanoparticles. Nat Commun 2018; 9(1): 1991.
[http://dx.doi.org/10.1038/s41467-018-04315-4] [PMID: 29777137]

[85] Zarschler K, Rocks L, Licciardello N, *et al.* Ultrasmall inorganic nanoparticles: State-of-the-art and perspectives for biomedical applications. Nanomedicine 2016; 12(6): 1663-701.
[http://dx.doi.org/10.1016/j.nano.2016.02.019] [PMID: 27013135]

[86] Singh N, Shi S, Goel S. Ultrasmall silica nanoparticles in translational biomedical research: Overview and outlook. Adv Drug Deliv Rev 2023; 192: 114638.
[http://dx.doi.org/10.1016/j.addr.2022.114638] [PMID: 36462644]

[87] Juthani R, Madajewski B, Yoo B, *et al.* Ultrasmall core-shell silica nanoparticles for precision drug delivery in a high-grade malignant brain tumor model. Clin Cancer Res 2020; 26(1): 147-58.
[http://dx.doi.org/10.1158/1078-0432.CCR-19-1834] [PMID: 31515460]

[88] Krishna KV, Saha RN, Dubey SK. Biophysical, biochemical, and behavioral implications of ApoE3 conjugated donepezil nanomedicine in a $A\beta_{1-42}$ induced Alzheimer's disease Rat Model. ACS Chem Neurosci 2020; 11(24): 4139-51.
[http://dx.doi.org/10.1021/acschemneuro.0c00430] [PMID: 33251785]

[89]   Pinzón-Daza ML, Garzón R, Couraud PO, *et al.* The association of statins plus LDL receptor-targeted liposome-encapsulated doxorubicin increases *in vitro* drug delivery across blood–brain barrier cells. Br J Pharmacol 2012; 167(7): 1431-47.
[http://dx.doi.org/10.1111/j.1476-5381.2012.02103.x] [PMID: 22788770]

[90]   Zhang B, Sun X, Mei H, *et al.* LDLR-mediated peptide-22-conjugated nanoparticles for dual-targeting therapy of brain glioma. Biomaterials 2013; 34(36): 9171-82.
[http://dx.doi.org/10.1016/j.biomaterials.2013.08.039] [PMID: 24008043]

[91]   Nikanjam M, Blakely EA, Bjornstad KA, Shu X, Budinger TF, Forte TM. Synthetic nano-low density lipoprotein as targeted drug delivery vehicle for glioblastoma multiforme. Int J Pharm 2007; 328(1): 86-94.
[http://dx.doi.org/10.1016/j.ijpharm.2006.07.046] [PMID: 16959446]

[92]   Nikanjam M, Gibbs AR, Hunt CA, Budinger TF, Forte TM. Synthetic nano-LDL with paclitaxel oleate as a targeted drug delivery vehicle for glioblastoma multiforme. J Control Release 2007; 124(3): 163-71.
[http://dx.doi.org/10.1016/j.jconrel.2007.09.007] [PMID: 17964677]

[93]   Jin J, Bae KH, Yang H, *et al. In vivo* specific delivery of c-Met siRNA to glioblastoma using cationic solid lipid nanoparticles. Bioconjug Chem 2011; 22(12): 2568-72.
[http://dx.doi.org/10.1021/bc200406n] [PMID: 22070554]

[94]   Floeth M, Elges S, Gerss J, *et al.* Low-density lipoprotein receptor (LDLR) is an independent adverse prognostic factor in acute myeloid leukaemia. Br J Haematol 2021; 192(3): 494-503.
[http://dx.doi.org/10.1111/bjh.16853] [PMID: 32511755]

[95]   Moradi A, Maleki M, Ghaemmaghami Z, *et al.* Mutational spectrum of *LDLR* and *PCSK9* genes identified in iranian patients with premature coronary artery disease and familial hypercholesterolemia. Front Genet 2021; 12: 625959.
[http://dx.doi.org/10.3389/fgene.2021.625959] [PMID: 33732287]

[96]   Huang L, Li H, Ye Z, *et al.* Berbamine inhibits Japanese encephalitis virus (JEV) infection by compromising TPRMLs-mediated endolysosomal trafficking of low-density lipoprotein receptor (LDLR). Emerg Microbes Infect 2021; 10(1): 1257-71.
[http://dx.doi.org/10.1080/22221751.2021.1941276] [PMID: 34102949]

[97]   Yu L, Wang C, Zhang D, *et al.* Exosomal circ_0008285 in follicle fluid regulates the lipid metabolism through the miR-4644/ LDLR axis in polycystic ovary syndrome. J Ovarian Res 2023; 16(1): 113.
[http://dx.doi.org/10.1186/s13048-023-01199-x] [PMID: 37322492]

[98]   Glickson JD, Lund-Katz S, Zhou R, *et al.* Lipoprotein nanoplatform for targeted delivery of diagnostic and therapeutic agents. Mol Imaging 2008; 7(2): 7290.2008.0012.
[http://dx.doi.org/10.2310/7290.2008.0012] [PMID: 18706292]

[99]   Glickson JD, Lund-Katz S, Zhou R, *et al.* Lipoprotein nanoplatform for targeted delivery of diagnostic and therapeutic agents. In: Advances in Experimental Medicine and Biology. 2009.
[http://dx.doi.org/10.1007/978-0-387-85998-9_35]

[100]   Zheng G, Chen J, Li H, Glickson JD. Rerouting lipoprotein nanoparticles to selected alternate receptors for the targeted delivery of cancer diagnostic and therapeutic agents. Proc Natl Acad Sci USA 2005; 102(49): 17757-62.
[http://dx.doi.org/10.1073/pnas.0508677102] [PMID: 16306263]

[101]   Huntosova V, Buzova D, Petrovajova D, *et al.* Development of a new LDL-based transport system for hydrophobic/amphiphilic drug delivery to cancer cells. Int J Pharm 2012; 436(1-2): 463-71.
[http://dx.doi.org/10.1016/j.ijpharm.2012.07.005] [PMID: 22814227]

[102]   Harisa GI, Alanazi FK. Low density lipoprotein bionanoparticles: From cholesterol transport to delivery of anti-cancer drugs. Saudi Pharm J 2014; 22(6): 504-15.
[http://dx.doi.org/10.1016/j.jsps.2013.12.015] [PMID: 25561862]

[103]   Ostrom QT, Cote DJ, Ascha M, Kruchko C, Barnholtz-Sloan JS. Adult glioma incidence and survival by race or ethnicity in the United States from 2000 to 2014. JAMA Oncol 2018; 4(9): 1254-62.
[http://dx.doi.org/10.1001/jamaoncol.2018.1789] [PMID: 29931168]

CHAPTER 3

Present Status and Prospects of Drug Delivery Approaches: Managing the Blood-Brain Barrier Treatment in Brain Tumors

Sakshi Soni[1], **Vandana Soni**[1] and **Sushil K. Kashaw**[1,*]

[1] Department of Pharmaceutical Sciences, Dr. Harisingh Gour Central University, Sagar, Madhya Pradesh-470003, India

Abstract: Developing effective treatments for CNS disorders remains a formidable challenge due to the existence of multiple physiological barriers, primarily the blood-brain barrier (BBB), which severely restricts medication invasion into the brain and consequently compromises therapeutic efficacy. Effective brain-targeted drug delivery, especially to diseased cells, requires overcoming these barriers to develop promising therapies for brain disorders. Current research focuses on diverse nanocarrier structures and surface-engineered, site-specific novel transporters to improve effectiveness and minimize the untoward effects of brain therapy. These methods aim to bypass the BBB or enhance its permeability, thereby increasing the absorption of medication in the brain. However, the effectiveness of innovative transporter systems is influenced by physiological factors such as Efflux-mediated excretion, Brain protein coating, Persistence, Cytotoxicity of the nanocarriers, and patient-specific factors. Thus, understanding the composition of the brain, the BBB, and related features is crucial for developing effective carrier systems. Additionally, alternative routes like direct nasal-to-brain drug transfer proposal promise revenue to contact the brain without the BBB barrier. This chapter discusses the characteristics of several biological barriers, as well as the BBB and BCSFB (blood-cerebrospinal fluid barrier), in drug treatment and the mechanisms of drug transport that cross the BBB. It additionally explores innovative approaches for brain-targeted drug delivery, as well as dendrimers, nanogels, inorganic nanoparticles, liposomes, polymeric nanoparticles, nanoemulsions, quantum dots, lipidic nanoparticles, and intranasal drug delivery. Features disturbing the drug-targeting efficacy of these innovative transporter systems are also illustrated.

Keywords: Brain, Blood-brain barrier, Drug targeting, Lipid vesicle, Nanoparticle, Pharmacotherapy, Trans nasal medication administration.

---

* **Corresponding author Sushil K. Kashaw:** Department of Pharmaceutical Sciences, Dr. Harisingh Gour Central University, Sagar, Madhya Pradesh-470003, India; E-mail: sushilkashaw@gmail.com

Prashant Tiwari, Pankaj Kumar Singh & Sunil Kumar Kadiri (Eds.)
All rights reserved-© 2025 Bentham Science Publishers

INTRODUCTION

The treatment of brain tumors is among the most formidable challenges in contemporary medical science [1, 2]. This is mainly due to the existence of a tightly regulated gateway, the blood-brain barrier, which segregates the circulating blood from the brain's extracellular environment. While the BBB is crucial for regulating the brain's internal milieu and protecting it from toxins and pathogens, it also significantly hinders the delivery of therapeutic agents [3, 4]. This dual role of the BBB as both a protector and an obstacle underscores the complexity of developing effective treatments for brain tumors, where the ability to deliver drugs directly to the tumor site within the brain is often severely limited. The BBB is composed of tightly joined endothelial cells, pericytes, astrocyte end-feet, and a basement membrane, all of which work in concert to regulate the passage of substances between the bloodstream and the brain [5, 6]. Tight junctions between endothelial cells restrict the passage of large and hydrophilic molecules, while efflux transporters, such as the P-glycoprotein pump, expel many drugs that manage to enter the endothelial cells [7, 8]. This highly regulated environment ensures the brain's protection, but also means that many conventional therapeutic agents cannot reach effective concentrations within the brain tissue when administered systemically. The current arsenal against brain tumors includes surgery, radiation, and medications that target cancer cells (chemotherapy) [9]. Surgical resection is often the first line of treatment; however, its success is highly dependent on the tumor's location and accessibility. Many brain tumors are situated in regions that are difficult to access without risking significant damage to vital brain structures [10, 11]. Radiation therapy is another common treatment that aims to destroy tumor cells with targeted radiation. However, it often cannot completely eradicate the tumor and can cause damage to surrounding healthy brain tissue, leading to long-term cognitive deficits. Chemotherapy, while a mainstay in cancer treatment, faces substantial barriers in treating brain tumors due to the BBB [12, 13]. Most chemotherapeutic agents are unable to cross the BBB in sufficient concentrations to be effective, and those that do often result in systemic toxicity. Given these limitations, innovative drug delivery therapeutic strategies are being developed and optimized to increase the distribution of healing agents beyond the BBB. These advanced approaches can be broadly classified into aggressive and non-aggressive approaches, each offering distinct advantages and facing unique challenges [14, 15]. Fig. (**1**) shows the medication delivery techniques to the CNS.

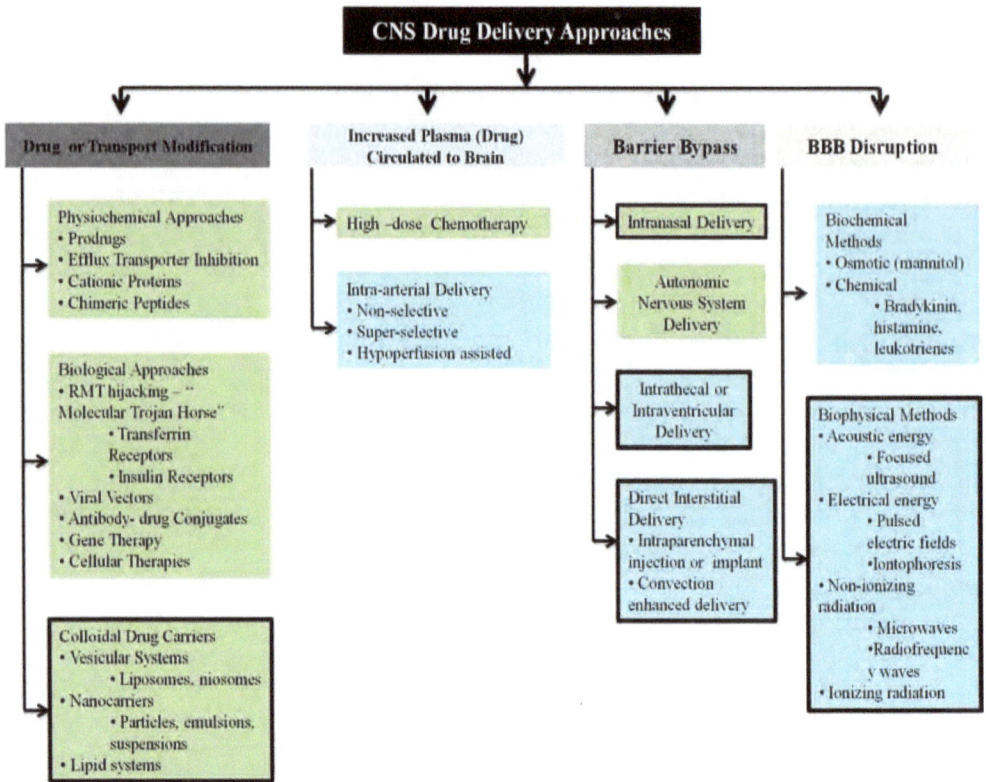

Fig. (1). A comparison of methods for distributing cargoes to the central nervous system (CNS), highlighting non-invasive (green boxes) and invasive (blue boxes) techniques.

One invasive approach is intracerebral implantation, where drug-releasing devices are placed directly into the brain. This method enables localized, high-concentration drug delivery, thereby minimizing systemic exposure and associated side effects. An example of this is the Gliadel® wafer, which releases the chemotherapeutic agent carmustine directly into the tumor resection cavity over time [16, 17]. Another promising invasive technique is CED, or convection-enhanced delivery, which precisely delivers medications within the brain by infusing them through a catheter driven by a pressure gradient. CED allows for the distribution of therapeutic agents over a larger volume and deeper into the brain tissue than would be possible with simple diffusion. Focused ultrasound (FUS) combined with microbubbles is another innovative approach that has shown promise in preclinical studies [16, 18]. This technique temporarily disrupts the BBB at targeted locations, allowing for the enhanced transport of medicines to brain tumors. The application of FUS can be precisely controlled, thereby minimizing damage to healthy brain tissue and improving drug penetration at the tumor site. Non-invasive methods of drug delivery also hold significant promise.

Nanoparticle-based delivery systems are at the forefront of this research. Nanocarriers can be customized to encapsulate drugs, preventing their breakdown and facilitating their passage across the BBB [19]. Several categories of nanoparticles, including vesicular carriers, synthetic carriers, and polymeric nanoparticles/carriers, are being explored for their potential to deliver cargoes primarily to brain tumors. These nanoparticles can be equipped with targeting moieties to enhance their targeting of tumor cells, thereby improving therapeutic efficacy and mitigating unwanted effects. Receptor-mediated transcytosis represents another non-invasive strategy that leverages the brain's natural transport mechanisms [20, 21]. Therapeutic agents are modified with ligands that have affinity for specific endothelial receptors, such as transferrin or insulin receptors. These ligand-receptor complexes undergo endocytosis and are transported across the BBB, enabling the drug to target the brain parenchyma. Cell-penetrating peptides (CPPs) offer yet another approach. These short peptides can facilitate the transport and delivery of multiple therapeutic molecules to cellular membranes, including the BBB. By conjugating drugs to CPPs, researchers aim to increase the brain bioavailability of these drugs, thereby overcoming one of the most significant barriers to effective brain tumor treatment [22, 23].

The current landscape of drug delivery methods for managing the BBB in brain tumor therapy is one of dynamic research and innovation. While significant progress has been made, many of these methods are still in various stages of clinical trials and preclinical research. The prospects for these advanced drug delivery systems are promising, offering the potential for more effective and targeted treatments for brain tumors. Ongoing investigation and advancement in this area are essential to overcoming the barriers posed by the BBB and improving outcomes for patients with brain tumors.

CNS BARRIER TO DRUG TARGETING

The CNS is shielded from the blood by three distinct protective layers: the arachnoid barrier, the BCSFB, and BBB. These hindrances act as both physical and functional separators, keeping the fluid surrounding the brain (extracellular fluid) distinct from the blood. All three barriers are essential for ensuring the proper functioning of the CNS internal equilibrium through controlled movement of materials in and out of the bloodstream. This control is achieved through physical mechanisms, including the limitation of movement between cells *via* tight junctions, the facilitation of molecular passage by specific transport proteins, and the breakdown of materials as they cross these barriers [24 - 26]. Fig. (2) shows the drug delivery transport pathways to the brain across the BBB.

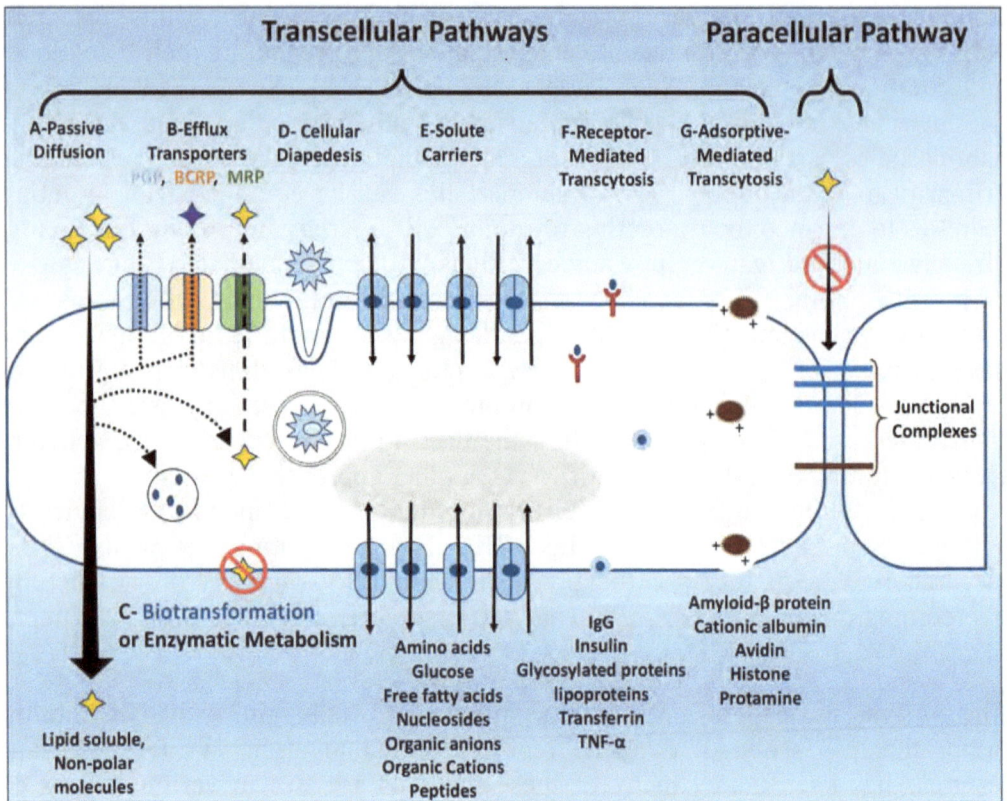

Fig. (2). Pathways for Substances Crossing BBB; Passive Diffusion (**A**); Efflux Pumps (**B**); Metabolism (**C**); Leukocyte Passage (**D**); Carrier-Mediated Transport (**E**); Transcytosis (**F, G**).

Arachnoid and BCSFD

The central nervous system has two guardians protecting it from harmful substances in the bloodstream: the arachnoid barrier and the blood-cerebrospinal fluid barrier (BCSFB). The arachnoid barrier, a sheet of tightly linked cells, separates the cerebrospinal fluid (CSF) from the dura mater; however, its lack of blood vessels and small surface area limit its role in overall exchange [24, 27]. The BCSFB, located within the brain's ventricles in structures called choroid plexuses, plays a more critical role. These choroid plexuses function like filters, featuring a dense network of capillaries surrounded by tightly sealed epithelial cells. This design allows essential nutrients to pass from blood to CSF while restricting harmful substances [26 - 28].

BCF Interface: Structure and Its Role in Brain Protection

The BBB acts as a highly discriminatory checkpoint,modulating what enters your brain. It is a complex structure lining the dense network of capillaries in your brain. This meshwork is formed by specialized endothelial cells, a basal membrane, pericytes, and finger-like astrocyte processes. Due to its dense design, the BBB strictly controls what substances can pass through, making it a foremost hurdle for distributing medications to the brain [24, 25].

Endothelial cells (BECs) of BBB

The roles of the BBB are first and foremost ascribed to the exclusive traits of brain endothelial cells (BECs), which are produced appreciably by the endothelium in further capillary networks and firmly adjust access routes to the central nervous system. Physiologically developed mammalian BECs require blood vessels and have a restricted ability to take in fluids, notably limiting the intracellular passage of molecules [25, 26]. Natural exchange of substances between blood and to brain barrier is typically restricted to molecules that dissolve easily in fats and are relatively small (around 400-500 Daltons) with few attractions to water (less than six hydrogen bonds) [29, 30]. The paracellular passage of water-loving molecules is impeded by the junctional complexes created amongst neighboring endothelial cells. These complexes comprise interlocking transmembrane integral proteins, including claudins, occludin, and JAM (junctional adhesion molecules), that form tight connections among the cells lining the blood vessels in the brain, cytoplasmic proteins (like zonula occludens), and components of the cytoskeleton (including actin). Tight junctions (TJs) connect the plasma membranes of near brain endothelial cells (BECs), involving claudins 1, 3, 5, and 12, as well as occludin plasma proteins, which contribute to the BBB's high transendothelial electrical resistance (TEER), representing the integrity of the barrier [24, 26, 29]. Zonula occludens proteins link claudins and occludin, acting as tethers that connect the tight junctions (TJs) to the actin cytoskeleton, providing structural support. Cell adhesion molecules (CAMs), belonging to the immunoglobulin superfamily, enhance brain endothelial cell (BEC) polarity, aid in the placement of zonula occludens proteins, reinforce TJs by interacting with cell architecture proteins, and facilitate leukocyte cell extravasation through transversely BBB. Adherens junctions (AJs) located at the basolateral membrane of BECs are composed of transmembrane cadherins, which are vital for cell-cell adhesion and the formation of TJs. These cadherins are secured by catenins, which attach them to cellular architecture proteins and modulate intracellular signal transduction [31]. The main adherens junction (AJ) protein in brain endothelial cells (BECs) is VE-cadherin. In addition, BECs possess a diverse array of active drug transport proteins, categorized within the ATP-binding cassette (ABC) and solute carrier (SLC) families, which are responsible for transporting organic anions, cations, and peptides [24, 25, 32]. Integral membrane proteins function as transporters, utilizing ATP hydrolysis to relocate cargos across cell membranes, and are critical for the circulation and removal of drugs from the brain. Xenobiotic transporters in the ATP-binding cassette (ABC) gene cluster, like P-glycoprotein (involved in general drug resistance), multidrug resistance protein (broad spectrum drug efflux), and protein associated with resistance in breast cancer, are necessary to the purpose of BBB by actively eliminate various endogenous and xenobiotic lipophilic compounds from brain endothelial cells (BECs) [24, 32, 33]. Moreover, BBBs have a variety of enzymes that metabolize neurotransmitters and inactivate drugs, such as HNMT (histamine N-methyltransferase), COMT (catechol O-methyl transferase), and cytochrome P450. These enzymes alter or inactivate drugs, therefore blocking their ability to enter the brain [34].

BBB-astrocytes

Astrocytes' terminal endfeet are intimately related to the luminal-facing surface of brain endothelial cells (BECs). These astrocytic end feet are enriched with molecules that regulate ionic concentrations at the BBB, such as Kir4.1 potassium channels and aquaporin-4 [24, 35]. Generation and stabilization of a BBB-like phenotype in BECs depend on the intimate direct interaction between astrocytes and BECs [36]. Studies have revealed that fenestrated capillary-rich tissues, which normally permit the passage of direct Blue 53 dye (BD) under normal conditions, become impermeable to BD and form a tightly regulated endothelial phenotype after being injected with enriched astrocyte fractions [37]. The stimulation of BBB characteristics in vascular cells can be commenced by soluble mediators released by astrocytes. For example, bovine vascular cells showed higher transendothelial electrical resistance (TEER) when cultured in a media environment supplemented with astrocytes. The secretion of factors such as vascular endothelial growth factor (VEGF), fibroblast growth factor (bFGF), glial cell line-derived neurotrophic factor (GDNF), and angiopoietin-1 from astrocytes is important for the formation of constricted connections in brain endothelial cells (BECs). These factors are also essential for BBB restore and revitalization in various central nervous system diseases [24, 25, 38].

BBB-pericytes

Perivascular cells are essential components of microvessels and post-capillary venules within the brain's vascular architecture [39]. Like brain endothelial cells (BECs), pericytes are enveloped by a continuous basal lamina, which serves as a structural barrier separating them from astrocytic end-feet. Pericytes are the closest cellular partners to BECs, forming direct physical connections through adherens junctions [40].

Pericytes play a pivotal role in maintaining the integrity of the blood-brain barrier (BBB). During development, they suppress the expression of genes in BECs that promote the formation of pinocytic vesicles—fluid-filled sacs that compromise barrier selectivity. In addition to contributing to the synthesis of basement membrane components, pericytes regulate proteins critical for the formation and maintenance of tight junctions (TJs).

Experimental studies have demonstrated that co-culturing pericytes with BECs enhances BBB function by increasing transendothelial electrical resistance (TEER) and boosting the activity of efflux transporters [41]. Furthermore, intercellular signaling between astrocytes and pericytes mediated by platelet-derived growth factor (PDGF) plays a significant role in preserving tight junction integrity and overall BBB stability [42]. **Basement membrane (BM) of BBB**.

The basement membrane (BM) is an exclusive, cell-free layer separating brain endothelial cells (BECs) from both the pericytes within the blood vessels and those residing in the surrounding brain tissue. This extremely prearranged, 50-100 nanometer-thick protein layer is composed of perlecan (a heparan sulfate proteoglycan), along with laminin, collagen IV, and nidogen. These components are produced pervasively by BECs, pericytes, and astrocytes. BM plays several vital roles, including supporting the BBB structurally, promoting cell adhesion, participating in signal transduction, and controlling cell relocation, survival, and differentiation. The basement membrane (BM) plays a critical role in maintaining BBB function. Mutations that inactivate genes responsible for BM components can significantly increase BBB permeability, leading to bleeding within the brain (intracerebral hemorrhage) and the development of atypical blood vessels and brain tissue (vascular and brain parenchymal malformations). However, our understanding of the specific methods by which the BM contributes to BBB maintenance and repair remains limited [43, 44].

MOLECULAR TRAFFICKING ACROSS THE BLOOD-BRAIN BARRIER

Molecules can cross the blood-brain barrier (BBB) through two main alleyways: the paracellular pathway involves molecules passing between the brain endothelial cells (BECs). In the transcellular pathway (transcytosis), substances cross the BEC itself. They first enter from the luminal side (bloodstream side) of the BEC plasma membrane, travel through the BEC cell cytoplasm, and then exit on the basolateral side (brain tissue side) into the brain's interstitial fluid. Under regular physiological conditions, the tight junctions of BECs obstruct paracellular molecular translocation through BBB. However, Transcytosis can be driven by energetic (active) or non-energetic (passive) mechanisms. Passive diffusion across BECs is considerably influenced by a molecule's lipid solubility, electrical charge, and size of the molecule. Essential polar molecules, such as scattered amino acids and glucose, which cannot diffuse across the BBB, are transported *via* carrier-mediated mechanisms. This protein-mediated transportation can be unidirectional, bi-directional (based on the concentration gradient), or engage substrate exchange. Large molecules, substances with poor fat solubility, and even whole cells can cross the BBB through facilitated transport. These mechanisms include cellular diapedesis, adsorptive-mediated transcytosis (AMT), or receptor-mediated transcytosis (RMT) [24]. In receptor-mediated transcytosis (RMT), Large molecule binders connect to particular receptors on brain endothelial cells (BECs). This binding begins endocytosis of the receptor-ligand complex, which is then transported across the cytoplasm of BECs within vesicles and released through exocytosis at the basolateral membrane. After binding to its receptor on the BEC surface, the signaling molecule (ligand) may detach during transport within vesicles or upon release from the cell (exocytosis). Adsorptive-mediated transcytosis (AMT) involves positively charged molecules (cations) interacting with binding sites on the cell sheath. This interaction triggers the cell to take up the molecule (endocytosis) and then transport it across the cell in a vesicle (vesicular transcytosis). White blood cells (mononuclear cells) can cross the intact BBB by squeezing through the BEC cytoplasm in a process called diapedesis [45, 46].

EMERGING STRATEGIES FOR BRAIN DRUG DELIVERY

The numerous approaches explored to transport drugs effectively to the brain highlight the complexities and challenges involved. Broadly, these strategies to improve CNS drug delivery can be categorized by their mechanism of action, which include: (1) designing the physicochemical properties of ingredients to enable their transcellular passage into the CNS; (2) bypassing the BBB using targeted delivery methods (locoregional drug delivery techniques). (3) Enhancing the movement of drugs among brain endothelial cells (paracellular transport) to increase delivery transversely through the BBB [30, 47].

Nanocarriers for Site-specific Drug Delivery

Nanocarriers for targeted drug delivery systems are a type of drug delivery method that uses microscopic particles to transport medication. These particles can be either suspensions or solid structures. One specific type of CDCS utilizes nanoparticles, which are extremely minute units ranging from 1 to 100 nanometers in dimensions, and are practical for carrying drugs across BBB for two reasons: their small size and their capacity to be decorated with target-specific ligands. These systems, which include dendrimers, micelles, liposomes, nanoparticles, emulsions, minicells, and microparticles, aim to improve drug efficacy and specificity while reducing toxicities. CDCS enhances bioavailability, protects drug payloads from degradation, and enables sustained delivery with targeted cell/tissue specificity. In brain tumors, where vasculature is often leaky, CDCS can be designed to take advantage of the leaky vasculature of tumors for better tumor targeting (Table **1**) [48, 49].

Liposomes

Liposomes are microscopic spherical vesicles with a distinctive bilayer structure [50, 51]. They resemble tiny bubbles, featuring a hydrophilic (water-filled) core surrounded by one or more layers of amphiphilic lipid molecules similar to the components of biological membranes [52 - 54]. These lipids can be derived from natural sources or synthesized, with common examples including phosphatidylcholine and cholesterol [55, 56]. By varying the lipid composition, scientists can modulate the flexibility, size, and surface characteristics of liposomes to suit specific therapeutic needs [57]. One of the most compelling features of liposomes is their ability to encapsulate both hydrophilic and lipophilic drugs. Hydrophilic drugs are housed within the aqueous core, while lipophilic drugs are embedded within the lipid bilayer—much like hiding two types of messages in a balloon: one in the water-filled center and the other in the rubbery wall. This dual-loading capability makes liposomes highly versatile carriers for drug delivery.

Despite their advantages, liposomes face certain limitations, including challenges in achieving high drug encapsulation efficiency and maintaining long-term stability. To address these issues, researchers are actively developing advanced formulations and surface modifications. One promising avenue is the use of liposomes for direct nasal-to-brain drug delivery. By decorating the liposome surface with targeting ligands or functional molecules, scientists can enhance their ability to cross the blood-brain barrier and reach specific brain regions. Preclinical studies have demonstrated encouraging results. For instance, H102 peptide-loaded liposomes have shown potential in treating Alzheimer's disease, while donepezil-loaded liposomes have exhibited improved brain targeting and therapeutic efficacy [58, 59].

Polymeric Nanoparticles

Polymeric nanoparticles are highly versatile drug delivery vehicles, typically ranging in size from 1 to 100 nanometers. Structurally, they can be categorized into two main types: nanospheres, which consist of a solid polymer matrix with the drug uniformly dispersed throughout, and nanocapsules, which feature a polymer shell encasing a drug-filled core. These microscopic carriers are engineered using biodegradable polymers such as polylactic acid (PLA), polyglycolic acid (PGA), or chitosan, with the choice of material influencing their biocompatibility, degradation rate, and drug release profile [60, 61].

Table 1. Summary of clinical applicability of colloidal drug delivery to the brain.

Entrance Mechanism	Targeting Effector/Ligand	NPs Type	Therapeutic Agent	Route of administration	Outcomes	Refs
Adsortive- transcytosis	CPPs peptide	Polymeric nanoparticles; Polymeric micelles; Quantum dots	siRNA for Raf-1/CPT	intravenous, intranasal, intra-arterial	- Suppression of tumor development in a rat model of aggressive glioma - Targeted delivery of quantum dots to the brain for theranostic applications	[86]
Transporter-mediated transcytosis	Mannose	Lipid nanoparticles; Cationic HSA nanoparticles	DT, DOX	intravenous	- Enhanced brain uptake of DT in mice - Improved DOX delivery across cell layer and increased uptake by U87MG glioblastoma cells - Reduced tumor volume in mice with glioblastoma	[87, 88]
	Amino acids	Liposomes; Lipid nanoparticles	DOX	intravenous	- Enhanced DOX uptake by C6 glioma cells and improved DOX accumulation in the brains of mice.	[89, 90]
Receptor transcytosis	Transferrin		DOX	intravenous	- Potent antitumor efficacy of DOX/PTX Tf-NPs against glioma in mice.	[91]
	Lactoferrin	Polymeric nanoparticles; Cyclodextrins; Dendrimers; Lipid nanoparticles; Liposomes	ETP, DOX, BCNU, pβ-Gal, GDNF, UCN, NAP, DOX	intravenous, intranasal,	- Improved brain uptake and biodistribution in mice - Increased BBB permeability and suppression of U87MG glioblastoma cell proliferation - Greater accumulation within tumor cells and extended lifespan in rats with glioma by DOX-NPs	[91-94]
	Angiopep-2	Liposomes, solid lipid nanoparticles.	AMB, DOX, TRAIL	intravenous	- Improved survival rates in glioma-bearing mice following treatment with DOX-loaded gold nanoparticles (Au NPs) and multi-walled nanotubes (MWNTs) - Enhanced brain accumulation and lifespan extension in tumor-bearing mice after administration of TRAIL-conjugated dendrimers	[95, 96]
	CDX	Liposomes; Polymeric nanoparticles	DOX, PTX	i/v injection	- Capacity to traverse the BBB and extended median survival in glioma-bearing mice following administration of DOX and PTX-loaded NPs-CDX.	[97, 98]
	RGD peptide	Magnetic nanoparticles; Liposomes	DT/QDs	i/v injection	- Localization within the tumor site, enabling MRI contrast for brain tumor visualization in mice - Improved brain delivery and imaging following administration of DT/QDs-loaded liposomes in mice	[99, 100]
	NGR peptide	Polymeric nanoparticles; Liposomes; Quantum dots	DT, DOX, siRNA-Luc	i/v injection	- Anti-angiogenic effects and extended lifespan in mice with intracranial glioma following administration of DT or DOX-loaded nanoparticles (NPs) - Suppression of reporter gene (Luciferase) expression in glioma cells of mice - Improved visualization of brain cancer in a rat glioma model following administration of QDs-NGR	[101, 102]

Surface modification plays a critical role in enhancing the targeting capabilities of polymeric nanoparticles. Functionalizing the surface with molecules like polysorbate 80 or proteins such as albumin serves as a biological "flag," directing the nanoparticles toward specific receptors or transport mechanisms in the brain. This strategy facilitates crossing the blood-brain barrier (BBB) and improves site-specific drug delivery.

Promising preclinical studies have demonstrated the efficacy of polymeric nanoparticles in brain-targeted therapies. For example, potent anticancer agents such as carmustine and doxorubicin have been encapsulated within nanoparticles and successfully delivered to brain tumors. These nanoparticles were surface-decorated with targeting ligands that enabled selective uptake by cancer cells. Similarly, nanoparticles coated with lactoferrin, a protein known to interact with brain endothelial cells, have been employed to deliver therapeutic agents for glioma treatment [62, 63].

Inorganic Nanoparticles

Although traditionally employed for diagnostic applications, inorganic nanoparticles are increasingly recognized for their potential in site-specific drug delivery to the brain. These nanoparticles are composed of materials such as metals (*e.g.*, gold, silver), metal oxides (*e.g.*, iron oxide), silica, and carbon-based structures (*e.g.*, carbon nanotubes). Unlike organic nanoparticles, inorganic variants offer a unique combination of uniform particle size, structural stability, and surface modifiability, making them attractive candidates for central nervous system (CNS) targeting. A notable example is mesoporous silica nanoparticles (MSNs), which possess several advantageous properties: High surface area, enabling substantial drug loading capacity. Large pore volume, allowing encapsulation of diverse therapeutic agents. Excellent biocompatibility, ensuring minimal toxicity and favorable interaction with biological systems [64].

Surface functionalization further enhances their brain-targeting capabilities. For instance, polyethylene glycol (PEG) conjugation facilitates BBB penetration *via* transcytosis, a cellular transport mechanism. This strategy has proven effective not only for silica nanoparticles but also for carbon nanotubes, which share similar physicochemical traits. Additionally, lactoferrin, a naturally occurring protein that binds to receptors on brain endothelial cells, has shown superior BBB permeability compared to PEG in some studies. Other targeting ligands are also being explored to improve specificity and uptake. Beyond drug delivery, inorganic nanoparticles hold promise in brain imaging and theranostics. Researchers have developed multi-modal imaging probes by combining gold and magnetic nanoparticles, enabling simultaneous diagnostic imaging and therapeutic guidance. These hybrid systems allow clinicians to visualize brain tissue with high resolution and potentially direct treatment in real time. Moreover, gold nanoparticles can be engineered to deliver therapeutic payloads and enhance radiation therapy for brain tumors [65].

Nanogels

Nanogels are microscopic, three-dimensional, crosslinked polymer networks that resemble tiny sponges capable of retaining large amounts of water. Their high water content contributes to excellent biocompatibility, while the porous structure facilitates efficient drug loading and release, making them particularly well-suited for brain-targeted drug delivery [66]. One of the key advantages of nanogels is their ability to allow drugs to diffuse in and out of the matrix with ease. Studies have shown that cationic (positively charged) nanogels exhibit enhanced cellular uptake due to favorable interactions with negatively charged cell membranes, further improving their therapeutic potential [67]. Innovative formulations have demonstrated promising results in preclinical models. For instance, a pH-sensitive nanogel loaded with doxorubicin was engineered to release the drug selectively in the acidic microenvironment of brain tumors, thereby minimizing systemic toxicity and off-target effects [68]. In another study, a biodegradable nanogel was used to deliver an anti-HIV drug to the brain, offering a strategy to reduce adverse effects commonly associated with systemic administration [69]. These examples underscore the versatility and promise of nanogels as smart, responsive carriers for CNS therapeutics, capable of overcoming biological barriers and enhancing drug efficacy through targeted release mechanisms.

Nanoemulsions

Nanoemulsions are microscopic oil droplets, less than 200 nanometers in size, suspended in water or *vice versa*. These tiny spheres can carry both water-soluble and fat-soluble drugs, making them versatile for drug delivery [70, 71]. They are often made from natural oils, such as peanut or sunflower oil, which enhances their biocompatibility. One exciting application of nanoemulsions is the delivery of drugs directly to the brain through the nose. A nanoemulsion loaded with an anti-tuberculosis drug to improve brain penetration through the nasal route [72, 73]. A mucoadhesive nanoemulsion carrying quercetin, an antioxidant, delivered nasally to treat stroke in rats. This approach provides a non-invasive method for delivering medication to the brain [74].

Dendrimers

Dendrimers are like tiny, bushy spheres made from repeatedly branched polymers. These complex structures offer a unique advantage for drug delivery: a hollow core for carrying medication and a reactive surface for customization [75, 76]. This enables scientists to tailor dendrimers for targeted delivery to the brain. The attachment of specific molecules to the dendrimer surface. These molecules act like flags, guiding the dendrimer towards brain cells or tumors. Dendrimers are modified with a protein that interacts with brain cells to deliver medication for the treatment of memory loss [77, 78]. This resulted in improved brain drug uptake and enhanced memory function in animal models. These studies highlight the promise of dendrimers for brain-targeted drug delivery. By fine-tuning their surface, scientists can create powerful tools to deliver medications directly to the brain, potentially alleviating symptoms for several neurological conditions [79].

Quantum Dots

Quantum dots (QDs) are nanoscale semiconductor crystals that function like tiny, highly luminous light sources. Their most distinctive property is size-dependent fluorescence, the ability to emit different colors of light based on their particle size [80]. This tunable emission makes quantum dots exceptionally useful as fluorescent imaging probes, particularly in applications related to brain health and neuro-oncology.

By conjugating quantum dots with targeting ligands, such as antibodies or peptides, researchers can direct these probes to bind specifically to brain tumor cells. This targeted imaging approach enables enhanced visualization of tumors during surgical procedures, potentially improving diagnostic accuracy and treatment outcomes [81].

Beyond imaging, quantum dots also encapsulate therapeutic agents within their core or attach them to their surface, offering a dual-function platform for theranostics — the integration of therapy and diagnostics. However, similar to other inorganic nanoparticles, quantum dots face significant challenges in their applications for drug delivery. These include concerns over cytotoxicity, particularly due to heavy metal content, and slow or uncontrolled drug release, which can limit their clinical utility [82]. Despite these limitations, ongoing research is focused on developing biocompatible quantum dot formulations, including carbon-based and silicon-based variants, to reduce toxicity and enhance therapeutic performance. Their unique optical properties continue to make them valuable tools in brain imaging, tumor tracking, and real-time monitoring of treatment responses.

Intranasal Delivery

Intranasal delivery is an auspicious method for direct and rapid drug delivery to the brain, avoiding BBB. The nasal cavity features a large, richly vascularized membrane system that envelops the brain and spinal cord, enabling quick absorption into the bloodstream and direct access to the brain. This route provides non-invasive administration, faster therapeutic onset, higher absorption by avoiding presystemic metabolism, and enhanced brain penetration without modifying the parent drug [83, 84]. In veterinary clinical neurology, intranasal deposition, particularly for anticonvulsant therapy in seizure management, has been in use for decades. Drugs can pass through the olfactory epithelium *via* paracellular routes, transcellular passage, or intracellular transport involving endocytosis and axonal transport. This method is favored for home use due to its ease of administration, safety, and preference over other routes [85].

Intrathecal Delivery

Intrathecal (IT) administration involves the direct inoculation of therapeutic agents into the spinal canal, delivering substances directly into the cerebrospinal fluid (CSF) within the central nervous system (CNS). This approach bypasses both the blood-brain barrier (BBB) and the blood-CSF barrier (BCSFB), enabling reduced systemic dosages, minimized toxicity, and elevated drug concentrations within the CNS. IT delivery can be performed *via* lumbar puncture, cerebellomedullary cisternal injection, or ventricular access. Despite its advantages, IT administration presents several limitations, including variable CSF distribution, which leads to inconsistent drug exposure. Rapid clearance of hydrophilic agents, reducing therapeutic duration. Limited diffusion of hydrophobic agents, restricting their efficacy. Poor saturation of brain parenchyma by macromolecules, limiting their therapeutic reach. To address these challenges, continuous drug delivery systems (CDCS) have been developed to enhance distribution and targeting of intrathecally administered drugs. In companion animals, IT administration requires general anesthesia and poses technical difficulties. Nevertheless, it is frequently employed in veterinary medicine for analgesia, particularly in surgical settings. For example, IT morphine, often combined with local anesthetics, has been shown to provide adequate pain relief in dogs undergoing surgery. However, adverse effects such as pruritus and urinary retention have been reported. In senior canines with cancer undergoing extensive procedures, IT morphine offers a safer alternative to intravenous analgesics, reducing systemic risks. In human medicine, IT injection is commonly used for managing spasticity, CNS neoplasms, and postoperative pain. It also plays a role in myelography, and in the treatment of hematologic malignancies, brain tumors with leptomeningeal dissemination, and other CNS disorders [50, 51].

Interstitial Delivery

Delivery of drugs directly into the interstitium of the brain offers a highly targeted approach that bypasses the BBB and BCSFB, allowing for the accurate delivery of both small and large particles for site-specific transfer to the CNS, thereby reducing distribution throughout the body. Techniques include interstitial bolus injections, implantation of Biodegradable implants, and advanced delivery methods. These methods are aggressive and normally limited to treating the disease, which emphasizes that it affects a specific location in the brain due to challenges in achieving efficient and safe global brain targeting. Bolus injections may lead to reflux or exposure of non-target areas, whereas biodegradable implants, such as PLGA microcylinders, have shown promise in delivering agents like temozolomide (TMZ) with controlled release profiles. Diffusion-dependent drug distribution within the brain parenchyma and procedural challenges underscore the current limitations of these approaches. Nonetheless, interstitial delivery holds the potential for enhancing therapeutic outcomes in localized brain disorders [52].

Convection-enhanced delivery (CED) involves inserting catheters into the brain to infuse agents under constant pressure, displacing extracellular fluid and promoting bulk fluid flow. This method utilizes Darcy's law to achieve a homogeneous distribution of infusates across larger tissue volumes compared to diffusion-based methods. While diffusion relies on concentration gradients, CED's pressure gradient ensures a uniform distribution, regardless of molecular weight, although it is constrained by the brain's extracellular space capacity. Challenges include optimizing catheter performance, infusate distribution in heterogeneous tissues, and managing infusion variables. Advances like reflux-preventing catheters and real-time MRI monitoring improve precision and safety, enhancing therapeutic outcomes in canine brain tumor treatments and potential applications in other neurologic disorders.

Focused Ultrasound BBB Disruption (FUS-BBBD)

Focused ultrasound (FUS) is a non-invasive technique that utilizes acoustic waves concentrated at a focal point to induce localized thermal or mechanical ablation within tissues. In therapeutic contexts, particularly MRI-guided FUS for blood-brain barrier disruption (MRg-FUS BBBD), low-intensity ultrasound combined with intravenously administered microbubbles enables transient and controlled opening of the blood-brain barrier (BBB). This approach, monitored in real-time *via* magnetic resonance imaging (MRI), facilitates the precise delivery of therapeutic agents, including large biomolecules and even intact cells, into targeted brain regions.

Mechanistically, FUS-BBBD enhances BBB permeability through several pathways, including the modulation of tight junctions, which temporarily loosens endothelial cell connections. Stimulation of vesicular transcytosis, increasing intracellular transport.Ultrasound-induced permeabilization of the vascular endothelium, allowing passage of therapeutic agents. Recent preclinical studies in elderly canines with cerebral amyloidosis have demonstrated the feasibility and safety of FUS-BBBD, suggesting its potential for treating neurological disorders in companion animals. However, several challenges remain: The transient nature of BBB opening necessitates repeated treatments for sustained therapeutic effect. Anatomical differences between humans and animals necessitate the adaptation of FUS systems, originally designed for human use, to suit veterinary applications. These findings underscore the translational promise of FUS-BBBD in both human and veterinary neurology, offering a platform for non-invasive, targeted CNS drug delivery.

Pulsed Electric Field BBB Penetration

Pulsed electrical field (PEF) therapies employ intermittent electric current pulses to induce controlled biophysical effects in tissues. These effects are governed by pulse parameters including amplitude, waveform, duration, and the electrical properties of the target tissue. One widely studied PEF modality is electroporation, which transiently creates nanopores in cell membranes, allowing for enhanced permeability and facilitating drug delivery. A specialized form of electroporation, known as irreversible electroporation (IRE), is used for non-thermal tumor ablation. IRE induces permanent damage to tumor cells while sparing adjacent critical structures, making it particularly valuable in anatomically sensitive regions such as the central nervous system (CNS). In addition to tumor ablation, IRE has been shown to transiently disrupt the blood-brain barrier (BBB), thereby enhancing drug delivery to infiltrative tumor margins.

An advanced variant, high-frequency irreversible electroporation (HFIRE), utilizes ultrashort electrical pulses to achieve more uniform tissue effects and controlled BBB opening. Both IRE and HFIRE have demonstrated feasibility in veterinary oncology, particularly in the treatment of canine brain tumors, offering promising translational insights.

However, several limitations persist: The invasive nature of electrode placement. The requirement for precise treatment planning to ensure optimal therapeutic outcomes and avoid collateral damage. Recent research has begun exploring non-invasive PEF techniques for transient BBB modulation, which may overcome current procedural challenges and expand the therapeutic potential of PEF-based interventions in neuro-oncology [55, 56].

CONCLUSION AND FUTURE PROSPECTS

Transporting therapeutic agents to the brain while circumventing the blood-brain barrier (BBB) and minimizing peripheral neurotoxicity remains one of the most formidable challenges in neurology. Over recent decades, significant advancements have been made, with targeted drug delivery systems emerging as promising strategies for central nervous system (CNS) therapeutics. Researchers have investigated a variety of mechanisms to facilitate drug transport across physiological barriers, including: Receptor-mediated transcytosis, Active efflux pump modulation, Transporter exploitation, Cell-mediated endocytosis, Adsorptive-mediated transcytosis. A wide array of nanoparticulate carriers—such as liposomes, dendrimers, nanogels, quantum dots, polymeric nanoparticles, and nanoemulsions—have demonstrated potential for brain-targeted delivery. Additionally, direct nasal drug transport has gained attention as a viable alternative route for bypassing the BBB and achieving rapid CNS access. Despite encouraging preclinical outcomes, none of these approaches have yet achieved widespread clinical adoption, mainly due to discrepancies between laboratory efficacy and clinical performance. Challenges include: Functional limitations in drug release and targeting. Toxicity concerns and poor biocompatibility. Suboptimal drug loading capacity and release kinetics. Instability and scalability issues. To overcome these barriers, future research must focus on: Optimizing physicochemical properties of drug carriers and enhancing targeting specificity to minimize off-target effects. Ensuring safety and biocompatibility for long-term use. The integration of artificial intelligence (AI) and machine learning (ML) offers a transformative opportunity to develop predictive models that can streamline formulation design, preclinical testing, and clinical translation. These technologies can accelerate the identification of optimal delivery parameters and improve decision-making throughout the drug development pipeline. Collaborative efforts among researchers, clinicians, and regulatory agencies will be crucial to advancing these innovative delivery systems from the bench to the bedside. By addressing current limitations and harnessing emerging technologies, there is substantial potential to improve therapeutic outcomes for patients with brain tumors, ultimately enhancing both quality of life and survival rates.

AUTHORS' CONTRIBUTIONS

Sakshi Soni was responsible for writing the original draft, conducting the investigation, gathering resources, and performing the formal analysis. Vandana Soni contributed to data curation, participated in the investigation, and was involved in writing, reviewing, and editing the manuscript. Sushil K. Kashaw was responsible for the conceptualization, supervision, provision of resources, and visualization.

REFERENCES

[1]   Vanbilloen WJF, Rechberger JS, Anderson JB, Nonnenbroich LF, Zhang L, Daniels DJ. Nanoparticle strategies to improve the delivery of anticancer drugs across the blood–brain barrier to treat brain tumors. Pharmaceutics 2023; 15(7): 1804.
[http://dx.doi.org/10.3390/pharmaceutics15071804] [PMID: 37513992]

[2]   Zhao Y, Yue P, Peng Y, *et al.* Recent advances in drug delivery systems for targeting brain tumors. Drug Deliv 2023; 30(1): 1-18.
[http://dx.doi.org/10.1080/10717544.2022.2154409] [PMID: 36597214]

[3]   Mitusova K, Peltek OO, Karpov TE, Muslimov AR, Zyuzin MV, Timin AS. Overcoming the blood–brain barrier for the therapy of malignant brain tumor: current status and prospects of drug delivery approaches. J Nanobiotechnology 2022; 20(1): 412.
[http://dx.doi.org/10.1186/s12951-022-01610-7] [PMID: 36109754]

[4]   Kumar PB, Kadiri SK, Khobragade DS, *et al.* Synthesis, characterization and biological investigations of some new Oxadiazoles: *In-vitro* and *In-Silico* approach. Results in Chemistry. 2024;7:101241.
[http://dx.doi.org/10.1016/j.rechem.2023.101241]

[5]   Gould A, Gonzales VA, Dmello CC, *et al.* Advances in blood-brain barrier disruption to facilitate drug delivery for infiltrative gliomas. Advances in oncology. 2023 May 1; 3(1): 77-86.
[http://dx.doi.org/10.1016/j.yao.2023.01.017]

[6]   Kianinejad N, Kwon YM. Dual-targeting of brain tumors with nanovesicles. Bioimpacts 2023; 13(1): 1-3.
[http://dx.doi.org/10.34172/bi.2022.26321] [PMID: 36816997]

[7]   Upton DH, Ung C, George SM, Tsoli M, Kavallaris M, Ziegler DS. Challenges and opportunities to penetrate the blood-brain barrier for brain cancer therapy. Theranostics 2022; 12(10): 4734-52.
[http://dx.doi.org/10.7150/thno.69682] [PMID: 35832071]

[8]   Bhunia S, Chaudhuri A. Crossing blood-brain barrier with nano-drug carriers for treatment of brain tumors: advances and unmet challenges. Brain Tumors. 2022 Apr 7.
[http://dx.doi.org/10.5772/intechopen.101925]

[9]   Bahadur S, Prakash A. A comprehensive review on nanomedicine: Promising approach for treatment of brain tumor through intranasal administration. Curr Drug Targets 2023; 24(1): 71-88.
[http://dx.doi.org/10.2174/1389450124666221019141044] [PMID: 36278468]

[10]   Hashizume R. New Therapeutic Strategies for the Treatment of Brain Tumor. InBrain Tumors-Current and Emerging Therapeutic Strategies 2011 Aug 23. IntechOpen.
[http://dx.doi.org/10.5772/23997]

[11]   Ahmad F, Varghese R, Panda S, *et al.* Smart nanoformulations for brain cancer theranostics: challenges and promises. Cancers (Basel) 2022; 14(21): 5389.
[http://dx.doi.org/10.3390/cancers14215389] [PMID: 36358807]

[12]   Dubey S, Suraj MR, Goni T, Gulalkai TV, Shankar SV, Gowda SS, Tiwari P. 3D QSAR studies of 3, 16 and 17 position modifications in steroidal derivatives for CNS anticancer activity. Curr Res Chem. 2023;15(1):1–4.
[http://dx.doi.org/10.3923/crc.2023.1.4]

[13]   Tashima T. Brain cancer chemotherapy through a delivery system across the blood-brain barrier into the brain based on receptor-mediated transcytosis using monoclonal antibody conjugates. Biomedicines 2022; 10(7): 1597.
[http://dx.doi.org/10.3390/biomedicines10071597] [PMID: 35884906]

[14]   Aryal M, Porter T. Emerging therapeutic strategies for brain tumors. Neuromolecular Med 2022; 24(1): 23-34.
[http://dx.doi.org/10.1007/s12017-021-08681-z] [PMID: 34406634]

[15]   Ganjeifar B, Morshed SF. Targeted drug delivery in brain tumors-nanochemistry applications and advances. Curr Top Med Chem 2021; 21(14): 1202-23.
[http://dx.doi.org/10.2174/1568026620666201113140258] [PMID: 33185163]

[16]   Pouliopoulos AN. Evaluating drug delivery enhancement following ultrasound treatment. Lancet Oncol 2023; 24(5): 420-2.
[http://dx.doi.org/10.1016/S1470-2045(23)00149-3] [PMID: 37142367]

[17]   Feldman L, Badie B. Novel cell delivery systems: Intracranial and intrathecal. InNK Cells in Cancer Immunotherapy: Successes and Challenges 2023 Jan 1 (pp. 263-280). Academic Press.

[18]   Mehkri Y, Woodford S, Pierre K, *et al.* Focused delivery of chemotherapy to augment surgical management of brain tumors. Curr Oncol 2022; 29(11): 8846-61.
[http://dx.doi.org/10.3390/curroncol29110696] [PMID: 36421349]

[19]   Duan L, Li X, Ji R, *et al.* Nanoparticle-based drug delivery systems: an inspiring therapeutic strategy for neurodegenerative diseases. Polymers (Basel) 2023; 15(9): 2196.
[http://dx.doi.org/10.3390/polym15092196] [PMID: 37177342]

[20]   Mohapatra TK, Nayak RR, Ganeshpurkar A, Tiwari P, Kumar D. Opportunities and difficulties in the repurposing of hdac inhibitors as antiparasitic agents. Drugs Drug Candidates 2024, 3, 70-101.
[http://dx.doi.org/10.3390/ddc3010006]

[21]   Qiu Z, Yu Z, Xu T, *et al.* Novel nano-drug delivery system for brain tumor treatment. Cells 2022; 11(23): 3761.
[http://dx.doi.org/10.3390/cells11233761] [PMID: 36497021]

[22]   Gu Y, Wu L, Hameed Y, Nabi-Afjadi M. Overcoming the challenge: cell-penetrating peptides and membrane permeability. Biomaterials and Biosensors 2023; 2(1): 7.
[http://dx.doi.org/10.58567/bab02010002]

[23]   Kotadiya DD, Patel P, Patel HD. Cell-penetrating peptides: a powerful tool for targeted drug delivery. Curr Drug Deliv 2024; 21(3): 368-88.
[http://dx.doi.org/10.2174/1567201820666230407092924] [PMID: 37026498]

[24]   Abbott NJ, Patabendige AAK, Dolman DEM, Yusof SR, Begley DJ. Structure and function of the blood–brain barrier. Neurobiol Dis 2010; 37(1): 13-25.
[http://dx.doi.org/10.1016/j.nbd.2009.07.030] [PMID: 19664713]

[25]   Villabona-Rueda A, Erice C, Pardo CA, Stins MF. The evolving concept of the blood brain barrier (BBB): from a single static barrier to a heterogeneous and dynamic relay center. Front Cell Neurosci 2019; 13: 405.
[http://dx.doi.org/10.3389/fncel.2019.00405] [PMID: 31616251]

[26]   Castro Dias M, Mapunda JA, Vladymyrov M, Engelhardt B. Structure and junctional complexes of endothelial, epithelial and glial brain barriers. Int J Mol Sci 2019; 20(21): 5372.
[http://dx.doi.org/10.3390/ijms20215372] [PMID: 31671721]

[27]   Nakada T, Kwee IL. Fluid dynamics inside the brain barrier: current concept of interstitial flow, glymphatic flow, and cerebrospinal fluid circulation in the brain. Neuroscientist 2019; 25(2): 155-66.
[http://dx.doi.org/10.1177/1073858418775027] [PMID: 29799313]

[28]   Wolburg H, Paulus W. Choroid plexus: biology and pathology. Acta Neuropathol 2010; 119(1): 75-88.
[http://dx.doi.org/10.1007/s00401-009-0627-8] [PMID: 20033190]

[29]   Pardridge WM. The blood-brain barrier: Bottleneck in brain drug development. NeuroRx 2005; 2(1): 3-14.
[http://dx.doi.org/10.1602/neurorx.2.1.3] [PMID: 15717053]

[30]   Partridge BR, Kani Y, Lorenzo MF, *et al.* High-frequency irreversible electroporation (H-FIRE) induced blood–brain barrier disruption is mediated by cytoskeletal remodeling and changes in tight junction protein regulation. Biomedicines 2022; 10(6): 1384.

[http://dx.doi.org/10.3390/biomedicines10061384] [PMID: 35740406]

[31]   Bazzoni G, Dejana E. Endothelial cell-to-cell junctions: molecular organization and role in vascular homeostasis. Physiol Rev 2004; 84(3): 869-901.
[http://dx.doi.org/10.1152/physrev.00035.2003] [PMID: 15269339]

[32]   Hartz AM, Bauer B. ABC transporters in the CNS - an inventory. Curr Pharm Biotechnol 2011; 12(4): 656-73.
[http://dx.doi.org/10.2174/138920111795164020] [PMID: 21118088]

[33]   Berntsson RPA, Smits SHJ, Schmitt L, Slotboom DJ, Poolman B. A structural classification of substrate-binding proteins. FEBS Lett 2010; 584(12): 2606-17.
[http://dx.doi.org/10.1016/j.febslet.2010.04.043] [PMID: 20412802]

[34]   Shawahna R, Declèves X, Scherrmann JM. Hurdles with using *In vitro* models to predict human blood-brain barrier drug permeability: a special focus on transporters and metabolizing enzymes. Curr Drug Metab 2013; 14(1): 120-36.
[http://dx.doi.org/10.2174/138920013804545232] [PMID: 23215812]

[35]   Nag S. Morphology and properties of brain endothelial cells. The Blood-Brain and Other Neural Barriers: Reviews and Protocols. 2011: 3-47.
[http://dx.doi.org/10.1007/978-1-60761-938-3_1]

[36]   Rubin LL, Barbu K, Bard F, *et al.* Differentiation of brain endothelial cells in cell culture. Ann N Y Acad Sci 1991; 633(1): 420-5.
[http://dx.doi.org/10.1111/j.1749-6632.1991.tb15631.x] [PMID: 1665033]

[37]   Janzer RC, Raff MC. Astrocytes induce blood–brain barrier properties in endothelial cells. Nature 1987; 325(6101): 253-7.
[http://dx.doi.org/10.1038/325253a0] [PMID: 3543687]

[38]   Alvarez JI, Katayama T, Prat A. Glial influence on the blood brain barrier. Glia 2013; 61(12): 1939-58.
[http://dx.doi.org/10.1002/glia.22575] [PMID: 24123158]

[39]   Dore-Duffy P, Cleary K. Morphology and properties of pericytes. The Blood-Brain and Other Neural Barriers: Reviews and Protocols. 2011:49-68.
[http://dx.doi.org/10.1007/978-1-60761-938-3_2]

[40]   Daneman R, Zhou L, Kebede AA, Barres BA. Pericytes are required for blood–brain barrier integrity during embryogenesis. Nature 2010; 468(7323): 562-6.
[http://dx.doi.org/10.1038/nature09513] [PMID: 20944625]

[41]   Nakagawa S, Deli MA, Nakao S, *et al.* Pericytes from brain microvessels strengthen the barrier integrity in primary cultures of rat brain endothelial cells. Cell Mol Neurobiol 2007; 27(6): 687-94.
[http://dx.doi.org/10.1007/s10571-007-9195-4] [PMID: 17823866]

[42]   Bonkowski D, Katyshev V, Balabanov RD, Borisov A, Dore-Duffy P. The CNS microvascular pericyte: pericyte-astrocyte crosstalk in the regulation of tissue survival. Fluids Barriers CNS 2011; 8(1): 8.
[http://dx.doi.org/10.1186/2045-8118-8-8] [PMID: 21349156]

[43]   Thomsen MS, Routhe LJ, Moos T. The vascular basement membrane in the healthy and pathological brain. J Cereb Blood Flow Metab 2017; 37(10): 3300-17.
[http://dx.doi.org/10.1177/0271678X17722436] [PMID: 28753105]

[44]   Hynes RO. The extracellular matrix: not just pretty fibrils. Science 2009; 326(5957): 1216-9.
[http://dx.doi.org/10.1126/science.1176009] [PMID: 19965464]

[45]   Sauer I, Dunay IR, Weisgraber K, Bienert M, Dathe M. An apolipoprotein E-derived peptide mediates uptake of sterically stabilized liposomes into brain capillary endothelial cells. Biochemistry 2005; 44(6): 2021-9.
[http://dx.doi.org/10.1021/bi048080x] [PMID: 15697227]

[46] Carman CV, Springer TA. Trans-cellular migration: cell–cell contacts get intimate. Curr Opin Cell Biol 2008; 20(5): 533-40.
[http://dx.doi.org/10.1016/j.ceb.2008.05.007] [PMID: 18595683]

[47] Pardridge WM. Blood-brain barrier and delivery of protein and gene therapeutics to brain. Front Aging Neurosci 2020; 11: 373.
[http://dx.doi.org/10.3389/fnagi.2019.00373] [PMID: 31998120]

[48] Barratt G. Colloidal drug carriers: achievements and perspectives. Cell Mol Life Sci 2003; 60(1): 21-37.
[http://dx.doi.org/10.1007/s000180300002] [PMID: 12613656]

[49] Allen TM, Brandeis E, Hansen CB, Kao GY, Zalipsky S. A new strategy for attachment of antibodies to sterically stabilized liposomes resulting in efficient targeting to cancer cells. Biochim Biophys Acta Biomembr 1995; 1237(2): 99-108.
[http://dx.doi.org/10.1016/0005-2736(95)00085-H] [PMID: 7632714]

[50] Pardridge WM. Drug transport in brain *via* the cerebrospinal fluid. Fluids Barriers CNS 2011; 8(1): 7.
[http://dx.doi.org/10.1186/2045-8118-8-7] [PMID: 21349155]

[51] Papisov MI, Belov VV, Gannon KS. Physiology of the intrathecal bolus: the leptomeningeal route for macromolecule and particle delivery to CNS. Mol Pharm 2013; 10(5): 1522-32.
[http://dx.doi.org/10.1021/mp300474m] [PMID: 23316936]

[52] Rossmeisl JH. New treatment modalities for brain tumors in dogs and cats. Vet Clin North Am Small Anim Pract 2014; 44(6): 1013-38.
[http://dx.doi.org/10.1016/j.cvsm.2014.07.003] [PMID: 25441624]

[53] Fishman PS, Frenkel V. Focused ultrasound: an emerging therapeutic modality for neurologic disease. Neurotherapeutics 2017; 14(2): 393-404.
[http://dx.doi.org/10.1007/s13311-017-0515-1] [PMID: 28244011]

[54] Meng Y, Hynynen K, Lipsman N. Applications of focused ultrasound in the brain: from thermoablation to drug delivery. Nat Rev Neurol 2021; 17(1): 7-22.
[http://dx.doi.org/10.1038/s41582-020-00418-z] [PMID: 33106619]

[55] Aycock KN, Davalos RV. Irreversible electroporation: background, theory, and review of recent developments in clinical oncology. Bioelectricity 2019; 1(4): 214-34.
[http://dx.doi.org/10.1089/bioe.2019.0029] [PMID: 34471825]

[56] Rossmeisl JH Jr, Garcia PA, Pancotto TE, *et al.* Safety and feasibility of the NanoKnife system for irreversible electroporation ablative treatment of canine spontaneous intracranial gliomas. J Neurosurg 2015; 123(4): 1008-25.
[http://dx.doi.org/10.3171/2014.12.JNS141768] [PMID: 26140483]

[57] Chaves MA, Esposto BS, Martelli-Tosi M, Pinho SC. Liposomes. 2021.

[58] Akbarzadeh A, Rezaei-Sadabady R, Davaran S, *et al.* Liposome: classification, preparation, and applications. Nanoscale Res Lett 2013; 8(1): 102.
[http://dx.doi.org/10.1186/1556-276X-8-102] [PMID: 23432972]

[59] Amin M, Seynhaeve ALB, Sharifi M, Falahati M, ten Hagen TLM. Liposomal drug delivery systems for cancer therapy: the rotterdam experience. Pharmaceutics 2022; 14(10): 2165.
[http://dx.doi.org/10.3390/pharmaceutics14102165] [PMID: 36297598]

[60] Khan S, Wong A, Zeb S, Mortari B, Villa JE, Sotomayor MD. Characterization of polymeric nanoparticles. InSmart Polymer Nanocomposites 2023 Jan 1 (pp. 141-163). Elsevier.
[http://dx.doi.org/10.1016/B978-0-323-91611-0.00003-7]

[61] Escudero AR. The potential application of Polymeric Nanoparticles in different cancer treatments, 2023.
[http://dx.doi.org/10.3390/mol2net-08-13910]

[62]   Gouthami K, Lakshminarayana L, Faniband B, *et al.* Introduction to polymeric nanomaterials. In Smart polymer nanocomposites 2023 Jan 1 (pp. 3-25). Elsevier.
[http://dx.doi.org/10.1016/B978-0-323-91611-0.00008-6]

[63]   Dristant U, Mukherjee K, Saha S, Maity D. An overview of polymeric nanoparticles-based drug delivery system in cancer treatment. Technol Cancer Res Treat 2023; 22: 15330338231152083.
[http://dx.doi.org/10.1177/15330338231152083] [PMID: 36718541]

[64]   Sharma S, Dang S. Nanocarrier-based drug delivery to brain: interventions of surface modification. Curr Neuropharmacol 2023; 21(3): 517-35.
[http://dx.doi.org/10.2174/1570159X20666220706121412] [PMID: 35794771]

[65]   Sabu C, Ameena Shirin VK, Sankar R, Pramod K. Inorganic Nanoparticles for Drug-delivery Applications. Nanomaterials and Nanotechnology in Medicine. 2022 Sep 30:367-99.
[http://dx.doi.org/10.1002/9781119558026.ch14]

[66]   Altuntaş E, Özkan B, Güngör S, Özsoy Y. Biopolymer-based nanogel approach in drug delivery: basic concept and current developments. Pharmaceutics 2023; 15(6): 1644.
[http://dx.doi.org/10.3390/pharmaceutics15061644] [PMID: 37376092]

[67]   Tiwari S, Singh S, Tripathi PK, Dubey CK. A Review-Nanogel Drug Delivery System. Asian Journal of Research in Pharmaceutical Science 2015; 5(4): 253-5.
[http://dx.doi.org/10.5958/2231-5659.2015.00037.5]

[68]   Jiao F. Mechanism of Cross-Linking, Self-Assembly, Controlled Release, and Applications of Nanogels in Drug Delivery System. Highlights in Science. Engineering and Technology 2023; 40: 326-31.

[69]   Attama AA, Nnamani PO, Onokala OB, Ugwu AA, Onugwu AL. Nanogels as target drug delivery systems in cancer therapy: A review of the last decade. Front Pharmacol 2022; 13: 874510.
[http://dx.doi.org/10.3389/fphar.2022.874510] [PMID: 36160424]

[70]   Talele C, Talele D, Shah N, *et al.* Intranasal Drug Delivery by Nanoemulsions, 2023; 151–157.
[http://dx.doi.org/10.9734/bpi/mono/978-81-19315-51-2/CH17]

[71]   Mishra N, Kaushik N, Sharma PK, Alam MA. Nano Emulsion Drug Delivery System: A Review. Current Nanomedicine (Formerly: Recent Patents on Nanomedicine). 2023 Mar 1;13(1):2-16.
[http://dx.doi.org/10.2174/2468187313666230213121011]

[72]   Dhumal N, Yadav V, Borkar S. Nanoemulsion as Novel Drug Delivery System: Development, Characterization and Application. Asian Journal of Pharmaceutical Research and Development 2022; 10(6): 120-7.
[http://dx.doi.org/10.22270/ajprd.v10i6.1205]

[73]   Kesavan K, Mohan P, Jain SK, Parra-Marín O, Murugesan S. Nanoemulsion: A Potential Carrier for Topical Drug Delivery. Nanoparticles and Nanocarriers Based Pharmaceutical Formulations. 2022 Dec 9:230.
[http://dx.doi.org/10.2174/9789815049787122010011]

[74]   Gaikwad R, Shinde A. Overview of nanoemulsion preparation methods, characterization techniques and applications. Asian Journal of Pharmacy and Technology 2022; 12(4): 329-36.
[http://dx.doi.org/10.52711/2231-5713.2022.00053]

[75]   Aravind M, Kumar SP, Begum AS. An overview of dendrimers as novel carriers in drug delivery. Research Journal of Pharmacy and Technology 2023; 16(4): 2051-6.

[76]   Kaurav M, Ruhi S, Al-Goshae HA, *et al.* Dendrimer: An update on recent developments and future opportunities for the brain tumors diagnosis and treatment. Front Pharmacol 2023; 14: 1159131.
[http://dx.doi.org/10.3389/fphar.2023.1159131] [PMID: 37006997]

[77]   Gorain B, Choudhury H, Nair AB, Al-Dhubiab BE. Dendrimers: an effective drug delivery and therapeutic approach. InDesign and Applications of Theranostic Nanomedicines 2023 Jan 1 (pp. 125-

142). Woodhead Publishing.
[http://dx.doi.org/10.1016/B978-0-323-89953-6.00002-7]

[78]   Li H, Zha S, Li H, Liu H, Wong KL, All AH. Polymeric dendrimers as nanocarrier vectors for neurotheranostics. Small 2022; 18(45): 2203629.
[http://dx.doi.org/10.1002/smll.202203629] [PMID: 36084240]

[79]   Choudhury H, Pandey M, Mohgan R, *et al.* Dendrimer-based delivery of macromolecules for the treatment of brain tumor. Biomater Adv 2022; 141: 213118.
[http://dx.doi.org/10.1016/j.bioadv.2022.213118] [PMID: 36182834]

[80]   Rajamanickam K. Application of quantum dots in bio-sensing, bio-imaging, drug delivery, anti-bacterial activity, photo-thermal, photo-dynamic therapy, and optoelectronic devices. InQuantum Dots-Recent Advances, New Perspectives and Contemporary Applications 2022 Sep 16. IntechOpen.

[81]   Armăşelu A, Jhalora P. Application of quantum dots in biomedical and biotechnological fields. InQuantum Dots 2023 Jan 1 (pp. 245-276). Elsevier.

[82]   Saravanan V, Krishna Tippavajhala V. Quantum Dots: Targeted and Traceable Drug Delivery System. Research Journal of Pharmacy and Technology 2022; 15(12): 5895-902.
[http://dx.doi.org/10.52711/0974-360X.2022.00994]

[83]   Hu X, Yue X, Wu C, Zhang X. Factors affecting nasal drug delivery and design strategies for intranasal drug delivery. Zhejiang da xue xue bao. Yi xue ban= Journal of Zhejiang University. Med Sci 2023; 52(3): 328-37.

[84]   Ainurofiq A, Prasetya A, Rahayu BG, Al Qadri MS, Kovusov M, Laksono OEP. Recent developments in brain-targeted drug delivery systems *via* the intranasal route. Farm Pol 2022; 78(12): 695-708.
[http://dx.doi.org/10.32383/farmpol/163334]

[85]   Patel D, Thakkar H. Formulation considerations for improving intranasal delivery of CNS acting therapeutics. Ther Deliv 2022; 13(7): 371-81.
[http://dx.doi.org/10.4155/tde-2022-0018] [PMID: 36416617]

[86]   Kumagai AK, Eisenberg JB, Pardridge WM. Absorptive-mediated endocytosis of cationized albumin and a beta-endorphin-cationized albumin chimeric peptide by isolated brain capillaries. Model system of blood-brain barrier transport. J Biol Chem 1987; 262(31): 15214-9.
[http://dx.doi.org/10.1016/S0021-9258(18)48160-4] [PMID: 2959663]

[87]   Broadwell RD. Transcytosis of macromolecules through the blood-brain barrier: a cell biological perspective and critical appraisal. Acta Neuropathol 1989; 79(2): 117-28.
[http://dx.doi.org/10.1007/BF00294368] [PMID: 2688350]

[88]   Schnitzer JE. Caveolae: from basic trafficking mechanisms to targeting transcytosis for tissue-specific drug and gene delivery *in vivo.* Adv Drug Deliv Rev 2001; 49(3): 265-80.
[http://dx.doi.org/10.1016/S0169-409X(01)00141-7] [PMID: 11551399]

[89]   Tuma PL, Hubbard AL. Transcytosis: crossing cellular barriers. Physiol Rev 2003; 83(3): 871-932.
[http://dx.doi.org/10.1152/physrev.00001.2003] [PMID: 12843411]

[90]   Gao X, Tao W, Lu W, *et al.* Lectin-conjugated PEG–PLA nanoparticles: Preparation and brain delivery after intranasal administration. Biomaterials 2006; 27(18): 3482-90.
[http://dx.doi.org/10.1016/j.biomaterials.2006.01.038] [PMID: 16510178]

[91]   Gao X, Wu B, Zhang Q, *et al.* Brain delivery of vasoactive intestinal peptide enhanced with the nanoparticles conjugated with wheat germ agglutinin following intranasal administration. J Control Release 2007; 121(3): 156-67.
[http://dx.doi.org/10.1016/j.jconrel.2007.05.026] [PMID: 17628165]

[92]   Bereczki E, Re F, Masserini ME, Winblad B, Pei JJ. Liposomes functionalized with acidic lipids rescue Aβ-induced toxicity in murine neuroblastoma cells. Nanomedicine 2011; 7(5): 560-71.
[http://dx.doi.org/10.1016/j.nano.2011.05.009] [PMID: 21703989]

[93]   Rousselle C, Smirnova M, Clair P, *et al.* Enhanced delivery of doxorubicin into the brain *via* a peptide-vector-mediated strategy: saturation kinetics and specificity. J Pharmacol Exp Ther 2001; 296(1): 124-31.
[http://dx.doi.org/10.1016/S0022-3565(24)29672-5] [PMID: 11123372]

[94]   Kanazawa T, Morisaki K, Suzuki S, Takashima Y. Prolongation of life in rats with malignant glioma by intranasal siRNA/drug codelivery to the brain with cell-penetrating peptide-modified micelles. Mol Pharm 2014; 11(5): 1471-8.
[http://dx.doi.org/10.1021/mp400644e] [PMID: 24708261]

[95]   Xia H, Gao X, Gu G, *et al.* Penetratin-functionalized PEG–PLA nanoparticles for brain drug delivery. Int J Pharm 2012; 436(1-2): 840-50.
[http://dx.doi.org/10.1016/j.ijpharm.2012.07.029] [PMID: 22841849]

[96]   Béduneau A, Saulnier P, Benoit JP. Active targeting of brain tumors using nanocarriers. Biomaterials 2007; 28(33): 4947-67.
[http://dx.doi.org/10.1016/j.biomaterials.2007.06.011] [PMID: 17716726]

[97]   Singh I, Swami R, Jeengar MK, Khan W, Sistla R. p-Aminophenyl-α-d-mannopyranoside engineered lipidic nanoparticles for effective delivery of docetaxel to brain. Chem Phys Lipids 2015; 188: 1-9.
[http://dx.doi.org/10.1016/j.chemphyslip.2015.03.003] [PMID: 25819559]

[98]   Li L, Di X, Zhang S, *et al.* Large amino acid transporter 1 mediated glutamate modified docetaxel-loaded liposomes for glioma targeting. Colloids Surf B Biointerfaces 2016; 141: 260-7.
[http://dx.doi.org/10.1016/j.colsurfb.2016.01.041] [PMID: 26859117]

[99]   Kharya P, Jain A, Gulbake A, *et al.* Phenylalanine-coupled solid lipid nanoparticles for brain tumor targeting. J Nanopart Res 2013; 15(11): 2022.
[http://dx.doi.org/10.1007/s11051-013-2022-6]

[100] Jefferies WA, Brandon MR, Hunt SV, Williams AF, Gatter KC, Mason DY. Transferrin receptor on endothelium of brain capillaries. Nature 1984; 312(5990): 162-3.
[http://dx.doi.org/10.1038/312162a0] [PMID: 6095085]

[101] Kuo YC, Wang LJ. Transferrin-grafted catanionic solid lipid nanoparticles for targeting delivery of saquinavir to the brain. J Taiwan Inst Chem Eng 2014; 45(3): 755-63.
[http://dx.doi.org/10.1016/j.jtice.2013.09.024]

[102] Chang J, Jallouli Y, Kroubi M, *et al.* Characterization of endocytosis of transferrin-coated PLGA nanoparticles by the blood–brain barrier. Int J Pharm 2009; 379(2): 285-92.
[http://dx.doi.org/10.1016/j.ijpharm.2009.04.035] [PMID: 19416749]

CHAPTER 4

Targeted Drug Delivery: Opportunities and Challenges in Brain Tumour

Meenakshi Attri[1], **Asha Raghav**[2] and **Mohit Agrawal**[1,*]

[1] *School of Medical & Allied Sciences, K. R. Mangalam University, Gurugram, Haryana-122103, India*

[2] *Department of Pharmaceutics, School of Health Sciences, Sushant University, Gurugram, Haryana, India*

Abstract: A promising approach to improving the treatment of brain tumors is the targeted delivery of drugs, which offers opportunities to increase drug efficacy while reducing systemic toxicity. This abstract provides an overview of the advantages and disadvantages of targeted medication delivery for brain tumors. Opportunities include minimising side effects by reducing systemic exposure, improving efficacy by accurately delivering therapeutic drugs to the tumor site, as well as being able to pass across the blood-brain barrier (BBB) to reach the brain tumor. Additionally, combination therapy approaches, and personalized medicine approaches catered to the molecular features of particular tumours are made possible by targeted drug delivery. To fully achieve the potential of targeted drug delivery against brain tumours, some challenges must be overcome. The complex nature and diversity of brain tumours, the BBB's impenetrable barrier, the development of resistance to targeted therapy, and the conversion of preclinical research into clinically effective treatments are some of these difficulties. Collaboration between researchers, physicians, engineers, and regulatory authorities will be necessary to address these issues. To improve the field of targeted drug delivery against brain tumours, novel approaches are required to target specific molecular pathways, get over the blood-brain barrier, and overcome drug resistance mechanisms. In conclusion, targeted drug delivery has great promise for improving patient outcomes and revolutionising the treatment of brain tumours with more research and development.

Keywords: Brain tumor, Blood-brain barrier, Targeted drug delivery.

---

* **Corresponding author Mohit Agrawal:** School of Medical & Allied Sciences, K. R. Mangalam University, Gurugram, Haryana-122103, India; E-mail mohitagrawalmohu@gmail.com

Prashant Tiwari, Pankaj Kumar Singh & Sunil Kumar Kadiri (Eds.)
All rights reserved-© 2025 Bentham Science Publishers

INTRODUCTION

According to the most recent worldwide cancer data provided by the World Health Organisation (WHO) in 2020 [1], brain tumours account for approximately 1.6% of incident cases and 2.5% of deaths from all tumours, respectively. In contrast, China has the highest rates of brain tumour morbidity and death worldwide, with rates as high as 32% and 26%, respectively, and an incidence rate that is still growing for younger people [2]. Gliomas account for approximately 30% of all brain tumours and are the most prevalent and invasive type of brain tumours. They have aggressive invasion, a high recurrence rate, and a poor prognosis.

The preferred course of treatment for brain tumours is surgery; however, its invasive nature and hazy boundaries make it challenging to eradicate the tumour entirely. Furthermore, almost 90% of surgical recurrences occur [3]. Furthermore, radiation and chemotherapy used after surgery are becoming the norm for treating brain tumours. Temozolomide (TMZ), an alkylating agent, acts as a first-line chemotherapeutic medication for brain tumours and introduces a methyl group to DNA purine bases to induce cell death [4]. However, due to its short half-life, the higher dosage has resulted in several side effects, including lymphopenia, neutropenia, and thrombocytopenia.

Furthermore, the deregulation of signalling pathways, DNA repair, autophagy, and other associated mechanisms may cause the tumour cells to develop resistance to TMZ [5]. In addition to TMZ, bevacizumab is a VEGFR inhibitor that the FDA has approved for the treatment of brain tumours. However, the usage of this anti-angiogenesis medication is still debatable because it has not been able to increase patients' overall survival [6]. It is challenging to develop specific therapeutic drugs for the treatment of brain tumours because of their poor efficacy and serious toxic and side effects. Examples of these drugs are nitrosoureas (carmustine, lomustine), anthracyclines (adriamycin), platinums (cisplatin, carboplatin, oxaliplatin), topoisomerase inhibitors (camptothecin, irinotecan, etoposide), integrin receptor inhibitors (cilengitide), EGFR inhibitors (erlotinib, gefitinib, afatinib), and histone deacetylase inhibitors (vorinostat, panobinostat) [7]. Novel treatments for gliomas, including gene therapy, angiogenesis inhibition, and immunotherapy, have shown promise but have had limited success [8 - 13]. Thus, the development of highly effective, low-toxic, and targeted medications for brain tumours is imperative.

OVERVIEW OF TARGETED DRUG DELIVERY IN CANCER TREATMENT

An advanced strategy for enhancing the effectiveness of cancer treatments while reducing adverse effects is targeted medication delivery. By targeting treatment drugs exclusively at cancer cells, this approach protects healthy organs and lowers systemic toxicity. There are two main techniques in use: passive targeting and active targeting. In order to specifically bind to receptors or antigens that are overexpressed on cancer cells, ligands or antibodies that are coupled to the drug delivery system are used in active targeting. By taking advantage of the increased permeability and retention (EPR) effect, which occurs when tumour tissues have leaky vasculature and inadequate lymphatic drainage, nanoparticles can accumulate there. Targeted drug delivery is achieved through a variety of methods, such as the use of nanoparticles (like liposomes, dendrimers, and polymeric nanoparticles), antibody-drug conjugates (ADCs), which combine powerful cytotoxic drugs with monoclonal antibodies specific to cancer cell antigens, and small molecule inhibitors that prevent the growth of cancer cells. By enhancing the delivery of therapeutic drugs to tumour locations, this targeted method reduces side effects associated with traditional chemotherapy and improves treatment results [14 - 17].

Importance of Targeting Tumors Specifically

To improve treatment success and minimise negative side effects, it is imperative that tumours be properly targeted during cancer treatment. Conventional radiation and chemotherapy treatments frequently target both healthy and malignant cells, resulting in severe side effects and substantial toxicity that can negatively influence a patient's quality of life. Targeted medication delivery minimises damage to normal tissues by directing therapeutic chemicals selectively to cancer cells. This reduces side effects, such as nausea, immunological suppression, and hair loss. Due to this accuracy, greater drug concentrations can be administered precisely to the tumour site, enhancing therapy efficacy and potentially improving clinical outcomes. Furthermore, targeted medicines offer a more successful and long-lasting approach to cancer therapy by circumventing the drug resistance mechanisms that frequently arise with conventional treatments [18 - 20].

TUMOUR HETEROGENEITY

In clinical settings, tumours are classified into several categories using histological and genetic analysis. For instance, there are 120 different types of central nervous system cancer [21]. Most significantly, this classification is insufficient to capture the heterogeneous character of intra-tumour cancer cells, which poses a challenge to efforts to develop targeted drugs for treating human disease. Research on the intra-tumoral heterogeneity in cancer cell population gained momentum in the early 1980s [23 - 27], even though the existence of several cellular populations within a tumour has been recognised since 1953 [22]. It is often known that because cancer cells are malleable, their characteristics can vary with a tumour throughout time and space [28]. As a result, a single tumour can yield a variety of human cell lines. It follows that studying specific medication delivery methods *in vitro* and *in vivo* on certain cell lines would not precisely replicate the genetic and phenotypic characteristics of the original tumour.

Theories of cancer stem cells and clonal evolution both lend credence to intra-tumoral heterogeneity resulting from genetic variation and epigenetic alteration [29, 30]. The cancer stem cell theory predicts the hierarchical heterogeneity of cell populations observed in blood cells derived from hematopoietic stem cells. It is reasonable to anticipate that longer-lived somatic stem cells are more likely than shorter-lived non-stem cells to accumulate an oncogenic repertoire of genetic alterations. Furthermore, it is believed that metastasis, recurrence, and resistance to chemotherapy and radiation are caused by cancer stem cells. In fact, some have proposed that cancer is a type of stem cell illness. However, any differentiated bulk cancer cells within a tumour may haphazardly acquire genetic mutations and/or epigenetic variations to transform into cancer stem-like cells (also called tumour initiating cells), according to the standard stochastic clonal evolution theory. Thus, the question of whether cancer is a stem cell disease remains unanswered. A single cell type's monoculture or aggregate bulk is not what constitutes a tumour. Even cultivated cell lines exhibit heterogeneity within the cell population, comprising cancer stem-like cells that differ from ordinary bulk cells in their surface markers and are resistant to common cytotoxic therapies [31]. It is significant to highlight that surface markers characterising a population of cancer stem-like cells may differ from isolated cell lines from that tumour [32], suggesting that cancer stem cells may not have the same origin.

One intriguing characteristic of cancer cells is their dynamic condition within the tumor's makeup, as noted in a 1985 review [28]. Tumour composition is dynamic due to the genetic instability and epigenetic diversity of cancer cells, which are selected to adapt to different tumour microenvironments in order to survive. The epithelial–mesenchymal transition is one of the most well-known transformations of differentiated cells into a more carcinogenic phenotype, which includes cancer stem-like cells in metastasis [33, 34]. In a form of dynamic equilibrium, distinct subpopulations of different phases of transition between these different cell phenotypes can exist in a tumour at any given time. For instance, recent research indicates that soluble factors, such as interleukin-6, may induce differentiated cancer cells (*i.e.*, non-cancerous stem cells) to revert to cells that resemble cancer stem cells [35]. Therefore, tumours can be viewed as a community of diverse subpopulations that interact with the tumour stroma's immune inflammatory cells, cancer-associated fibroblasts, pericytes, non-cancerous endothelial cells, and stem and progenitor cells. They can also respond to environmental cues brought on by soluble factors. These cells have the ability to combine to form an extremely complex tumour microenvironment [36]. When combined, targeting cancer cells with a single surface marker corresponds to targeting a single population within a mixed population that is ever-changing and in motion. A single surface marker's ability to identify and diagnose a specific cancer cell type may lead to an overestimation of malignancies because that surface marker shares characteristics with normal cells found inside tumours. Therefore, it is commonly accepted that single surface marker techniques are "outdated." The numerous surface marker strategy is thought to be a superior method for identifying and separating cancer cells. A lot of research is being conducted to investigate novel targeting moieties using cutting-edge methods, such as phase display and aptamer screening, employing a primary tumour sample or a specific cell line that expresses multiple surface markers. It is anticipated that using cell-specific techniques will lead to more selective tumour targeting. Over the past few decades, a noticeable surge has occurred in the number of research articles on nanoparticles that specifically target cancer cells [37]. The methods discussed have not resulted in notable advancements in clinical practice. Cell-specific targeted drug delivery was considered as far back as 1987 [38], and it may not be a practical alternative for treating solid tumours in clinics. The "magic bullet" proposed by Paul Erlich is still merely a desirable idea.

BIOLOGICAL BARRIERS

Compared to peripheral cancers, brain tumours are more difficult to treat (Fig. **1**). Physiological barriers, such as the blood-brain tumour barrier (BBTB), the overexpressed efflux pumps, and the blood-brain barrier (BBB), prevent drugs from passing through the central nervous system (CNS) and reaching the cancer site. However, the intrinsic characteristics of brain tumours, which are further limited in their therapeutic effects and raise the risk of failure and recurrence, include immune escape, drug resistance, invasion, infiltration, and high heterogeneity caused by the tumour microenvironment (TME) and cancer stem cells (CSC). Patients with brain tumours receiving standard care had a miserable 2- and 5-year survival rate of 27% and 10%, respectively, and a median survival of only about 20 months.

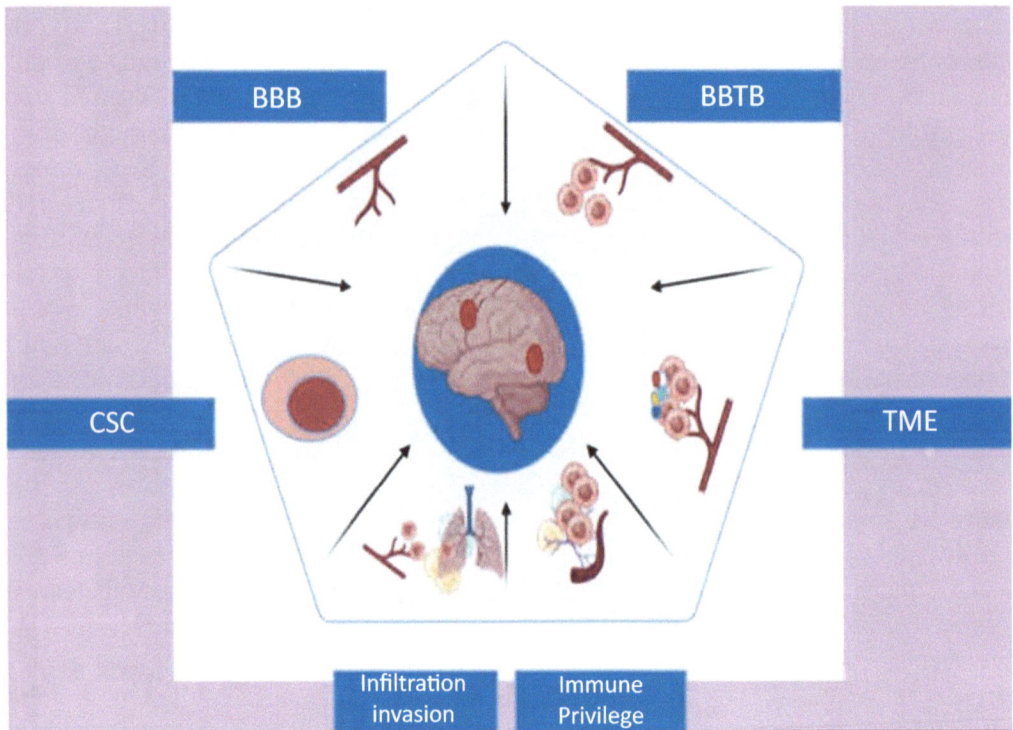

Fig. (1). Treatment difficulties for brain cancer.

Blood-Brain Barrier and Blood-Tumour Barrier

Both endogenous and foreign substances must pass through the blood-brain barrier (BBB), which is composed of neurons, astrocytes, pericytes, brain capillary endothelial cells (BCEC), and basement membranes. The BBB prevents about 98% of small molecule drugs and all macromolecular medications from entering the central nervous system [39]. It does this by operating under the following principles.

Paracellular Barrier: Only lipophilic compounds and hydrophilic small molecules are permitted to reach the brain, and this is due to the tight connections between BCEC cells, which severely restrict the passive diffusion of medicines into the central nervous system [40].

Transcellular Barrier: The lower endocytosis activity of BCEC cells in comparison to other brain cells significantly restricts the transcellular transport of medications across the blood-brain barrier [41].

Enzyme Barrier: BCEC dramatically up-expresses the enzymes nucleotidase, peptidase, phosphatase, esterase, and cytochrome P450, which enhances the ability of BBB cells to hydrolyze drugs.

Immunologic Barrier: Mast cells, macrophages, and microglia work together to create an immunologic barrier that speeds up the removal of drugs [42].

Efflux Proteins: The BBB's overexpression of efflux proteins, which actively pump out medicines and restrict permeability, includes ATP-binding cassette transporters (P-gp, BCRP, MRPs), as well as solute carrier transporters. Furthermore, it is a primary cause of medication resistance in brain tumours. The growth of angiogenesis compromises the integrity and normal functions of the BBB when a brain tumour exceeds 2 mm^3, creating the BBTB [43]. The best method for nanoparticle accumulation has long been thought to be passive tumour targeting *via* the enhanced permeability and retention (EPR) effect. On the other hand, brain tumours have significantly smaller vascular pores (7 ~ 100 nm) and a much weaker EPR. As a result, medications still have trouble getting to brain tumour sites *via* the EPR effect [44]. Because of this, BBTB is thought to constitute yet another significant barrier to drug transport in the treatment of brain tumours, severely limiting the ability of medications to reach tumour tissues.

Infiltration and Invasion

The cells that make up brain tumours have an aggressive behaviour towards the surrounding tissues. Through the following processes, a single brain tumour cell can penetrate healthy tissues and develop into a tumour. The brain tumour cells migrate, gather at the surrounding arteries, and produce glioma-derived substances that interfere with normal endothelial cell-capillary membrane interaction. These factors include TGF-β2, reactive oxygen species (ROS), and proinflammatory peptides. The factors then cause and activate matrix metalloproteinases (MMPs), which downregulate claudin proteins and cause more tight junction breakdown. Due to VEGF overexpression, these mechanisms aid in the development of aberrant new blood vessels, the migration of endothelial cells, and the breakdown of the extracellular matrix and vascular basement membrane [45 - 47]. Consequently, the abnormally rapid expansion of the vasculature causes the BBB to be destroyed, which in turn causes the tight junctions to cease functioning, allowing the tumour to infiltrate and spread.

Brain Cancer Stem Cells

Brain tumour cell subsets exhibit traits of stem cells and display markers characteristic of stem cells, such as CD133, A2B5, and EGFRvIII [48, 49]. Following are the traits displayed by the stem cells:

- **Aggressiveness:** This includes resistance to chemoradiotherapy and a highly migratory and invasive nature.
- **Self-renewal and Differentiation:** Like normal stem cells or progenitor cells, CSCs have the ability to self-renew and differentiate into various cancer cell lineages within tumor tissues.
- **Drug Resistance:** CSCs exhibit multidrug resistance (MDR), characterized by their enhanced capacity to repair damaged DNA and eliminate toxic compounds [50].

Furthermore, CSC can increase angiogenesis and the production of efflux transporters and anti-apoptotic genes. Because of their invasiveness, resistance, self-renewal, and differentiation, stem cells that have penetrated the brain parenchyma may eventually cause disease recurrence even while normal therapy kills the majority of tumour cells [51, 52]. Thus, eliminating tumour stem cells is a crucial area of research to overcome multidrug resistance and boost the effectiveness of tumour treatment.

Immune Escape

The BBB keeps out the majority of dangerous substances, keeping the brain in a comparatively safe environment and preventing immune system attacks from happening too frequently. There are dire repercussions when autoimmunity targets brain cells. As a result, the CNS immune system is typically suppressed. However, it is a well-established dogma that the central nervous system (CNS) lacks healthy functioning dendritic and lymphatic cells necessary for antigen presentation [52]. As a result, there is rarely active immune surveillance in the central nervous system (CNS), which creates a favourable environment for tumour formation [53]. It has been demonstrated that immunotherapy offers therapeutic potential for treating various solid tumours, including non-small cell lung cancer and melanoma [54]. It has not been demonstrated that the present immunotherapy significantly increases the survival rate of brain tumour patients in clinical settings in any case. This is primarily due to the BBB's inability to allow immunological components, including antibodies and immune cells, to penetrate the central nervous system [55].

Tumor Microenvironment

Tumour cells, tumour stem cells, blood vessels, lymphatics, immune cells, fibroblasts, and the extracellular matrix comprise the tumour microenvironment, which provides an ideal setting for tumour cell development, division, angiogenesis, and metastasis [56]. Additionally, TME primarily uses the following strategies to shield tumour cells. High microvessel proliferation is caused by enhanced vascular endothelial growth factor activity. Drug resistance is the outcome of tumour cells interacting with growth factors or cytokines to obtain nutrients from aberrant blood vessels. This leads to the proliferation and invasion of fibroblasts and macrophages. Medications cannot enter tumour cells through the microenvironment due to the cross-linking structure of the extracellular matrix, which is composed of fibrous collagen, proteoglycans, stromal cell proteins, and hyaluronic acid. As a result, the tumour cells reject the treatment of the medications. Tumour initiation and progression are promoted by extracellular matrix, which not only provides essential structure but also aids in the transportation of nutrients and oxygen [57].

PRECISION MEDICINE AND PERSONALIZED THERAPY

The field of precision oncology is complex and requires a growing number of diagnostic and therapy prediction data to guide treatment choices and clinical trial enrollment. Together with clinically validated genetic studies, other omics have begun to appear in clinical labs, adding to the complexity and demand for data processing, with machine learning emerging as a viable method. Creative study designs are essential for utilizing the new therapies and incorporating additional data [58].

An evolving strategy for tumour prevention and therapy that considers the genetic variations both inside and between tumours, the immunological milieu surrounding the tumour, the patient's lifestyle, and any associated morbidities is known as personalised medicine, or precision medicine in oncology. PM has the capacity to modify the tumour immune environment and target treatment towards the oncogenic drivers of the tumour. Moreover, PM seeks to maximise tumour response while accounting for the toxicities brought on by therapy for each individual patient. In this method, the maintenance of organ function and, consequently, quality of life, is paired with optimised tumour response. Furthermore, the goal of this is, of course, to provide better patient care.

Firstly, Gambardella *et al.* [59] summarise the history of the advancement of targeted treatments for tumours with a distinct oncogenic driver during the last few decades. They stress the importance of utilising a multi-omics strategy that integrates changes in DNA and RNA to enhance comprehension of tumour biology, intra-tumor heterogeneity, and the formation of immune-defense systems. This makes it possible to find tumour-specific biomarkers and to create cutting-edge treatment plans that maximise response rates and prevent medication resistance. Additionally, Gambarella *et al.* discussed the developing field of liquid biopsies, a non-invasive method of early treatment resistance development and cancer detection and diagnosis.

One popular and widely utilised model where tumour-tailored treatment has already been applied is colorectal cancer. The recognised process of how genetic and epigenetic events gradually accumulate to generate carcinomas in colorectal tumours is well understood. As a result, focused therapy decisions in the current standard of care are guided by the identification of prognostic and predictive biomarkers, such as microsatellite instability (MSI) and KRAS (Kirsten Rat Sarcoma virus). Koulis *et al.* [60] provided an overview of new liquid biopsy platforms and developing biomarkers in this Special Issue. These developments could lead to novel combination treatment approaches that target both the tumour microenvironment and tumour cells.

Furthermore, driver mutations in non-small cell lung malignancies (NSCLCs) are being found, and effective treatment plans have been created for these alterations. Known oncogenic drivers in NSCLC, as well as various targeted therapeutic approaches that have been demonstrated to increase overall survival in these patients, are described by Ferrara *et al.* [61]. The scientists also conclude that further investigation is required to identify new molecular targets, as only a small percentage of NSCLC express programs presently understood are oncogenic. Mazo *et al.* [62] present an *in silico* analysis to optimise tumour- and patient-tailored treatments in breast cancer patients. They show that in patients with human epidermal growth factor receptor (HER)-2-negative and estrogen receptor (ER)-positive breast cancer, four well-known numeric risk scores (OncoMasTR, EndoPredict, OncotypeDX, and tumor-infiltrating lymphocytes) were significantly predictive of pathological complete remission to neoadjuvant chemotherapy. PM can be used in oncology to customise supportive care for patients with cancer diagnoses in addition to optimising anticancer medicines. Sufficient management of cancer-related pain is critical to improving the quality of life for all patients receiving anticancer treatment, especially when they approach the end of their lives and no longer have access to antitumor medications. The main purpose of opioids is to treat chronic pain brought on by cancer. The best possible pain management was hampered by interpatient variation in opioid-related adverse effects and treatment efficacy. In order to optimise analgesic treatment techniques and fine-tune opioid prescriptions, Bugada *et al.* [10] evaluated the pharmacogenomics data on genetic variations and opioid responsiveness [63].

Role of Target Drug Delivery in Personalized Medicine

Because targeted drug delivery allows doctors to customise treatments to each patient's unique tumor's genetic and molecular profile, it is essential to personalised medicine, especially in the treatment of cancer. With this method, side effects are reduced and treatment success is increased because it guarantees that the particulars of each patient's malignancy are targeted. Utilising cutting-edge diagnostic techniques like genome sequencing and biomarker analysis, personalised therapy pinpoints particular targets within cancer cells. These medicines can more effectively target these targets and improve therapeutic outcomes by adding targeted medication delivery. Patients will have a better overall treatment experience thanks to this precision, which also lowers the risk of adverse responses. Targeted delivery methods, such as antibody-drug conjugates and nanoparticles, can also be tailored to each patient's unique requirements, providing a more efficient and customised course of treatment. The combination of personalised medicine and targeted drug delivery is a major breakthrough in the fight against cancer, opening the door to more efficient and individualised treatment choices [64 - 66].

IMMUNOTHERAPY AND COMBINATION THERAPIES

Understanding the immune system's function in cancer is crucial in order to comprehend the reasoning behind combining immunotherapy with other modalities. Three essential stages comprise the immune response to cancer: escape, balance, and elimination [67]. First, early lesions can be eliminated by elements of the innate and adaptive immune systems upon detection of antigens specific to a tumour, which triggers an immune response. Tumour cells that are not removed during the second phase are kept in an equilibrium, preventing the tumour from growing but also preventing it from being completely eradicated.

On the other hand, cancer cells that can evade, thwart, or suppress the body's natural defences may proliferate if the immune system and tumour are not in balance [68, 69]. Tumour cells can evade immune monitoring due to a variety of variables present in the tumour microenvironment [69]. For instance, interleukin-6 (IL-6), tumour necrosis factor-α, and IL-23 are produced from myeloid-derived suppressive cells (MDSC) and exhibit dominant cancer-promoting activity. These elements, along with others present in the tumour microenvironment, may cause regulatory T cells (Tregs) to be recruited and/or induced. Tregs utilise a variety of methods, including TGF-β and IL-10, to suppress the immune response against cancer [69 - 71]. Enhancing the immune system's innate antitumor properties is the aim of cancer immunotherapy against this immunosuppressive background. Remarkably, however, several traditional cancer therapies, including radiation and chemotherapy, also possess immune-stimulating modes of action (Fig. **1**) [72]. Tumour antigens released by treatment-induced cell death may be ingested by antigen-presenting cells, processed, and presented to naïve T lymphocytes, rendering the cells vulnerable to lysis. Likewise, immunosuppressive cells, such as Tregs and MDSC, can be depleted by anticancer therapy to bolster a latent antitumor immune response [72 - 74].

In this context, physical tumor burden reduction is one way of potentiating the immune response that is frequently overlooked. For instance, when a cancer antigen is repeatedly presented to mice, cytotoxic T cells that were previously reactive to the antigen develop tolerance, and the tumor grows. It is interesting to note that T cells can return to full activity in an antigen-free environment [75]. These findings show that the removal of tumor antigen may contribute to the development of a successful immune response. They also imply that cancer therapies that are successful in lowering the burden of a primary tumor, such as androgen ablation for prostate cancer [76], may be useful adjuncts to immunotherapy [77].

ENGINEERING TARGETED DRUG DELIVERY SYSTEMS

Because most therapeutic agents are restricted by the blood-brain barrier (BBB), designing targeted drug delivery systems for brain tumours has particular problems. Technological developments in biomedical engineering and nanotechnology have enabled the creation of highly targeted delivery systems that can penetrate the blood-brain barrier and specifically target tumour cells. Among the most promising carriers are polymeric micelles, liposomes, and nanoparticles, which are designed to encapsulate medications and deliver them directly into the tumor. Targeting ligands, including peptides or antibodies, can be used to functionalize these carriers by recognising and binding to particular receptors that are overexpressed on brain tumour cells. Furthermore, stimuli-responsive systems improve the accuracy of drug administration by releasing their payload in reaction to variations in the tumor's specific microenvironment, such as pH or enzyme activity. With these advances, systemic toxicity and side effects are reduced while also improving drug accumulation within the tumour. Brain tumour patients may now benefit from more effective and individualised treatment plans thanks to the combination of targeted drug delivery systems and cutting-edge imaging methods, which also offer real-time monitoring of drug distribution and therapy response [78 - 81].

BIOMARKER IDENTIFICATION AND VALIDATION

The advancement of targeted medicines and personalised medicine depends on the identification and validation of biomarkers. High-throughput methods including transcriptomics, proteomics, metabolomics, and genomics are used in the identification of biomarkers. These methods examine biological samples and identify chemicals associated with specific diseases or disorders. In order to evaluate and interpret the enormous amount of data produced and find possible biomarkers by looking for patterns and correlations, bioinformatics tools are essential.

Biomarkers are rigorously validated after they are discovered to ensure their validity and applicability. The biomarker's specificity, sensitivity, and repeatability in identifying the target condition are evaluated by analytical validation. Assessing the biomarker's performance in clinical settings is a crucial step in clinical validation to ensure it accurately represents disease states and can predict treatment responses. This thorough procedure is essential for improving disease diagnosis, prognosis, and the creation of individualised treatment plans, as well as for converting biomarkers into clinically beneficial instruments [82 - 85]. Cancer biomarkers, also known as tumor-associated antigens (TAA), are important targets for cancer treatments. Antibody-based therapies that target cancer biomarkers include monoclonal antibodies (MoAbs), radiolabeled MoAbs, antibody-drug conjugates (ADCs), and bispecific T cell engagers (BiTEs) [86 - 91]. CAR-T cells, also known as chimeric antigen receptor-engineered T cells, have recently advanced the field of cancer immunotherapy [92 - 95]. Apart from enhancing the creation and production of these targeted agents, it is equally important to identify novel indicators of cancer. More drugs targeting the following key biomarkers are rapidly advancing from the bench to the patient's bedside in cancer treatment.

FUTURE PROSPECTIVE

Anticipated advancements and emerging technologies in targeted drug delivery for tumours have great potential to enhance the outcomes of cancer treatments. Technological developments in nanotechnology are opening doors for the creation of multifunctional nanoparticles that can carry drugs, monitor therapy response, and diagnose conditions simultaneously. These so-called "theranostic" nanoparticles can be designed to specifically target tumour markers, improve contrast on imaging scans, and transport therapeutic medicines to cancer cells. Furthermore, gene editing technologies—such as CRISPR-Cas9—offer the opportunity to correct genetic abnormalities that accelerate cancer development or make tumors more susceptible to current treatments. Bioengineering innovations like 3D printing are making it possible to create customised medication delivery systems that are suited to the distinct anatomical and clinical characteristics of particular tumours. Additionally, real-time treatment regimen adjustments and patient-specific response prediction are made possible by the integration of artificial intelligence and machine learning into targeted delivery systems, which optimises their performance and design. It is anticipated that these developments will greatly increase the accuracy and effectiveness of cancer treatments, thereby reducing side effects and enhancing patient outcomes [96 - 100].

CONCLUSION

In conclusion, because of their invasiveness, high recurrence rates, and dismal prognosis, brain tumors—gliomas in particular—present considerable therapy problems. Due to problems including medication resistance, significant side effects, and the inability to fully eliminate or target tumour cells, current therapeutic techniques, such as surgery, chemotherapy, and radiation therapy, frequently fail to manage these tumours effectively. One potential way to maximise therapeutic benefit and reduce side effects is through the development of targeted medication delivery systems. These systems can increase drug delivery precision and guarantee that higher quantities of therapeutic drugs reach the tumour location by utilising both passive and active targeting mechanisms. However, because brain tumours are so complex—due to their heterogeneity and the blood-brain barrier—high-tech, multimodal techniques are required. Important challenges that need to be addressed include tumour heterogeneity, the dynamic nature of the tumour microenvironment, and biological barriers, including the blood-brain barrier and blood-brain tumour barrier.

Novel therapeutics including gene therapy, immunotherapy, and personalised medicine have promise, but they still need to be further investigated and proven before they can be used in clinical settings. Precision medicine seeks to customise treatments to the unique genetic and molecular characteristics of individual tumours by utilising targeted drug delivery and biomarker detection. By improving drug targeting and lowering systemic toxicity, this strategy promises to improve treatment outcomes. Therefore, it is imperative that innovative and interdisciplinary efforts continue to focus on developing targeted, highly efficient, and low-toxicity medicines for the management and treatment of brain tumours.

AUTHORS' CONTRIBUTIONS

Meenakshi Attri was responsible for investigation, providing resources, and performing formal analysis. Asha Raghav contributed to data curation, investigation, and the review and editing of the manuscript. Mohit Agrawal handled the conceptualization of the study, supervision, provision of resources, visualization, and writing of the original draft.

REFERENCES

[1]   Sung H, Ferlay J, Siegel RL, *et al.* Global cancer statistics 2020: GLOBOCAN estimates of incidence and mortality worldwide for 36 cancers in 185 countries. CA Cancer J Clin 2021; 71(3): 209-49.
        [http://dx.doi.org/10.3322/caac.21660] [PMID: 33538338]

[2]   Patel AP, Fisher JL, Nichols E, *et al.* Global, regional, and national burden of brain and other CNS cancer, 1990–2016: a systematic analysis for the Global Burden of Disease Study 2016. Lancet Neurol 2019; 18(4): 376-93.
        [http://dx.doi.org/10.1016/S1474-4422(18)30468-X] [PMID: 30797715]

[3]   Reifenberger G, Wirsching HG, Knobbe-Thomsen CB, Weller M. Advances in the molecular genetics of gliomas — implications for classification and therapy. Nat Rev Clin Oncol 2017; 14(7): 434-52.
[http://dx.doi.org/10.1038/nrclinonc.2016.204] [PMID: 28031556]

[4]   Lin S, Xu H, Zhang A, *et al.* Prognosis analysis and validation of m6A signature and tumor immune microenvironment in glioma. Front Oncol 2020; 10: 541401.
[http://dx.doi.org/10.3389/fonc.2020.541401] [PMID: 33123464]

[5]   Ganz JC. Low grade gliomas. Prog Brain Res 2022; 268(1): 271-7.
[PMID: 35074085]

[6]   Zhang J, Stevens MFG, Bradshaw TD. Temozolomide: mechanisms of action, repair and resistance. Curr Mol Pharmacol 2012; 5(1): 102-14.
[http://dx.doi.org/10.2174/1874467211205010102] [PMID: 22122467]

[7]   Yan Y, Xu Z, Dai S, Qian L, Sun L, Gong Z. Targeting autophagy to sensitive glioma to temozolomide treatment. J Exp Clin Cancer Res 2016; 35(1): 23.
[http://dx.doi.org/10.1186/s13046-016-0303-5] [PMID: 26830677]

[8]   Ozdemir-Kaynak E, Qutub AA, Yesil-Celiktas O. Advances in glioblastoma multiforme treatment: new models for nanoparticle therapy. Front Physiol 2018; 9: 170.
[http://dx.doi.org/10.3389/fphys.2018.00170] [PMID: 29615917]

[9]   Aparicio-Blanco J, Sanz-Arriazu L, Lorenzoni R, Blanco-Prieto MJ. Glioblastoma chemotherapeutic agents used in the clinical setting and in clinical trials: Nanomedicine approaches to improve their efficacy. Int J Pharm 2020; 581: 119283.
[http://dx.doi.org/10.1016/j.ijpharm.2020.119283] [PMID: 32240807]

[10]   Sousa F, Dhaliwal HK, Gattacceca F, Sarmento B, Amiji MM. Enhanced anti-angiogenic effects of bevacizumab in glioblastoma treatment upon intranasal administration in polymeric nanoparticles. J Control Release 2019; 309: 37-47.
[http://dx.doi.org/10.1016/j.jconrel.2019.07.033] [PMID: 31344424]

[11]   Weenink B, French PJ, Sillevis Smitt PAE, Debets R, Geurts M. Immunotherapy in glioblastoma: current shortcomings and future perspectives. Cancers (Basel) 2020; 12(3): 751.
[http://dx.doi.org/10.3390/cancers12030751] [PMID: 32235752]

[12]   Chelliah SS, Paul EAL, Kamarudin MNA, Parhar I. Challenges and perspectives of standard therapy and drug development in high-grade gliomas. Molecules 2021; 26(4): 1169.
[http://dx.doi.org/10.3390/molecules26041169] [PMID: 33671796]

[13]   Conniot J, Talebian S, Simões S, Ferreira L, Conde J. Revisiting gene delivery to the brain: silencing and editing. Biomater Sci 2021; 9(4): 1065-87.
[http://dx.doi.org/10.1039/D0BM01278E] [PMID: 33315025]

[14]   Singh D, Tiwari P, Nagdev S. Particulate vaccine dispersions emerge as a novel carrier for deep pulmonary immunization. Curr Nanomedicine. 2023;13(2):71–4.
[http://dx.doi.org/10.2174/2468187313666230714124009]

[15]   Dubey S, Suraj MR, Goni T, *et al.* 3D QSAR studies of 3, 16 and 17 position modifications in steroidal derivatives for CNS anticancer activity. Curr Res Chem. 2023;15(1):1–4.
[http://dx.doi.org/10.3923/crc.2023.1.4]

[16]   Kim J, Cho H, Lim DK, Joo MK, Kim K. Perspectives for improving the tumor targeting of nanomedicine *via* the EPR effect in clinical tumors. Int J Mol Sci. 2023;24(12):10082.
[http://dx.doi.org/10.3390/ijms241210082]

[17]   Sutradhar KB, Amin ML. Nanoemulsions: Increasing possibilities in drug delivery. Eur J Nanomed 2014; 6(3): 157-70.

[18]   Kumar PB, Kadiri SK, Khobragade DS, *et al.* Synthesis, characterization and biological investigations of some new Oxadiazoles: *In-vitro* and *In-Silico* approach. Results in Chemistry. 2024;7:101241.

[http://dx.doi.org/10.1016/j.rechem.2023.101241]

[19]   Thomas A, Teicher BA, Hassan R. Antibody–drug conjugates for cancer therapy. Lancet Oncol 2016; 17(6): e254-62.
[http://dx.doi.org/10.1016/S1470-2045(16)30030-4] [PMID: 27299281]

[20]   Danhier F. To exploit the tumor microenvironment: Since the EPR effect fails in the clinic, what is the future of nanomedicine? J Control Release 2016; 244(Pt A): 108-21.
[http://dx.doi.org/10.1016/j.jconrel.2016.11.015] [PMID: 27871992]

[21]   Louis DN, Ohgaki H, Wiestler OD, *et al.* The 2007 WHO Classification of tumours of the central nervous system. Acta Neuropathol 2007; 114: 97-109. [PMC free article]. [PubMed]. [Google Scholar].

[22]   Levan A, Hauschka TS. Endomitotic reduplication mechanisms in ascites tumors of the mouse. J Natl Cancer Inst 1953; 14(1): 1-43. [PubMed]. [Google Scholar].
[PMID: 13097135]

[23]   Dexter DL, Calabresi P. Intraneoplastic diversity. Biochim Biophys Acta 1982; 695: 97-112. [PubMed]. [Google Scholar].

[24]   Fidler IJ, Hart IR. Biological diversity in metastatic neoplasms: origins and implications. Science 1982; 277: 998-1003. [PubMed]. [Google Scholar].

[25]   Heppner GH, Miller BE. Tumor heterogeneity: biological implications and therapeutic consequences. Cancer Metastasis Rev 1983; 2(1): 5-23.
[http://dx.doi.org/10.1007/BF00046903] [PMID: 6616442]

[26]   Miller FR. Intratumor heterogeneity. Cancer Metastasis Rev 1982; 1: 319-34. [PubMed]. [Google Scholar].
[http://dx.doi.org/10.1007/BF00124215]

[27]   Poste G, Greig R. On the genesis and regulation of cellular heterogeneity in malignant tumors. Invasion Metastasis 1982; 2(3): 137-76. [PubMed]. [Google Scholar].
[PMID: 6765249]

[28]   Welch DR, Tomasovic SP. The implication of tumor progression on clinical oncology. Clin Exp Matast 1985; 3: 151-88. [PubMed]. [Google Scholar].
[http://dx.doi.org/10.1007/BF01786761]

[29]   Adams JM, Strasser A. Is tumor growth sustained by rare cancer stem cells or dominant clones? Cancer Res 2008; 68(11): 4018-21.
[http://dx.doi.org/10.1158/0008-5472.CAN-07-6334] [PMID: 18519656]

[30]   Shackleton M, Quintana E, Fearon ER, *et al.* Heterogeneity in cancer: cancer stem cells *versus* clonal evolution. Cell 2009; 138(5): 822-9. [PubMed]. [Google Scholar].
[http://dx.doi.org/10.1016/j.cell.2009.08.017]

[31]   Charafe-Jauffret E, Ginestier C, Iovino F, *et al.* Breast cancer cell lines contain functional cancer stem cells with metastatic capacity and a distinct molecular signature. Cancer Res 2009; 69: 1302-13. [PMC free article]. [PubMed]. [Google Scholar].

[32]   Hwang-Verslues WW, Kuo W-H, Chang P-H, *et al.* Hwang-Verslues WW, Kuo W-H, Chang P-H, Pan C-C, Wang H-H, Tsai S-T, Jeng Y-M, Shew J-Y, Kung JT, Chen C-H, Lee EY-HP, Chang K-J, Lee W-H. Multiple lineages of human breast cancer stem/progenitor cells identified by profiling with stem cell markers. PLoS One 2009; 4(12): e8377. [PMC free article]. [PubMed]. [Google Scholar].
[http://dx.doi.org/10.1371/journal.pone.0008377]

[33]   Mani SA, Guo W, Liao M-J, *et al.* The epithelial–mesenchymal transition generates cells with properties of stem cells. Cell 2008; 133: 704-15. [PMC free article]. [PubMed]. [Google Scholar].

[34]   Gupta PB, Onder TT, Jiang G, *et al.* Identification of selective inhibitors of cancer stem cells by high-throughput screening. Cell 2009; 138(4): 645-59. [PMC free article]. [PubMed]. [Google Scholar].

[http://dx.doi.org/10.1016/j.cell.2009.06.034]

[35] Iliopoulos D, Hirsch HA, Wang G, Struhl K. Inducible formation of breast cancer stem cells and their dynamic equilibrium with non-stem cancer cells *via* IL6 secretion. Proc Natl Acad Sci USA 2011; 108(4): 1397-402.
[http://dx.doi.org/10.1073/pnas.1018898108] [PMID: 21220315]

[36] Hanahan D, Weinberg RA. Hanahan D, Weinberg RA. Hallmarks of cancer: the next generation. Cell 2011; 144(5): 646-74. [PubMed]. [Google Scholar].
[http://dx.doi.org/10.1016/j.cell.2011.02.013]

[37] Vicent MJ, Ringsdorf H, Duncan R. Polymer therapeutics: Clinical applications and challenges for development. Adv Drug Deliv Rev 2009; 61(13): 1117-20.
[http://dx.doi.org/10.1016/j.addr.2009.08.001] [PMID: 19682516]

[38] Welch DR. Biologic considerations for drug targeting in cancer patients. Cancer Treat Rev 1987; 14(3-4): 351-8.
[http://dx.doi.org/10.1016/0305-7372(87)90029-6] [PMID: 3326669]

[39] Zhao Y, Peng Y, Yang Z, *et al.* pH-redox responsive cascade-targeted liposomes to intelligently deliver doxorubicin prodrugs and lonidamine for glioma. Eur J Med Chem 2022; 235: 114281.
[http://dx.doi.org/10.1016/j.ejmech.2022.114281] [PMID: 35344903]

[40] Azarmi M, Maleki H, Nikkam N, Malekinejad H. Transcellular brain drug delivery: A review on recent advancements. Int J Pharm 2020; 586: 119582.
[http://dx.doi.org/10.1016/j.ijpharm.2020.119582] [PMID: 32599130]

[41] Alexander JJ. Blood-brain barrier (BBB) and the complement landscape. Mol Immunol 2018; 102: 26-31.
[http://dx.doi.org/10.1016/j.molimm.2018.06.267] [PMID: 30007547]

[42] Saidijam M, Karimi Dermani F, Sohrabi S, Patching SG. Efflux proteins at the blood–brain barrier: review and bioinformatics analysis. Xenobiotica 2018; 48(5): 506-32.
[http://dx.doi.org/10.1080/00498254.2017.1328148] [PMID: 28481715]

[43] Mojarad-Jabali S, Farshbaf M, Walker PR, *et al.* An update on actively targeted liposomes in advanced drug delivery to glioma. Int J Pharm 2021; 602: 120645.
[http://dx.doi.org/10.1016/j.ijpharm.2021.120645] [PMID: 33915182]

[44] Caro C, Avasthi A, Paez-Muñoz JM, Pernia Leal M, García-Martín ML. Passive targeting of high-grade gliomas *via* the EPR effect: a closed path for metallic nanoparticles? Biomater Sci 2021; 9(23): 7984-95.
[http://dx.doi.org/10.1039/D1BM01398J] [PMID: 34710207]

[45] Ishihara H, Kubota H, Lindberg RLP, *et al.* Endothelial cell barrier impairment induced by glioblastomas and transforming growth factor β2 involves matrix metalloproteinases and tight junction proteins. J Neuropathol Exp Neurol 2008; 67(5): 435-48.
[http://dx.doi.org/10.1097/NEN.0b013e31816fd622] [PMID: 18431253]

[46] Dubois LG, Campanati L, Righy C, *et al.* Gliomas and the vascular fragility of the blood brain barrier. Front Cell Neurosci 2014; 8: 418.
[http://dx.doi.org/10.3389/fncel.2014.00418] [PMID: 25565956]

[47] Oishi T, Koizumi S, Kurozumi K. Molecular mechanisms and clinical challenges of glioma invasion. Brain Sci 2022; 12(2): 291.
[http://dx.doi.org/10.3390/brainsci12020291] [PMID: 35204054]

[48] Ishii H, Mimura Y, Zahra MH, *et al.* Isolation and characterization of cancer stem cells derived from human glioblastoma. Am J Cancer Res 2021; 11(2): 441-57.
[PMID: 33575080]

[49] Smiley SB, Yun Y, Ayyagari P, *et al.* Development of CD133 targeting multi-drug polymer micellar nanoparticles for glioblastoma – *in vitro* evaluation in glioblastoma stem cells. Pharm Res 2021; 38(6):

1067-79.
[http://dx.doi.org/10.1007/s11095-021-03050-8] [PMID: 34100216]

[50] Phi LTH, Sari IN, Yang YG, *et al.* Cancer stem cells (CSCs) in drug resistance and their therapeutic Implications in cancer treatment. Stem Cells Int 2018; 2018: 1-16.
[http://dx.doi.org/10.1155/2018/5416923] [PMID: 29681949]

[51] Alcantara Llaguno S, Parada LF. Cancer stem cells in gliomas: evolving concepts and therapeutic implications. Curr Opin Neurol 2021; 34(6): 868-74.
[http://dx.doi.org/10.1097/WCO.0000000000000994] [PMID: 34581301]

[52] D'Agostino PM, Gottfried-Blackmore A, Anandasabapathy N, Bulloch K. Brain dendritic cells: biology and pathology. Acta Neuropathol 2012; 124(5): 599-614.
[http://dx.doi.org/10.1007/s00401-012-1018-0] [PMID: 22825593]

[53] Rustenhoven J, Kipnis J. Bypassing the blood-brain barrier. Science 2019; 366(6472): 1448-9.
[http://dx.doi.org/10.1126/science.aay0479] [PMID: 31857468]

[54] Waldman AD, Fritz JM, Lenardo MJ. A guide to cancer immunotherapy: from T cell basic science to clinical practice. Nat Rev Immunol 2020; 20(11): 651-68.
[http://dx.doi.org/10.1038/s41577-020-0306-5] [PMID: 32433532]

[55] Desbaillets N, Hottinger AF. Immunotherapy in glioblastoma: a clinical perspective. Cancers (Basel) 2021; 13(15): 3721.
[http://dx.doi.org/10.3390/cancers13153721] [PMID: 34359621]

[56] Petrova V, Annicchiarico-Petruzzelli M, Melino G, Amelio I. The hypoxic tumour microenvironment. Oncogenesis 2018; 7(1): 10.
[http://dx.doi.org/10.1038/s41389-017-0011-9] [PMID: 29362402]

[57] Zhao Y, Yue P, Peng Y, *et al.* Recent advances in drug delivery systems for targeting brain tumors. Drug Deliv 2023; 30(1): 1-18.
[http://dx.doi.org/10.1080/10717544.2022.2154409] [PMID: 36597214]

[58] Edsjö A, Holmquist L, Geoerger B, *et al.* Precision cancer medicine: Concepts, current practice, and future developments. J Intern Med 2023; 294(4): 455-81.
[http://dx.doi.org/10.1111/joim.13709] [PMID: 37641393]

[59] Gambardella V, Tarazona N, Cejalvo JM, *et al.* Personalized Medicine: Recent Progress in Cancer Therapy. Cancers (Basel) 2020; 12(4): 1009. [PMC free article]. [PubMed]. [CrossRef]. [Google Scholar].
[http://dx.doi.org/10.3390/cancers12041009]

[60] Koulis C, Yap R, Engel R, *et al.* Personalized Medicine—Current and Emerging Predictive and Prognostic Biomarkers in Colorectal Cancer. Cancers (Basel) 2020; 12(4): 812.
[http://dx.doi.org/10.3390/cancers12040812]

[61] Ferrara MG, Di Noia V, D'Argento E, *et al.* Oncogene-addicted non-small-cell lung cancer: Treatment opportunities and future perspectives. Cancers. 2020;12:1196.
[http://dx.doi.org/10.3390/cancers12051196]

[62] Mazo C, Barron S, Mooney C, Gallagher WM. Multi-gene prognostic signatures and prediction of pathological complete response to neoadjuvant chemotherapy in ER-positive, HER2-negative breast cancer patients. Cancers (Basel) 2020; 12(5): 1133.
[http://dx.doi.org/10.3390/cancers12051133] [PMID: 32369904]

[63] Hoeben A, Joosten EAJ, van den Beuken-van Everdingen MHJ. Personalized Medicine: Recent Progress in Cancer Therapy. Cancers (Basel) 2021; 13(2): 242.
[http://dx.doi.org/10.3390/cancers13020242] [PMID: 33440729]

[64] Collins FS, Varmus H. A new initiative on precision medicine. N Engl J Med 2015; 372(9): 793-5.
[http://dx.doi.org/10.1056/NEJMp1500523] [PMID: 25635347]

[65] Liu Y, Cao X. Characteristics and application of targeted drug delivery systems. Int J Mol Sci 2016; 17(7): 1130.

[66] Yin Y, Yao J, Chen X. Targeted drug delivery systems as personalized medicine for cancer. Curr Mol Med 2013; 13(5): 697-713.

[67] Schreiber RD, Old LJ, Smyth MJ. Cancer immunoediting: integrating immunity's roles in cancer suppression and promotion. Science 2011; 331(6024): 1565-70.
[http://dx.doi.org/10.1126/science.1203486] [PMID: 21436444]

[68] Dunn GP, Old LJ, Schreiber RD. Dunn GP, Old LJ, Schreiber RD. The three Es of cancer immunoediting. Annu Rev Immunol 2004; 22(1): 329-60. [PubMed]. [Google Scholar].
[http://dx.doi.org/10.1146/annurev.immunol.22.012703.104803]

[69] Swann JB, Smyth MJ. Immune surveillance of tumors. J Clin Invest 2007; 117(5): 1137-46. [PMC free article]. [PubMed]. [Google Scholar].
[http://dx.doi.org/10.1172/JCI31405]

[70] Zamarron BF, Chen W. Dual roles of immune cells and their factors in cancer development and progression. Int J Biol Sci 2011; 7(5): 651-8.
[http://dx.doi.org/10.7150/ijbs.7.651] [PMID: 21647333]

[71] Zou W. Immunosuppressive networks in the tumour environment and their therapeutic relevance. Nat Rev Cancer 2005; 5(4): 263-74.
[http://dx.doi.org/10.1038/nrc1586] [PMID: 15776005]

[72] Zou W. Immunosuppressive networks in the tumour environment and their therapeutic relevance. Nat Rev Cancer. 2005;5:263–274.

[73] Formenti SC, Demaria S. Systemic effects of local radiotherapy. Lancet Oncol 2009; 10(7): 718-26. [PMC free article]. [PubMed]. [Google Scholar].
[http://dx.doi.org/10.1016/S1470-2045(09)70082-8]

[74] Menard C, Martin F, Apetoh L, *et al.* Cancer chemotherapy: not only a direct cytotoxic effect, but also an adjuvant for antitumor immunity. Cancer Immunol Immunother. 2008;57:1579–1587.

[75] den Boer AT, van Mierlo GJ, Fransen MF, *et al.* The tumoricidal activity of memory CD8+ T cells is hampered by persistent systemic antigen, but full functional capacity is regained in an antigen-free environment. J Immunol. 2004;172:6074–6079.

[76] Drake CG, Doody AD, Mihalyo MA, *et al.* Androgen ablation mitigates tolerance to a prostate/prostate cancer-restricted antigen. Cancer Cell. 2005;7:239–249.

[77] North RJ. Cyclophosphamide-facilitated adoptive immunotherapy of an established tumor depends on elimination of tumor-induced suppressor T cells. J Exp Med 1982; 155(4): 1063-74.
[http://dx.doi.org/10.1084/jem.155.4.1063] [PMID: 6460831]

[78] Pardridge WM. Drug transport across the blood-brain barrier. J Cereb Blood Flow Metab 2012; 32(11): 1959-72.
[http://dx.doi.org/10.1038/jcbfm.2012.126] [PMID: 22929442]

[79] Blanco E, Shen H, Ferrari M. Principles of nanoparticle design for overcoming biological barriers to drug delivery. Nat Biotechnol 2015; 33(9): 941-51.
[http://dx.doi.org/10.1038/nbt.3330] [PMID: 26348965]

[80] Sumer B, Gao J. Theranostic nanomedicine for cancer. Nanomedicine (Lond) 2008; 3(2): 137-40.
[http://dx.doi.org/10.2217/17435889.3.2.137] [PMID: 18373419]

[81] Chen Y, Liu L. Modern methods for delivery of drugs across the blood–brain barrier. Adv Drug Deliv Rev 2012; 64(7): 640-65.
[http://dx.doi.org/10.1016/j.addr.2011.11.010] [PMID: 22154620]

[82] Rifai N, Gillette MA, Carr SA. Protein biomarker discovery and validation: the long and uncertain

path to clinical utility. Nat Biotechnol 2006; 24(8): 971-83.
[http://dx.doi.org/10.1038/nbt1235] [PMID: 16900146]

[83] Sawyers CL. The cancer biomarker problem. Nature 2008; 452(7187): 548-52.
[http://dx.doi.org/10.1038/nature06913] [PMID: 18385728]

[84] Diamandis EP. Cancer biomarkers: can we turn recent failures into success? J Natl Cancer Inst 2010;
102(19): 1462-7.
[http://dx.doi.org/10.1093/jnci/djq306] [PMID: 20705936]

[85] Benson EE, Jaiswal JK, Srinivasan S. Omics technologies and bioinformatics: Identifying biomarkers
for personalized medicine. Curr Mol Med 2013; 13(4): 471-84.

[86] Goede V, Fischer K, Robrecht S, Giza A, Dyer MJS, Eckart MJ. Long-term outcomes of
chemoimmunotherapy with obinutuzumab/chlorambucil in chronic lymphocytic leukemia. Blood Adv.
2025;9(10):2431–5.
[http://dx.doi.org/10.1182/bloodadvances.2024014875]

[87] Goede V, Fischer K, Busch R, *et al.* Obinutuzumab plus chlorambucil in patients with CLL and
coexisting conditions. N Engl J Med 2014; 370(12): 1101-10.
[http://dx.doi.org/10.1056/NEJMoa1313984] [PMID: 24401022]

[88] Kantarjian HM, DeAngelo DJ, Stelljes M, *et al.* Inotuzumab Ozogamicin *versus* standard therapy for
acute lymphoblastic leukemia. N Engl J Med 2016; 375(8): 740-53.
[http://dx.doi.org/10.1056/NEJMoa1509277] [PMID: 27292104]

[89] Maury S, Chevret S, Thomas X, *et al.* for G. Rituximab in B-lineage adult acute lymphoblastic
leukemia. N Engl J Med 2016; 375(11): 1044-53.
[http://dx.doi.org/10.1056/NEJMoa1605085] [PMID: 27626518]

[90] Verma S, Miles D, Gianni L, *et al.* Trastuzumab emtansine for HER2-positive advanced breast cancer.
N Engl J Med 2012; 367(19): 1783-91.
[http://dx.doi.org/10.1056/NEJMoa1209124] [PMID: 23020162]

[91] Yu S, Li A, Liu Q, *et al.* Recent advances of bispecific antibodies in solid tumors. J Hematol Oncol
2017; 10(1): 155.
[http://dx.doi.org/10.1186/s13045-017-0522-z] [PMID: 28931402]

[92] Zhang C, Liu J, Zhong JF, Zhang X. Engineering CAR-T cells. Biomark Res 2017; 5(1): 22.
[http://dx.doi.org/10.1186/s40364-017-0102-y] [PMID: 28652918]

[93] Yu S, Li A, Liu Q, *et al.* Chimeric antigen receptor T cells: a novel therapy for solid tumors. J
Hematol Oncol 2017; 10(1): 78.
[http://dx.doi.org/10.1186/s13045-017-0444-9] [PMID: 28356156]

[94] Maude SL, Laetsch TW, Buechner J, *et al.* Tisagenlecleucel in children and Young adults with B-cell
lymphoblastic leukemia. N Engl J Med 2018; 378(5): 439-48.
[http://dx.doi.org/10.1056/NEJMoa1709866] [PMID: 29385370]

[95] Porter DL, Levine BL, Kalos M, Bagg A, June CH. Chimeric antigen receptor-modified T cells in
chronic lymphoid leukemia. N Engl J Med 2011; 365(8): 725-33.
[http://dx.doi.org/10.1056/NEJMoa1103849] [PMID: 21830940]

[96] Peer D, Karp JM, Hong S, Farokhzad OC, Margalit R, Langer R. Nanocarriers as an emerging
platform for cancer therapy. Nat Nanotechnol 2007; 2(12): 751-60.
[http://dx.doi.org/10.1038/nnano.2007.387] [PMID: 18654426]

[97] Chen F, Ehlerding EB, Cai W. Theranostic nanoparticles. J Nucl Med 2017; 58(12): 2044-52.
[PMID: 25413134]

[98] Doudna JA, Charpentier E. The new frontier of genome engineering with CRISPR-Cas9. Science
2014; 346(6213): 1258096.
[http://dx.doi.org/10.1126/science.1258096] [PMID: 25430774]

[99]   Murphy SV, Atala A. 3D bioprinting of tissues and organs. Nat Biotechnol 2014; 32(8): 773-85.
        [http://dx.doi.org/10.1038/nbt.2958] [PMID: 25093879]

[100]   Esteva A, Kuprel B, Novoa RA, *et al.* Dermatologist-level classification of skin cancer with deep
        neural networks. Nature 2017; 542(7639): 115-8.
        [http://dx.doi.org/10.1038/nature21056] [PMID: 28117445]

CHAPTER 5

Receptor-Ligand-based Targeting Approaches in Brain Tumors

Ekta Singh[1] and **Sonal Dubey**[2,*]

[1] *Aditya Bangalore Institute of Pharmacy Education and Research, Bengaluru, Karnataka, India*

[2] *College of Pharmaceutical Science, Dayananda Sagar University, Bengaluru-562112, Karnataka, India*

Abstract: Brain tumors pose a significant therapeutic challenge due to their heterogeneity, invasive properties, and limited availability of treatment options. Targeted therapies offer a promising approach to address the complexity of brain tumors by selectively inhibiting molecular pathways critical for tumor growth and survival. Among these targeted approaches, receptor-ligand based targeting strategies have emerged as a promising avenue for precision therapy. This chapter provides a comprehensive overview of receptor-ligand based targeting approaches for brain tumors, focusing on the molecular interactions between receptors and their cognate ligands, the expression profiles of key receptors in different tumor subtypes, and the development of targeted therapeutics. The diverse range of receptors and ligands is implicated in brain tumor biology. Epidermal growth factor receptor (EGFR), vascular endothelial growth factor receptor (VEGFR), and human epidermal growth factor receptor 2 (HER2) are important in the context of this discussion.

Additionally, this chapter examines the challenges associated with delivering targeted therapeutics permeating the blood-brain barrier (BBB) and explores innovative strategies to enhance the delivery of drugs to brain tumors. Promising outcomes and areas for further investigation are highlighted based on a review of preclinical and clinical studies that have evaluated the efficacy and safety of receptor-ligand-based targeting approaches. A discussion on the challenges and future directions in this field, including strategies to overcome resistance mechanisms, enhance treatment specificity, and advance personalized medicine approaches, is incorporated. Overall, this chapter offers valuable insights into the current state and future prospects of receptor-ligand-based targeting approaches for brain tumors. This chapter therefore provides a roadmap for the development of innovative and operational therapies in the fight against this disease.

---

* **Corresponding author Sonal Dubey:** College of Pharmaceutical Science, Dayananda Sagar University, Bengaluru-562112, Karnataka, India; E-mail: drsonaldubey@gmail.com

Prashant Tiwari, Pankaj Kumar Singh & Sunil Kumar Kadiri (Eds.)
All rights reserved-© 2025 Bentham Science Publishers

Keywords: Brain tumor, Clinical studies, Chemotherapy, Drug repurposing, *In-silico* interactions, Multi-target approaches, Malignancy, Preclinical studies, Receptor-ligand interaction, Small molecules, Targeted approaches.

INTRODUCTION

Brain tumors represent a diverse group of neoplasms arising from abnormal growth of cells within the brain and central nervous system (CNS). They pose significant challenges in neuro-oncology due to their heterogeneity, location, and often aggressive nature. Understanding the molecular characteristics of brain tumors is quintessential for the development of effective treatment strategies. Hence, it is noteworthy that targeted therapies have materialized as a favorable approach to address these challenges [1]. Fig. (**1**) illustrates the process of tumorigenesis.

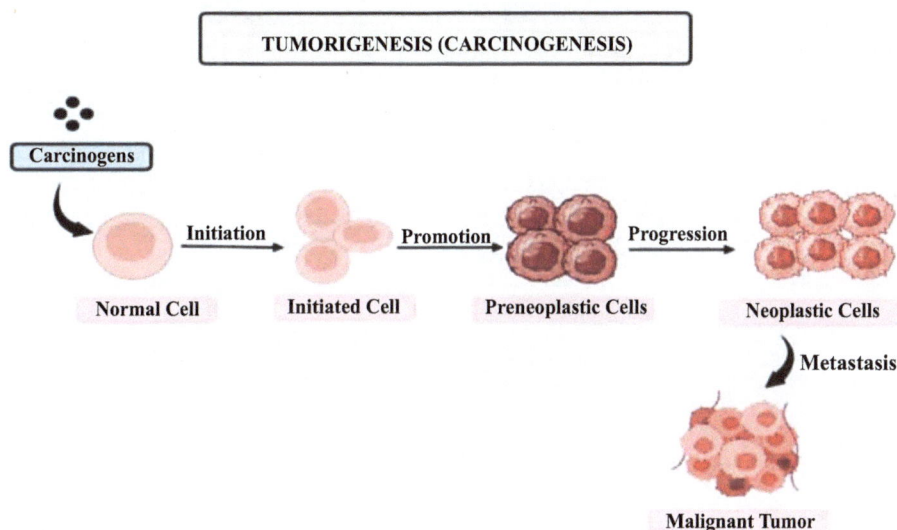

Fig. (1). Brain tumorigenesis.

Brain tumors are relatively rare compared to other cancers, comprising approximately 1.4% of all new cancer cases worldwide. However, they are associated with significant morbidity and mortality, particularly due to their propensity to invade surrounding brain tissue and cause neurological dysfunction.

Primary and secondary can be the two main categories of brain tumors. The origin of primary brain tumors is within the brain or the structures surrounding the brain. Primary brain tumors can be further classified depending on their cell of origin and histological features. Common primary brain tumors comprise gliomas (*e.g.*, glioblastoma (GBM), astrocytoma, oligodendroglioma), meningiomas, and medulloblastomas. Secondary brain tumors are metastatic tumors, and they arise from cancer cells that have spread (metastasized) to the brain from other parts of the body, such as the lung, breast, or colon.

The classification of brain tumors is evolving with the exploration of their site of origin, the cells involved, the parts of the brain affected, the rate of growth, the age of occurrence, and the grade of the tumors. The latest type of classification, which has been widely used, is based on the genes/proteins/biomarkers involved in the brain tumor [1].

Receptor Expression Profiles in Brain Tumors

The exploration of biomarkers involved in brain tumors has facilitated the nomenclature of the tumors based on the genes/ proteins involved. The information about biomarkers involved in brain tumors serves as the basis for receptor-ligand-based targeting approaches in brain tumors. Diagnostic pathology identifies the receptors overexpressed in the cancer, which may serve as potential targets for treatment [2].

The characterization of tumors based on overexpressed or suppressed biomarkers contributes to their specificity in identification. Some examples are compiled in Table **1**.

Table 1. Examples of tumor nomenclature.

Location	Classification of Tumor	Biomarker Modified
Glial cell	Glioma, glioneuroma, ganglioglioma	*IDH* wild type, *EGFR*
Neuronal cell	Neuroma, astrocytoma, gangliocytoma	*IDH* mutant, *ATRX, TP53*
Ependymal cells	Ependymoma	*NF2, MYCN*
Choroid plexus	Choroid plexus papilloma, carcinoma	*PRKCA*
Embryonal cells	Medulloblastoma, neuroblastoma	*WNT* Activated (CTNNB, APC) and SHH activated (*TP53, PTCH1)*
Pineal gland	Pineocytoma	*SMARCB1* mutant
Cranium and paraspinal nerves	Schwannoma, neurofibroma, perineurioma, paraganglioma	*NF1*
Meninges	Meningioma	*SMARCE1, BAP1, KLF4/TRAF7, TERT, CDKN2A/B, H3K27*
Vascular tissue	Hemangioma, hemangioblastoma	*VEGF*
Skeletal tissue	Rhabdomyosarcoma	*PAX3-FOXO1, PAX7-FOXO1*

(Table 1) cont.....

Location	Classification of Tumor	Biomarker Modified
Sellar region	Craniophyringioma, pituitary adenoma, pituicytoma	*DICER1*

Rationale for Targeted Therapies in Brain Tumors

The advent of molecular profiling techniques has provided valuable insights into the genetic and molecular alterations that drive the development and progression of brain tumors. These discoveries have underscored the heterogeneity of brain tumors and highlighted the need for personalized treatment approaches that target specific molecular pathways involved in tumor growth and survival.

Targeted therapies offer several advantages over conventional treatments, including increased specificity, reduced toxicity to normal tissues, and the potential to overcome resistance mechanisms. By selectively targeting key molecules and pathways involved in tumor progression, targeted therapies aim to disrupt essential cellular processes, like cell proliferation, angiogenesis, and DNA repair, thereby inhibiting tumor growth and metastasis [3].

Receptor-Ligand-based Targeting Approaches

Receptor-ligand based targeting approaches represent a promising strategy for precision therapy in brain tumors. These approaches utilize the interaction between specific receptors expressed on the surface of tumor cells and their corresponding ligands to deliver therapeutic agents directly to cancer cells, thereby minimizing off-target effects.

Key receptors targeted in brain tumors include receptor tyrosine kinases (RTKs), such as the epidermal growth factor receptor (EGFR), vascular endothelial growth factor receptor (VEGFR), platelet-derived growth factor receptor (PDGFR), and the Hepatocyte Growth Factor Receptor (HGF). By using ligands that selectively bind to these receptors, targeted therapies can deliver cytotoxic drugs, monoclonal antibodies, or other therapeutic agents directly to tumor cells, leading to enhanced anti-tumor efficacy and reduced systemic toxicity [4].

Receptor-Ligand Interactions in Brain Tumors

Receptor-ligand interactions play a crucial role in the neurobiology of brain tumors, influencing tumor development, progression, and response to therapy. Understanding these interactions is essential for elucidating the mechanisms that contribute to tumorigenesis and, subsequently, the development of targeted therapies aimed at disrupting these molecular pathways.

Receptor-ligand interactions involve the binding of a specific receptor protein on the surface of a cell to its corresponding ligand, typically a signaling molecule or another protein. This binding triggers intracellular signaling pathways that entail in cascading mechanisms that regulate various cellular processes, including proliferation, differentiation, and survival.

In the context of brain tumors, aberrant receptor-ligand interactions can drive oncogenic signaling pathways, promoting the growth of tumors and metastasis. Dysregulated signaling pathways often result from genetic mutations, amplifications, or overexpression of receptors or ligands, leading to sustained proliferative signaling and evasion of growth suppressors —hallmark features of cancer.

Understanding the molecular pathways underlying receptor-ligand interactions in brain tumors is critical for identifying potential therapeutic targets and developing strategies to disrupt tumor-promoting signaling pathways. For example, small molecule inhibitors that target RTKs inhibit the transphosphorylation of the protein, and another signaling cascade that is selective towards EGFR is considered an EGFR inhibitor. The inhibitors may be reversible or irreversible and lead to the inhibition of EGFR activation. Some of the FDA-approved drugs that work through this pathway are Gefitinib, Erlotinib, Osimertinib, Vandetanib, and Simotinib [5].

Epidermal Growth Factor Receptor (EGFR)

EGFR, a tyrosine kinase receptor, belongs to various families, including EGFR/ERBB1/HER1, NEU/ERBB2/HER2, ERBB3/HER3, and ERBB4/HER4. It is composed of a transmembrane receptor that is responsible for the activation of multiple proteins. EGF is a ligand found in bile, serum, and breast milk, among other sources. Transforming growth factor-α (TGF-α), amphiregulin (AREG), epiregulin (EREG), betacellulin (BTC), heparin-binding EGF-like growth factor (HB-EGF), and epigen (EPI) are other EGFR ligands. EGFR undergoes dimerization followed by autophosphorylation and promotes proliferation, differentiation, and cell division. When there is an overexpression of EGFR, abnormal changes follow in the pathway. This is one of the underlying pathways involved in the formation of GBM. There are various therapeutic approaches involved in treating cancer. Notably, monoclonal antibodies bind to the EGFR binding site, and tyrosine kinase inhibitors bind to the receptor kinase pocket, thereby preventing signal transduction. These drugs are used in head and neck cancers, as a monotherapy or in a combination therapy, even in a high dose of radiation. Anti-EGFR agents and tyrosine kinase inhibitors can generally improve the survival rate and are effective against cancer cells [6].

Vascular Endothelial Growth Factor Receptor (VEGFR)

VEGFR has three subtypes: VEGFR-1, VEGFR-2, and VEGFR-3. These receptors are activated by natural ligands known as vascular endothelial growth factors (VEGF-A to VEGF-E). VEGFRs exhibit relatively weak intrinsic tyrosine kinase activity. Upon activation, they initiate distinct intracellular signaling pathways. VEGFR signaling plays a critical role in angiogenesis—the formation of new blood vessels that support tumor growth. It also promotes endothelial cell proliferation and increases vascular permeability, thereby facilitating tumor angiogenesis and metastasis.

Platelet-derived Growth Factor Receptor (PDGFR)

PDGF Receptors alpha and beta are activated by platelet-derived growth factors (PDGF A, PDGF B, PDGF C, and PDGF D). These ligands are not active in monomeric form. They dimerize in different combinations to stimulate the kinase behavior of PDGFR. Ligand binding activates PDGFR, leading to signaling pathways involved in tumor cell growth and survival. This contributes to progression of glioma by promoting cell proliferation, migration, and invasion.

Hepatocyte Growth Factor Receptor (HGFR)

HGFR, also known as c-Met, is a type of tyrosine kinase receptor encoded by the *MET* gene. Its natural ligand is hepatocyte growth factor (HGF), which activates the receptor upon binding. Activation of c-Met/HGFR promotes tumor cell proliferation, invasion, and metastasis in glioblastoma multiforme (GBM) and other brain tumors.

EGFR Pathway

EGFR dimerization and autophosphorylation activate multiple biomolecules which promote the cell cycle, cell growth, cell survival, and phagocytosis. Fig. (**2**) illustrates the EGFR (Epidermal Growth Factor Receptor) signaling pathway and the effects of inhibitors, such as monoclonal antibodies and tyrosine kinase inhibitors.

VEGFR Pathway

VEGFR stimulation causes endothelial cell proliferation and increased vessel permeability, thereby facilitating tumor angiogenesis and metastasis, as elaborated in Fig. (**3**).

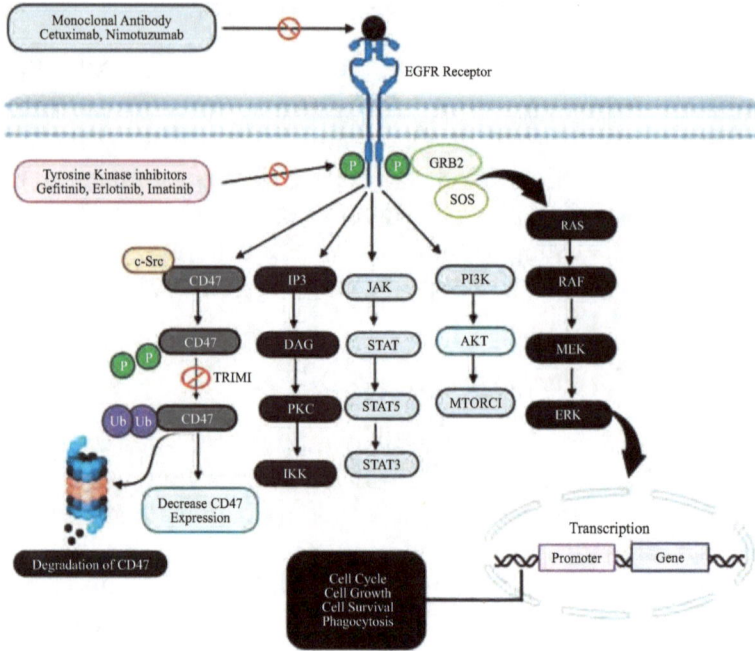

Fig. (2). EGFR Pathway.

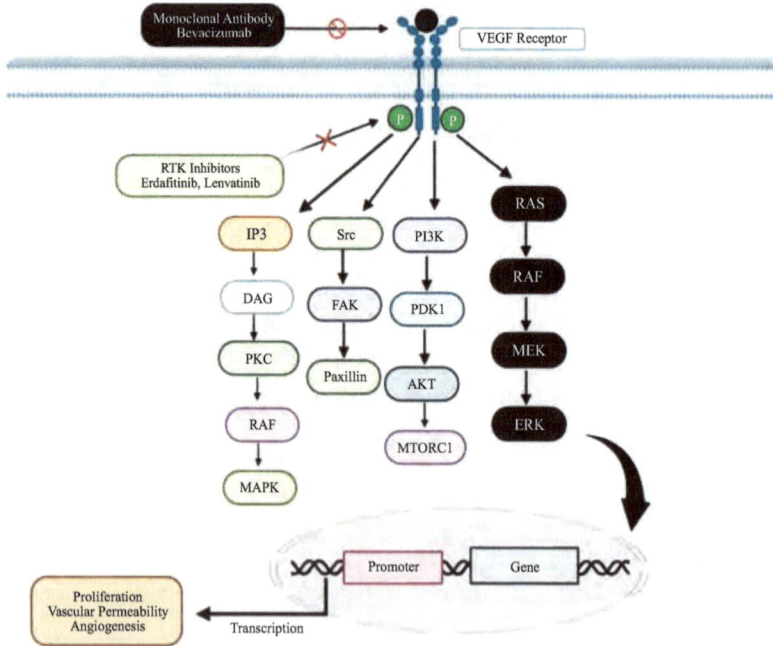

Fig. (3). VEGFR pathway.

PDGFR Pathway

PDGFR signaling contributes to the progression of glioma by promoting cell proliferation, migration, and invasion, as explained in Fig. (**4**).

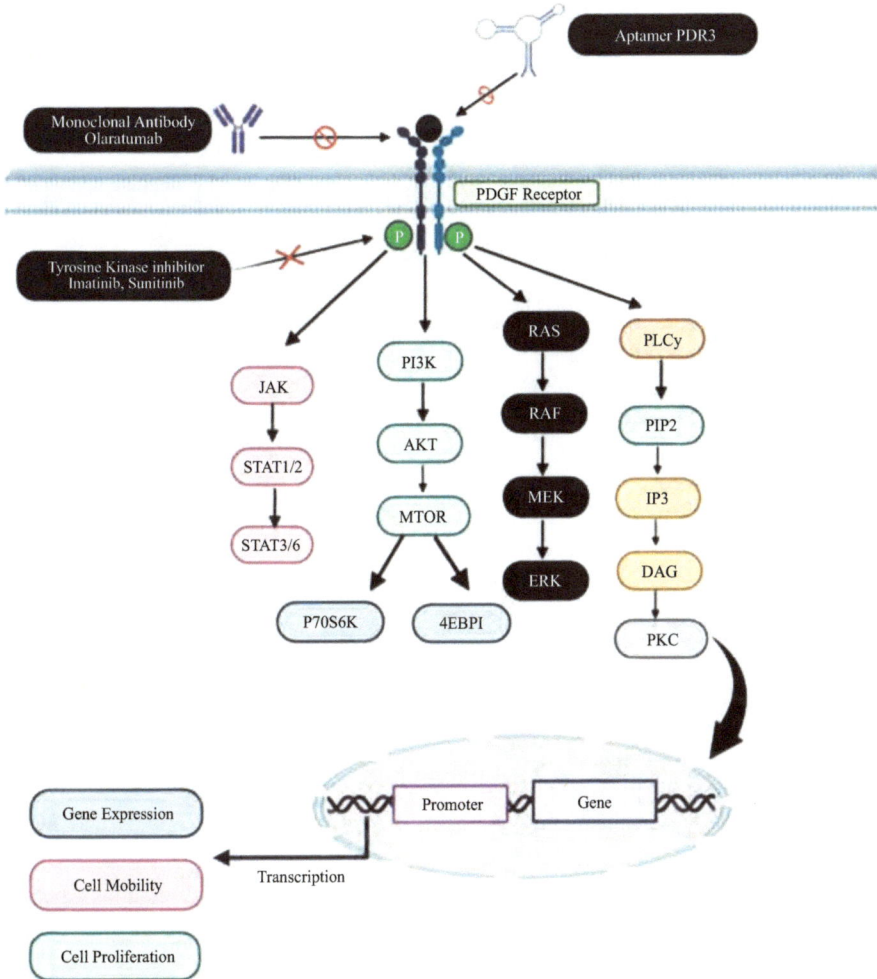

Fig. (4). PDGF pathway.

HGFR Pathway

Hepatocyte growth factor (HGF) is the ligand of the HGFR c-Met receptor. Activation of this receptor promotes tumor cell proliferation, invasion, and metastasis in GBM and other brain tumors, as shown in Fig. (**5**).

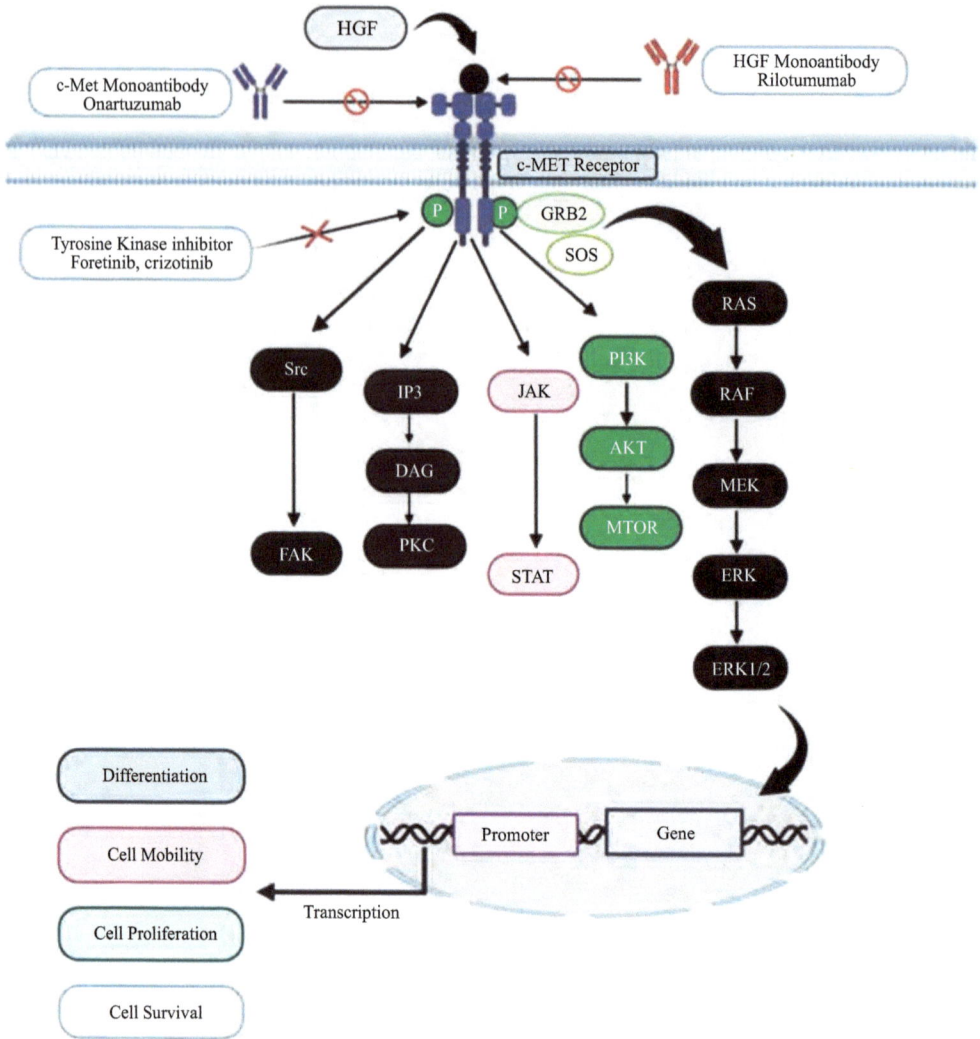

Fig. (5). HGF/c-MET pathway.

Chemoresistance

Receptor-ligand interactions play a multifaceted role in promoting the progression of tumor in neuro-oncology. Activation of oncogenic signaling pathways by the binding of ligands to their cognate receptors stimulates tumor cell proliferation, enhances cell survival, and promotes resistance to apoptosis. Temozolomide (TMZ) is an approved drug for GBMs that acts by DNA damage mechanism. The metabolites of the drug mercaptopurine methylate DNA purine bases, leading to the breaking of DNA strands and the cessation of the cell cycle. A combination of

6-Mercaptopurine and TMZ shows evidence of overcoming chemoresistance in a mouse model [7]. Resistance to TMZ in GBM is correlated with an increase in the Dopamine Receptor D2 (DRD2). A combination of TMZ with DRD2 antagonists like haloperidol has been reported to be effective in chemoresistance towards TMZ by inducing ferroptosis in GBM cells [8].

The DNA repair enzyme MGMT is used as a target in GBM cells because it is involved in one of the pathways that confers resistance to TMZ. Quercetin has been reported to work through different pathways and inhibit GBM. If it is given along with TMZ, it addresses the problem of chemoresistance. An *in-silico* study conducted on the selected disease targets reveals quercetin binding interactions in the docking studies and provides a rationale for combination therapy [9].

Chemoresistance, which involves evading the immune system, poses challenges to chemotherapy. The tumor cells follow Indoleamine-pyrrole-2,3-dioxygenase (IDO) pathway for their survival. The immune response towards the tumor is compromised mainly through the enzyme IDO. A phase I clinical trial on a pediatric population with glioma showed that indoximod (an inhibitor of IDO) works well with chemotherapy (TMZ) and radiation. The overall survival rate differed for various types of tumors. This combination therapy had better overall survival in the patients who continued with indoximod and full dose of repeated irradiation at the site of tumor than those who did not receive full dose re-irradiation with indoximod treatment.

Moreover, receptor-ligand interactions contribute to tumor cell invasion and metastasis by modulating cellular processes, such as cytoskeletal dynamics, cell adhesion, and extracellular matrix remodeling. By promoting tumor cell motility and invasiveness, aberrant receptor-ligand signaling facilitates the dissemination of the tumor cells from the primary site to distant organs, contributing to the aggressiveness and poor prognosis associated with advanced-stage brain tumors.

Preclinical Studies of Receptor Ligand-based Targeting Approaches

Receptor-ligand interactions have also been exploited in gene therapy, where viral vectors are engineered to bind to specific receptors on tumor cells, facilitating targeted gene delivery. The binding interaction of aldoxorubicin with the target gene provides the basis for gene therapy strategy.

Ligand-based imaging techniques have been utilized for the detection and characterization of brain tumors [10]. Tumor necrosis factor-related apoptosis-inducing ligand (TRAIL) is a molecule that targets GBM cells. However, this faces the challenge of chemoresistance. If TRAIL is administered along with an antineoplastic agent, such as cisplatin, which targets DNA intercalation and arrests the G2 phase of the cell cycle, the combination is effective in GBM cells. The on-site delivery of TRAIL *via* engineered stem cells in a mouse model, with co-administration of cisplatin, was studied. There was a significant decrease in tumor growth and an increase in survival in the mouse population [11]. A recombinant of TRAIL and oncolytic Herpes Simplex Virus (oHSV) in mice with resistant intracranial tumors showed decreased tumor growth and increased survival of the population. The mechanistic study of recombinant TRAIL-oHSV on resistant GBM cell lines revealed downregulation of MAPK and upregulation of JNK and p38 MAPK [12]. The poly ADP-ribose polymerase (PARP) trapping profile of AZD9574, as determined by fluorescence anisotropy, demonstrates its target specificity in a mouse model. The overall survival was better in AZD9574-TMZ combination than TMZ alone in the glioma model in mice [13].

Novel Approaches to Receptor Ligand-Based Targeting

Targeting more than one receptor is always advantageous over single-target therapy. SYHA1813 has been reported to co-target VEGFR and Colony stimulating factor 1 receptor (CSF1R). It exhibits strong antitumor activity against GBM [14]. A Phase I trial on a small group of IDH wild-type GBM patients showed that radiation concurrent with medium-frequency alternating electric fields has better overall survival when given post-surgery, along with TMZ chemotherapy. Such low-intensity fields that disrupt spindle fibers and cause cell cycle arrest are referred to as Tumor-treating fields (TTF). The evaluation of biomarkers such as O6-methylguanine-DNA methyltransferase (MGMT), Telomerase Reverse Transcriptase (TERT), EGFR, Tumor protein (TP53), and PTEN gene was done to explore the effect on overall survival. It was reported that MGMT methylation is favorable for disease prognosis. In this study, mutations in EGFR, TP53, and PTEN genes were associated with cell survival [15].

Immunotherapy and Vaccine

Different types of vaccines have been studied for GBM, including peptide vaccines, dendritic cell vaccines, mRNA vaccines, and viral vector vaccines [16]. Monoclonal antibodies targeting overexpressed EGFR are referred to as anti-EGFR antibodies. These comprise Cetuximab (C225; Erbitux), Panitumumab (ABX-EGF; Vectibix), and Nimotuzumab. These antibodies prevent ligand binding and/or dimerization of EGFR, and enhance the killing of tumor cells. A clinical trial that studied the intravenous administration of the radiolabeled antibody 125I-MAb 425, either alone or in combination with standard of care treatment, reported significantly improved median survival, along with the inhibition of EGFR signaling.

No benefit to patient survival was reported with Rindopepimut in a Phase III clinical trial of the vaccine. There are advanced therapeutic approaches, such as combination gene therapy, that are more effective than anti-tumor agents. The gene therapy is currently under clinical trials, but it is proposed to enhance the inhibitory effect on cell invasion and angiogenesis of mono-therapies targeting EGFR or EGFRvIII [17]. DC of Transtuzumab-Deruxtecan has shown good activity against solid tumors where HER2 is overexpressed, such as breast cancer [18] and lung cancer [19]. Since this combination is working well as a precision medicine in relapse and metastatic tumors, it can potentially be used for metastatic brain tumors expressing HER 2 [20]. B7-H3 is one of the immune response-related proteins that is overexpressed in brain stem tumors. It is modulated by a chimeric antigen receptor T cell (CART) specific to B7-H3. *In-vitro, in-vivo,* and clinical trials support the effectiveness of CAR T cells in counteracting brain stem tumor and diffuse intrinsic pontine glioma [21]. Several studies and clinical trials on oncolytic virus confirm its effectiveness in brain tumors. The delivery of o-HSV at the site of the tumor and stimulation of immune response is one of the novel strategies to address recurrent and resistant brain tumors [22, 23]. Numerous research studies have been conducted to investigate the application of immunotherapy and vaccines in neuro-oncology [24].

Nanoparticles have the advantage of site-specific delivery. This property of AGuIX nanoparticle has been reported in a clinical trial to radio-sensitize gliomas for radiation therapy in combination with TMZ [25].

RNA-based Therapy

Ribonucleic acid (RNA) is a genetic material involved in multiple functions within a cell, including gene expression. A small stand of RNA that interferes with the expression of genes is called interfering RNA (iRNA). Such small iRNAs are used specifically as an anti-cancer agent.

In-silico Studies on Brain Tumor Targets

In-silico studies provide a strong background for exploring the targets [26] involved in tumorigenesis [27] and targets responsible for drug resistance [28, 29]. They can provide a rationale for performing *in vitro* [30] and *in vivo* studies [31] for the targeted approach [32 - 34]. Additionally, *in-silico* studies are utilized to envisage the binding interactions of the active molecules [35]. The results of *in-silico* studies are used as complementary data to support the *in-vitro* studies [36]. *In-silico* studies have expanded the horizon to predict the target therapy response [37]. The genetic correlation of brain tumors has been studied to explore the therapeutic targets of some specific types of tumors. For example, DUSP6 and SOX2 genes are upregulated in GBM. These genes may serve as criteria for the prognosis of GBM and could be utilized in a target-based approach for the treatment of brain tumors [38].

The *in-silico* binding study of BCI (2-Benzylidine-3-(cylohexamino)-3H-indene-1-one) with ERK2 binding cavity of the gene DUSP6 exhibits good interactions. BCI has been reported to work in other cancers [39, 40]. Thus, in-silico studies provide a rationale for drug repurposing. Another example of drug repositioning is a study on antiretroviral therapy (ART), in which ART drugs were screened against 40 glioma cell lines. The drugs alone and in combination were tested for their anti-cancer activity against GBM [41]. An EGFR-targeted study demonstrated that sitravatinib induces ferroptosis and pyroptosis, in addition to other mechanisms, to inhibit GBM cells. The probable binding interactions of sitravatinib with EGFR were predicted by a docking study, which supports the *in vitro* study [42].

Drug Repurposing

Drug repurposing is a promising approach to treating the brain tumors. It reduces the time required by a drug to reach the phase III clinical trial stage. Dihydroartemisinin (an antimalarial drug) along with a photosensitizer in a self-assembly nanoplatform produces ferroptosis. This is an efficient way to manage apoptosis-resistant GBM. The drug repurposing has demonstrated prolonged survival in a rodent model [43]. The prevalent physicochemical property of drugs repurposed for brain tumors is brain penetrance. Based on this, numerous central nervous system drugs have been investigated for their anti-tumor effects in various cancers [44].

Challenges and Future Directions

The receptor-ligand strategy is not only used to inhibit overexpressed proteins, but it also has immense potential in surgical interventions. The receptor can be used for the detection and molecular fluorescence-guided surgery of meningiomas. A Phase I clinical trial targets the overexpression of VEGFα in meningioma using Bevacizumab and IRDye800CW for fluorescence-guided surgery [45]. The challenges associated with the receptor ligand approach stem from the involvement of multiple receptors in brain tumors and their recurrence. In any type of brain cancer/tumors, there is upregulation/downregulation of multiple biomolecules/genes. These multiple pathways provide advantages for tumor cells to develop an alternative mechanism of survival. Multi-target approach for the development of more efficient chemotherapeutic agents is desirable to address the challenges posed by drug resistance. Though dual target therapies are under study, chemotherapy alone cannot be a choice for the treatment of brain tumors. Instead, surgical interventions and radiation therapy need to support chemotherapy for better overall and progression-free survival. Moreover, personalized medicine, based on the nature of the tumor and other complications observed in the patient, is the most effective way to provide treatment.

CONCLUDING REMARKS

The survival rate of patients with brain tumors remains limited to a few years, despite advancements in surgery, targeted drug delivery, chemotherapy, and radiation therapy. In neuro-oncology, surgery and radiation therapy often have limited flexibility due to the sensitive and complex nature of the brain. The main area for therapeutic variation lies in the development and optimization of chemotherapeutic agents. The discovery and development of more selective, target-based anti-tumor molecules have the potential to significantly extend disease-free survival in tumor survivors. A thorough understanding of receptor expression, including their upregulation or downregulation in tumor cells, is essential for designing novel therapeutic agents. The most critical factor in target-based therapy is the accurate identification of tumor-specific molecular targets. Accurate selection of therapeutic targets significantly enhances treatment outcomes. Furthermore, a multi-targeted therapy will always be more advantageous than a monotherapy.

AUTHORS' CONTRIBUTIONS

Dr. Sonal Dubey was responsible for the conceptualization of the study, supervision of the project, providing resources, visualization, and writing the original draft. Dr. Ekta Singh carried out the investigation, contributed resources, performed formal analysis and data curation, and was involved in reviewing and editing the manuscript.

REFERENCES

[1]   Louis DN, Perry A, Wesseling P, *et al.* The 2021 WHO classification of tumors of the central nervous system: a summary. Neuro-oncol 2021; 23(8): 1231-51.
       [http://dx.doi.org/10.1093/neuonc/noab106] [PMID: 34185076]

[2]   Mendez FM, Núñez FJ, Zorrilla-Veloz RI, Lowenstein PR, Castro MG. Native chromatin immunoprecipitation using murine brain tumor neurospheres. J Vis Exp 2018; 131(131): 57016.
       [PMID: 29443090]

[3]   Arnold MA, Barr FG. Molecular diagnostics in the management of rhabdomyosarcoma. Expert Rev Mol Diagn 2017; 17(2): 189-94.
       [http://dx.doi.org/10.1080/14737159.2017.1275965] [PMID: 28058850]

[4]   Sundblom J, Skare TP, Holm O, *et al.* Central nervous system hemangioblastomas in von Hippel-Lindau disease: Total growth rate and risk of developing new lesions not associated with circulating VEGF levels. PLoS One 2022; 17(11): e0278166.
       [http://dx.doi.org/10.1371/journal.pone.0278166] [PMID: 36441756]

[5]   Wee P, Wang Z. Epidermal growth factor receptor cell proliferation signaling pathways. Cancers (Basel) 2017; 9(5): 52.

[http://dx.doi.org/10.3390/cancers9050052] [PMID: 28513565]

[6]   Zubair T, Bandyopadhyay D. Small molecule EGFR inhibitors as anti-cancer agents: discovery, mechanisms of action, and opportunities. Int J Mol Sci. 2023; 24: 2651.

[7]   Yin J, Wang X, Ge X, *et al.* Hypoxanthine phosphoribosyl transferase 1 metabolizes temozolomide to activate AMPK for driving chemoresistance of glioblastomas. Nat Commun 2023; 14(1): 5913.
[http://dx.doi.org/10.1038/s41467-023-41663-2] [PMID: 37737247]

[8]   Shi L, Chen H, Chen K, *et al.* The DRD2 antagonist haloperidol mediates autophagy-induced ferroptosis to increase temozolomide sensitivity by promoting endoplasmic reticulum stress in glioblastoma. Clin Cancer Res 2023; 29(16): 3172-88.
[http://dx.doi.org/10.1158/1078-0432.CCR-22-3971] [PMID: 37249604]

[9]   Wang W, Yuan X, Mu J, *et al.* Quercetin induces $MGMT^+$ glioblastoma cells apoptosis *via* dual inhibition of Wnt3a/β-Catenin and Akt/NF-κB signaling pathways. Phytomedicine 2023; 118: 154933.
[http://dx.doi.org/10.1016/j.phymed.2023.154933] [PMID: 37451151]

[10]   Marrero L, Wyczechowska D, Musto AE, *et al.* Therapeutic efficacy of aldoxorubicin in an intracranial xenograft mouse model of human glioblastoma. Neoplasia 2014; 16(10): 874-82.
[http://dx.doi.org/10.1016/j.neo.2014.08.015] [PMID: 25379024]

[11]   Redjal N, Zhu Y, Shah K. Combination of systemic chemotherapy with local stem cell delivered S-TRAIL in resected brain tumors. Stem Cells 2015; 33(1): 101-10.
[http://dx.doi.org/10.1002/stem.1834] [PMID: 25186100]

[12]   Tamura K, Wakimoto H, Agarwal AS, *et al.* Multimechanistic tumor targeted oncolytic virus overcomes resistance in brain tumors. Mol Ther 2013; 21(1): 68-77.
[http://dx.doi.org/10.1038/mt.2012.175] [PMID: 22929661]

[13]   Staniszewska AD, Pilger D, Gill SJ, *et al.* Preclinical characterization of AZD9574, a blood–brain barrier penetrant inhibitor of PARP1. Clin Cancer Res 2024; 30(7): 1338-51.
[http://dx.doi.org/10.1158/1078-0432.CCR-23-2094] [PMID: 37967136]

[14]   Liu Y, Zhan Z, Kang Z, *et al.* Preclinical and early clinical studies of a novel compound SYHA1813 that efficiently crosses the blood–brain barrier and exhibits potent activity against glioblastoma. Acta Pharm Sin B 2023; 13(12): 4748-64.
[http://dx.doi.org/10.1016/j.apsb.2023.09.009] [PMID: 38045044]

[15]   Cappelli L, Khan MM, Kayne A, *et al.* Differences in clinical outcomes based on molecular markers in glioblastoma patients treated with concurrent tumor-treating fields and chemoradiation: exploratory analysis of the SPARE trial. Chin Clin Oncol 2023; 12(3): 23.
[http://dx.doi.org/10.21037/cco-22-123] [PMID: 37417289]

[16]   Xiong Z, Raphael I, Olin M, Okada H, Li X, Kohanbash G. Glioblastoma vaccines: past, present, and opportunities. EBioMedicine 2024; 100: 104963.
[http://dx.doi.org/10.1016/j.ebiom.2023.104963] [PMID: 38183840]

[17]   Keller S, Schmidt M. EGFR and EGFRvIII promote angiogenesis and cell invasion in glioblastoma: combination therapies for an effective treatment. Int J Mol Sci 2017; 18(6): 1295.
[http://dx.doi.org/10.3390/ijms18061295] [PMID: 28629170]

[18]   Goto K, Goto Y, Kubo T, *et al.* Trastuzumab deruxtecan in patients with *HER2* -mutant metastatic non–small-cell lung cancer: Primary results from the randomized, phase II DESTINY-Lung02 trial. J Clin Oncol 2023; 41(31): 4852-63.
[http://dx.doi.org/10.1200/JCO.23.01361] [PMID: 37694347]

[19]   Mosele F, Deluche E, Lusque A, *et al.* Trastuzumab deruxtecan in metastatic breast cancer with variable HER2 expression: the phase 2 DAISY trial. Nat Med 2023; 29(8): 2110-20.
[http://dx.doi.org/10.1038/s41591-023-02478-2] [PMID: 37488289]

[20]   Coy S, Lee JS, Chan SJ, *et al.* Systematic characterization of antibody–drug conjugate targets in central nervous system tumors. Neuro-oncol 2024; 26(3): 458-72.

[http://dx.doi.org/10.1093/neuonc/noad205] [PMID: 37870091]

[21]   Vitanza NA, Wilson AL, Huang W, *et al.* Intraventricular B7-H3 CAR T cells for diffuse intrinsic pontine glioma: Preliminary first-in-human bioactivity and safety. Cancer Discov 2023; 13(1): 114-31.
[http://dx.doi.org/10.1158/2159-8290.CD-22-0750] [PMID: 36259971]

[22]   Singh D, Tiwari P, Nagdev S. Particulate vaccine dispersions emerge as a novel carrier for deep pulmonary immunization. Curr Nanomedicine. 2023;13(2):71–4.
[http://dx.doi.org/10.2174/2468187313666230714124009]

[23]   Ling AL, Solomon IH, Landivar AM, *et al.* Clinical trial links oncolytic immunoactivation to survival in glioblastoma. Nature 2023; 623(7985): 157-66.
[http://dx.doi.org/10.1038/s41586-023-06623-2] [PMID: 37853118]

[24]   Huang B, Li X, Li Y, Zhang J, Zong Z, Zhang H. Current Immunotherapies for Glioblastoma Multiforme. Front Immunol 2021; 11: 603911.
[http://dx.doi.org/10.3389/fimmu.2020.603911] [PMID: 33767690]

[25]   Ning J, Wakimoto H. Oncolytic herpes simplex virus-based strategies: toward a breakthrough in glioblastoma therapy. Front Microbiol 2014; 5: 303.
[http://dx.doi.org/10.3389/fmicb.2014.00303] [PMID: 24999342]

[26]   Thivat E, Casile M, Moreau J, *et al.* Phase I/II study testing the combination of AGuIX nanoparticles with radiochemotherapy and concomitant temozolomide in patients with newly diagnosed glioblastoma (NANO-GBM trial protocol). BMC Cancer 2023; 23(1): 344.
[http://dx.doi.org/10.1186/s12885-023-10829-y] [PMID: 37060055]

[27]   Brlek P, Kafka A, Bukovac A, Pećina-Šlaus N. Integrative cbioportal analysis revealed molecular mechanisms that regulate EGFR-PI3K-AKT-mTOR pathway in diffuse gliomas of the brain. Cancers (Basel) 2021; 13(13): 3247.
[http://dx.doi.org/10.3390/cancers13133247] [PMID: 34209611]

[28]   Zheng X, Chen W, Yi J, *et al.* Apolipoprotein C1 promotes glioblastoma tumorigenesis by reducing KEAP1/NRF2 and CBS-regulated ferroptosis. Acta Pharmacol Sin 2022; 43(11): 2977-92.
[http://dx.doi.org/10.1038/s41401-022-00917-3] [PMID: 35581292]

[29]   Roura AJ, Gielniewski B, Pilanc P, *et al.* Identification of the immune gene expression signature associated with recurrence of high-grade gliomas. J Mol Med (Berl) 2021; 99(2): 241-55.
[http://dx.doi.org/10.1007/s00109-020-02005-7] [PMID: 33215304]

[30]   Sumera, Anwer F, Waseem M, Fatima A, *et al.* Molecular docking and molecular dynamics studies reveal secretory proteins as novel targets of temozolomide in glioblastoma multiforme. Molecules. 2022; 27(21): 7198.

[31]   Hussein D, Saka M, Baeesa S, *et al.* Structure-based virtual screening and molecular docking approaches to identify potential inhibitors against KIF2C to combat glioma. J Biomol Struct Dyn 2023; 1-14.
[PMID: 37942622]

[32]   Balaji E V, Satarker S, Kumar BH, *et al.* In-silico lead identification of the pan-mutant IDH1 and IDH2 inhibitors to target glioblastoma. J Biomol Struct Dyn 2024; 42(7): 3764-89.
[http://dx.doi.org/10.1080/07391102.2023.2215884] [PMID: 37227789]

[33]   Swati K, Srivastava R, Agrawal K, *et al.* Structure-based virtual screening identifying novel FOXM1 inhibitors as the lead compounds for glioblastoma. Recent Pat Anti-cancer Drug Discov 2024.

[34]   Roy PK, Majumder R, Mandal M. *In-silico* identification of novel DDI2 inhibitor in glioblastoma *via* repurposing FDA approved drugs using molecular docking and MD simulation study. J Biomol Struct Dyn 2024; 42(5): 2270-81.
[http://dx.doi.org/10.1080/07391102.2023.2204371] [PMID: 37139547]

[35]   Feng Y, Zhu P, Wu D, Deng W. A network pharmacology prediction and molecular docking-based strategy to explore the potential pharmacological mechanism of *Astragalus membranaceus* for glioma.

Int J Mol Sci 2023; 24(22): 16306.
[http://dx.doi.org/10.3390/ijms242216306] [PMID: 38003496]

[36]   Rizvi SMD, Almazni IA, Moawadh MS, *et al.* Targeting NF-κB signaling cascades of glioblastoma by a natural benzophenone, garcinol, *via in vitro* and molecular docking approaches. Front Chem 2024; 12: 1352009.
[http://dx.doi.org/10.3389/fchem.2024.1352009] [PMID: 38435669]

[37]   Arora R, Cao C, Kumar M, *et al.* Spatial transcriptomics reveals distinct and conserved tumor core and edge architectures that predict survival and targeted therapy response. Nat Commun 2023; 14(1): 5029.
[http://dx.doi.org/10.1038/s41467-023-40271-4] [PMID: 37596273]

[38]   Caglar HO, Duzgun Z. Identification of upregulated genes in glioblastoma and glioblastoma cancer stem cells using bioinformatics analysis. Gene 2023; 848: 146895.
[http://dx.doi.org/10.1016/j.gene.2022.146895] [PMID: 36122609]

[39]   Marciniak B, Kciuk M, Mujwar S, *et al. In vitro* and *in silico* investigation of BCI Anti-cancer properties and. Cancers (Basel) 2023; 15(18): 4442.
[http://dx.doi.org/10.3390/cancers15184442] [PMID: 37760412]

[40]   Shin JW, Kwon SB, Bak Y, Lee SK, Yoon DY. BCI induces apoptosis *via* generation of reactive oxygen species and activation of intrinsic mitochondrial pathway in H1299 lung cancer cells. Sci China Life Sci 2018; 61(10): 1243-53.
[http://dx.doi.org/10.1007/s11427-017-9191-1] [PMID: 29524123]

[41]   Rivas SR, Mendez Valdez MJ, Chandar JS, *et al.* Antiretroviral drug repositioning for glioblastoma. Cancers (Basel) 2024; 16(9): 1754.
[http://dx.doi.org/10.3390/cancers16091754] [PMID: 38730705]

[42]   Lu H, Zhang B, Xie Y, *et al.* Sitravatinib is a potential EGFR inhibitor and induce a new death phenotype in Glioblastoma. Invest New Drugs 2023; 41(4): 564-78.
[http://dx.doi.org/10.1007/s10637-023-01373-4] [PMID: 37322389]

[43]   Liang J, Li L, Tian H, *et al.* Drug Repurposing-Based Brain-Targeting Self-Assembly Nanoplatform Using Enhanced Ferroptosis against Glioblastoma. Small 2023; 19(46): 2303073.
[http://dx.doi.org/10.1002/smll.202303073] [PMID: 37460404]

[44]   Aroosa M, Malik JA, Ahmed S, Bender O, Ahemad N, Anwar S. The evidence for repurposing anti-epileptic drugs to target cancer. Mol Biol Rep 2023; 50(9): 7667-80.
[http://dx.doi.org/10.1007/s11033-023-08568-1] [PMID: 37418080]

[45]   Dijkstra BM, Cordia QCF, Nonnekens J, *et al.* Bevacizumab-IRDye800CW for tumor detection in fluorescence-guided meningioma surgery (LUMINA trial): a single-center phase I study. J Neurosurg 2024; 141(6): 1655-66.
[http://dx.doi.org/10.3171/2024.4.JNS232766] [PMID: 38968617]

CHAPTER 6

Therapeutic Interference and Signaling Pathways in Brain Tumors

Rahul Kumar[1], **Santosh Kumar Guru**[1], **Prashant Tiwari**[3] and **Pankaj Kumar Singh**[2,*]

[1] *Department of Biological Science, National Institute of Pharmaceutical Education and Research (NIPER), Hyderabad–500037, India*

[2] *Department of Pharmaceutics, National Institute of Pharmaceutical Education and Research (NIPER), Hyderabad–500037, India*

[3] *College of Pharmaceutical Sciences, Dayananda Sagar University, Bengaluru-562112, India*

Abstract: Despite their rarity, brain tumors are associated with significant morbidity and mortality across all age groups. Although therapeutic options remain limited, the prognosis for individuals with brain tumors has markedly improved due to advances in immunotherapies, targeted treatments, and a deeper understanding of tumor biology. However, further progress in treating brain tumors such as gliomas, meningiomas, and brain germ cell tumors is hindered by low response rates and predictable drug resistance associated with currently approved therapies. Evidence from previous studies indicates that brain tumors dysregulate several distinct signaling pathways. Importantly, a more comprehensive understanding of the molecular mechanisms driving the malignant behavior of brain tumor cells could facilitate the development of novel targeted therapies. Therefore, an in-depth exploration of the pathophysiology of these tumors is urgently needed, as it holds the potential to significantly enhance therapeutic strategies. Glioblastoma, in particular, is a primary brain tumor characterized by high morbidity and poor responsiveness to conventional treatments. Recently, large-scale genome sequencing initiatives have intensified research efforts, providing new insights into the cellular signaling networks and genomic alterations underlying brain tumor pathogenesis. Current knowledge of molecular markers and tumorigenic pathways may prove instrumental in identifying new therapeutic avenues for brain cancers. Multiple signaling pathways including pRB, p53, NF-κB, RAS/MAPK, STAT3, ZIP3, and WNT are implicated in the development of various brain tumor types. This chapter explores the therapeutic interventions and signaling pathways involved in brain tumor progression.

Keywords: Chemotherapy resistance, Combination therapy, Drug delivery systems, Molecular inhibitors, Nanotechnology in therapy, Targeted therapy.

* Corresponding author Pankaj Kumar Singh: Department of Pharmaceutics, National Institute of Pharmaceutical Education and Research (NIPER), Hyderabad–500037, India; E-mail: pankajksingh3@gmail.com

Prashant Tiwari, Pankaj Kumar Singh & Sunil Kumar Kadiri (Eds.)
All rights reserved-© 2025 Bentham Science Publishers

INTRODUCTION

Brain and central nervous system (CNS) cancers represent a significant global public health challenge, characterized by high mortality rates, substantial financial burdens on patients and healthcare systems, low survival rates, and profound impacts on patients' quality of life [1]. Although brain tumors are relatively rare, they contribute disproportionately to morbidity and mortality across all age groups. Despite limited treatment options, advances in molecular biology and the development of targeted therapies and immunotherapies have led to notable improvements in patient outcomes [2]. Nevertheless, low therapeutic response rates and pronounced drug resistance remain major obstacles to further progress in brain tumor management [3].

According to data from the U.S. Central Brain Tumor Registry, the incidence of malignant brain tumors declined by approximately 0.8% annually between 2008 and 2017 across all age groups. In contrast, the prevalence of nonmalignant tumors increased during the same period [4, 5]. Globally, brain and CNS cancers were the 12th leading cause of cancer-related deaths in 2020, accounting for 2.5% of all cancer deaths, as reported by the Global Cancer Observatory (GLOBOCAN) [4, 6]. In Iraq, these cancers ranked as the fourth leading cause of death among both genders in 2020 [2]. Furthermore, only 35.1% of brain cancer patients in the United States survived five years following a local-stage diagnosis, despite 76.9% of brain and CNS cancer cases being confirmed between 2012 and 2018 [2, 4]. The incidence of primary brain tumors varies significantly by age, gender, and race. Adults over the age of 20 exhibit a higher occurrence of both malignant and nonmalignant brain tumors, with most histological subtypes being the most prevalent in individuals aged 65 and older. In this age group, malignant tumors were 1.5 to 8 times more frequent, while nonmalignant tumors occurred 2 to 9 times more often. However, in younger patients, the incidence of malignant tumors decreases with age, while nonmalignant tumors increase. Notably, the overall occurrence rate is higher in adults aged 0-4 and 15-19 compared to children aged 5-14 [3, 7].

Primary brain tumors form when neural cells divide uncontrollably in the brain parenchyma [8]. These tumors can be identified based on their location, histological features, or the presence of particular mutations [8]. The World Health Organization (WHO) classifies gliomas, the most prevalent type of primary tumor of the adult brain parenchyma, into four categories ranging from 1 to 4. Grade 4 glioma, known as glioblastoma (GBM), is characterized by an aggressive nature, treatment resistance, and short overall survival of the patients [8].

Meningiomas are the most common primary intracranial tumors in adults; however, the understanding of their carcinogenesis remains limited compared to other intracranial tumors, such as gliomas [9].

Increasing evidence underscores the pivotal roles of growth factors, particularly platelet-derived growth factor (PDGF) and epidermal growth factor (EGF), in initiating the molecular pathways that drive brain tumor development [10]. These growth factors activate critical signaling cascades, including the Ras/mitogen-activated protein kinase (MAPK) and phosphoinositide 3-kinase (PI3K-AKT) pathways, as well as secondary signaling routes such as mTOR and phospholipase C-γ/protein kinase C [9, 11, 12]. Similar to other malignancies, glioblastoma is characterized by disruptions in fundamental biological mechanisms that regulate cell proliferation and survival [13, 14]. The major molecular alterations that are commonly observed in glioblastoma include: Activation of MAPK and RTK/PI3K pathways present in approximately 90% of cases [14]; Inactivation of the p53 tumor suppressor pathway observed in 86% of cases [15]; Inactivation of the RB tumor suppressor pathway found in 79% of cases [16, 17]. Notably, 74% of glioblastomas exhibit concurrent alterations in all three pathways, highlighting the complexity and aggressiveness of this tumor type [18].

To improve preventive and therapeutic strategies, it is essential to increase public awareness of brain and CNS cancers, considering recent estimates of their incidence, mortality, and links to global socioeconomic factors. Estimating the signaling pathways involved in brain tumorigenesis and progression is critical for identifying therapeutic targets and managing brain tumors. This chapter provides an overview of different brain tumors and the signaling pathways driving their progression by regulating cell proliferation and cell cycle.

EPIDEMIOLOGY OF BRAIN TUMORS

As of 2023, more than one million individuals in the United States are living with primary brain tumors. Among these, approximately 72% are classified as benign, while 28% are malignant [18]. The adolescent and young adult (AYA) population in the U.S., defined as individuals aged 15 to 39, comprises over 110 million people, with an estimated 208,620 affected by a primary brain or spinal cord tumor. Meningiomas are the most common form of primary benign or non-malignant brain tumors, accounting for 39.7% of all brain tumors and 55.4% of non-malignant cases. In contrast, glioblastoma represents the most prevalent primary malignant brain tumor, comprising 14.2% of all brain tumors and 50.1% of malignant cases [18]. Malignant brain tumors, commonly referred to as brain cancers, are projected to cause approximately 18,990 deaths in the United States in 2023. Brain cancer ranks as the tenth leading cause of cancer-related mortality across all age groups for both men and women [4, 19]. Aligned with global health priorities, the United Nations Sustainable Development Goals (SDGs) and the World Health Organization (WHO) have set a target to reduce premature cancer mortality by one-third by 2030 [4, 20]. However, WHO projections suggest that the European Region will experience approximately 85,000 new cases and 70,000 deaths from brain and central nervous system cancers by 2030, an increase from 82,000 cases and 58,000 deaths reported in 2015 for both sexes combined [20].

BRAIN TUMOR

Brain tumor is a mass of abnormal cells in the brain that can be malignant (cancers) or benign (noncancerous). A brain tumor results from uncontrolled cell division in the brain tissue or its surroundings, including nerves, the pituitary gland, the pineal gland, and the meningitis membrane [21]. The brain's architecture is highly intricate, with several regions controlling various nervous system processes. Brain tumors can form in any portion of the brain or skull, including the protective lining, the underside of the brain (skull base), the brainstem, the sinuses and nasal cavity, and numerous other locations [22]. About 120 different forms of tumors can grow in the brain, depending on which tissue they originate. A tumor reported in the brain is called a primary brain tumor, while cancer that spreads from other parts of the body to the brain is defined as a secondary brain tumor or metastatic brain tumor.

Brain tumors, while uncommon, cause significant death and illness at every age. The prognosis of patients with brain malignancies has dramatically improved due to the limited number of available therapy options, growing biological knowledge, and previously unheard-of developments in targeted therapies and immunotherapies [3]. However, glioma, meningioma, brain germ cell tumors, and brain lymphoma are among the brain tumors for which there is currently no significant progress due to the low response rates and inevitable drug resistance of the available therapeutic options [19].

CLASSIFICATION OF BRAIN TUMORS

Several types of brain tumors have been reported that are classified based on the cell type of the cancer. Over 150 different kinds of brain tumors are known to exist, and they can be divided into two major groups: the first is a primary brain tumor, and the second is a secondary brain tumor [16, 23]. Primary brain tumors originate in the brain from brain cells (neurons and glial cells), meningitis membrane surrounding the brain, nerve cells, and glands such as the pituitary. Primary brain tumors are benign and malignant [2, 24]. Gliomas and meningiomas are the most common types of primary brain tumors. Primary brain tumors can be categorized as either benign or malignant and as glial (originating from glial cells) or non-glial (arising from brain structures like nerves, blood vessels, and glands). In contrast, secondary brain tumors, which make up the majority of brain tumors, originate from other parts of the body such as the lungs, breast, kidney, or skin and metastasize to the brain, typically as malignant cancers [24]. Table 1 provides an overview of the classification of brain tumors.

Table 1. Classification of brain tumors and their respective pathways involved in pathophysiology of brain tumor.

-	Types of Brain Tumor	Description	Cells Involved or Sites	Symptoms	Signalling Pathways	References
Benign brain tumour	Chordomas	Notochordal tumor of the skull base, mobile spine, and sacrum.	Notochord Cells	Skull may have headaches Double vision, Blurry vision, Facial numbness.	PI3K/Akt/mTOR pathway RAS/ERK/MAPK pathway Wnt/β-catenin pathway	[180, 181]
	Craniopharyngiomas	Craniopharyngiomas are a rare, slow-growing intracranial non-cancerous tumour reported in sellar-suprasellar region.	Epithelial of sellar–suprasellar region	Headaches, Visual impairment, Fatigue, Gastrointestinal abnormalities, Weight disturbance	Wnt/β-catenin pathway MAPK/ERK pathway	[182, 183]
	Gangliocytomas	Gangliocytomas are a slow-growing benign tumor of CNS, with the most common site being the temporal lobe.	Ganglion cells	Symptoms depend upon the tumor location and include seizures, endocrine disorders; and focal symptoms.	PI3K pathway	[184, 185]
	Meningiomas	A meningioma is a tumour that develops in the membranes covering the brain and spinal cord, known as the meninges.	Arachnoid cap cells that cover the whole brain and spinal cord.	Hearing loss, memory loss, smell loss, seizures, trouble speaking.	PI3K/Akt/mTOR pathway RAS/ERK/MAPK pathway Wnt/β-catenin pathway	[186 - 189]
	Pineocytomas	Pineocytoma is an uncommon benign tumour that develops gradually in the pineal gland, a tiny brain structure that is involved in the production of melatonin.	Pineal parenchymal cells	Memory impairment, headaches, difficulties in body balance in eye movement, and sleep disturbance.	p53 and Rb pathway	[190, 191]

(Table 1) cont.....

-	Types of Brain Tumor	Description	Cells Involved or Sites	Symptoms	Signalling Pathways	References
-	Pituitary adenomas	Benign tumours called pituitary adenomas develop in the pituitary gland, which is situated near the base of the brain.	Pit-1 lineage cells of the anterior pituitary	• Hormonal disbalance, vision loss, weight gain, menstrual irregularities, lactation, easy bleeding or bruising.	PI3K/Akt/mTOR pathway Raf/MEK/ERK pathway	[192]
	Schwannomas	A rare kind of tumour that develops in the nerve system is called a schwannoma.	Schwann cells	Pain, numbness, and tingling, facial weakness, loss of taste, Lumps or swollen areas.	Hippo pathway, MAPK signaling, and PI3K pathway	[193 - 195]
Malignant brain tumour	Astrocytomas	One kind of brain tumor is an astrocytoma. A particular type of star-shaped glial cell in the cerebrum known as an astrocyte is the source of astrocytomas, also known as astrocytomas.	Astrocytes glial cells	Behavior and mood changes; changes in personality, difficulty speaking, difficulty with balance, dizziness, headaches not alleviated by painkillers, worse in the morning, may cause nausea/vomiting, memory loss	PI3K/Akt/mTOR pathway RAS/ERK/MAPK pathway Wnt/β-catenin pathway	[196, 197]
	Ependymomas	Glial cell tumours known as ependymomas typically develop in the ventricular system's lining cells, although they can also occur intracranially or extraneurally in the brain. pericardium	Radial glial cells	Neck stiffness, Neck weakness, leg muscle weakness, and bladder and bowel dysfunction.	PI3K pathway, Wnt/β-catenin pathway, ErbB signaling, Notch signaling.	[198]
	Glioblastoma multiform	This is the most aggressive and prevalent kind of brain cancer, with an extremely low chance of survival.	Microglial cells	Persistent headaches, nausea, vomiting, blurred vision, changes in cognitive abilities, memory loss, personality changes	PI3K/Akt/mTOR pathway RAS/ERK/MAPK pathway Wnt/β-catenin pathway NFkB pathway	[115, 199]

(Table 1) cont.....

-	Types of Brain Tumor	Description	Cells Involved or Sites	Symptoms	Signalling Pathways	References
-	Medulloblastomas	Tumours are growths of cells that are the precursor to medulloma. Rapid cell growth and potential for brain metastasis are features of these cells.	Since they develop in foetal cells that survive after birth, medulloblastomas are often referred to as embryonal neuroepithelial tumours.	• Head bobbing, dizziness, nausea and vomiting, walking problem or ataxia •	Shh/Ptch signaling pathway, PI3K pathway, Wnt/β-catenin pathway, ErbB signaling, Notch signaling.	[200, 201]
	Oligodendrogliomas	The oligodendrocytes in the brain or a glial precursor cell are thought to be the source of oligodendrogliomas, a particular kind of glioma.	Oligodendrocytes glial cells	Seizures, headaches, and weakness or disability	p53 pathway and Notch signalling	[202,203]

Meningioma

Meningiomas are the most common primary intracranial tumors in adults; however, our understanding of their carcinogenesis remains limited compared to research on other intracranial tumors, such as gliomas [25]. These tumors arise from arachnoid cap cells and are generally slow-growing. Despite being historically categorized as benign, up to 35% of meningiomas are now classified as malignant, atypical, or anaplastic based on recent histological grading techniques [26, 27]. Various signaling pathways influence key cellular processes essential for angiogenesis, proliferation, and apoptosis, with models now emerging to illustrate the dysregulated signaling cascades involved in meningioma development (Nassiri et al., 2021). Increasing evidence shows that growth factors, particularly platelet-derived growth factor (PDGF) and epidermal growth factor (EGF), are critical for activating these cascades (28). Two primary anti-apoptotic cell signaling pathways, the Ras/mitogen-activated protein kinase (MAPK) pathway and the phosphoinositide 3-kinase (PI3K)- AKT pathway, are associated with these growth factors, along with secondary pathways such as phospholipase C-γ protein kinase C and the mammalian target of rapamycin (mTOR) (29). Tumor angiogenesis, often driven by growth factors or oncogene-induced deregulation, is primarily mediated by vascular endothelial growth factor (VEGF) [28, 29].

Glioblastoma

Glioblastoma multiforme (GBM) is the most prevalent malignant glioma and the most aggressive form of primary brain tumor, with no established curative treatment. It is characterized by a poor prognosis, limited therapeutic responsiveness, and a short median survival time for affected patients [20]. The etiology of GBM is complex, involving a multitude of genetic alterations and dysregulated biological processes that contribute to tumor angiogenesis, cellular migration, survival, and proliferation. Key molecular changes frequently observed in GBM include aberrations in the epidermal growth factor receptor (EGFR) and the phosphatidylinositol 3-kinase (PI3K)/Akt/mTOR signaling pathways. Approximately 88% of GBM cases exhibit genetic abnormalities, including EGFR overexpression, activating mutations in PIK3CA (p110) or PIK3R1 (p85), and loss of PTEN expression [30, 31]. Additionally, GBM pathogenesis is associated with altered or elevated expression of several critical proteins, including transforming growth factor β (TGF-β), vascular endothelial growth factor (VEGF), EGFR, cyclin-dependent kinase inhibitor 2A (CDKN2A), nuclear factor-κB (NF-κB), and components of the PI3K/Akt/mTOR pathway [32]. These molecular disruptions underscore the aggressive nature of GBM and highlight potential targets for therapeutic intervention.

BRAIN TUMOR PATHOPHYSIOLOGY

Brain tumor symptoms can differ based on the location of the tumor because different areas of the brain regulate distinct activities. For instance, movement, walking, balance, and coordination issues may result from a brain tumor in the cerebellum, which is the region near the rear of the head [33]. Vision alterations could happen if the tumor damages the optic pathway in charge of vision [34]. Primary CNS cancers that arise partly from alterations in cellular growth pathways are known as neural and glioneuronal tumors [35, 36]. Owing to their scarcity, it is still challenging to precisely characterize many of these cancers molecularly, but continuous research is still finding the genetic abnormalities that cause them to arise. The tumor site greatly influences the pathogenesis of most of these diseases, as these lesions are usually WHO grade 1. WHO grade 2 lesions include cerebellar liponeurocytoma, extraventricular neurocytoma, and central neurocytoma. The only WHO grade 3 lesions in the cohort are anaplastic gangliogliomas. Certain recently described tumors, like DLGNT and MGT, lack categorization at this time due to their rarity. As mentioned, although some cancers are more likely to occur in a particular area, the supratentorial or infratentorial area is a potential location for these malignancies [37]. For instance, in the context of Cowden syndrome, dysplastic cerebellar gangliocytoma arises in the cerebellum in a manner akin to Lhermitte-Duclos disease [38].

The temporal lobe is the most common site for certain brain tumors, primarily due to the frequent occurrence of dysembryoplastic neuroepithelial tumors (DNETs) and gangliogliomas in this region [39]. Nonetheless, these tumors can also arise in extra-temporal areas [40]. Seizures are the most prevalent clinical manifestations, typically originating from the temporal lobe or other epileptogenic zones [41, 42]. Owing to their resistance to optimal medical management for seizures, these tumors are often classified as long-term epilepsy-associated tumors (LEATs). Intra-tumoral electroencephalographic (EEG) recordings have demonstrated that both DNETs and gangliogliomas possess intrinsic epileptogenic activity. This is likely attributable to abnormal neuronal cells and structures within the tumors, indicating that their seizure-inducing potential is not solely dependent on anatomical location [43]. Furthermore, research suggests that the overexpression of glutamate receptors (GluRs) and decreased expression of gamma-aminobutyric acid receptors (GABA-Rs) contribute to the hyperexcitable state observed in these tumors [43, 44]. In addition to seizures, patients may experience other symptoms such as focal neurological deficits [45], hydrocephalus, papilledema, headache, nausea, and vomiting. Signs including headache, vomiting, and papilledema hallmarks of space-occupying lesions are indicative of elevated intracranial pressure [46]. These symptoms often precede diagnosis, as many of these tumors grow slowly and may reach substantial size before becoming clinically evident.

A headache is the most frequent symptom of a brain tumor. The majority of people who seek medical attention for them do not have a significant underlying cause, such as a brain tumor. Although brain tumors are rare, many of their patients do have headaches [47, 48]. These headaches typically accompany other neurological symptoms and signs. Whenever additional neuroimaging studies are being considered to rule out a severe underlying illness, a thorough clinical screening for warning signs should be conducted [46, 49].

Additionally, in patients with frontal lobe tumors, neurocognitive impairment is a significant issue. While the degree of tumor excision, progression-free survival, and overall survival are generally used to gauge the theoretical success of surgery for these types of malignancies, neurocognitive impairment is currently receiving much attention as a significant yet beneficial outcome metric [45]. Early on, neurocognitive function declines are often mild and may go undiagnosed for various reasons, including a lack of awareness and appropriate, straightforward tests [50]. Unfortunately, this impairs the patient's social, professional, and personal skills, lowering their quality of life and capacity for self-management. This, in turn, makes it more difficult for them to adhere to the therapy [51].

THERAPEUTIC INTERFERENCE IN BRAIN TUMOR

Significant interpatient variability and intratumoral heterogeneity pose major challenges to the effective treatment of brain tumors. These tumors rank among the most lethal malignancies, with glioblastoma being the most common primary brain tumor in adults. Despite advances in research, the prognosis for glioblastoma remains dismal, with a median survival of approximately 15 months post-diagnosis and a five-year survival rate of only 10%. Recurrence rates exceed 90% [52]. Unfortunately, therapeutic outcomes have seen slight improvement over the past several decades. This stagnation is primarily attributed to the high rate of primary therapeutic resistance, which continues to undermine treatment efficacy. The lack of clinical progress reflects the multi-scale and omics-wide complexity inherent in many cancers, particularly brain tumors [53]. The emergence of modern omics technologies has uncovered a vast array of molecular pathways involved in brain tumor pathogenesis. Omics-based studies across diverse patient populations have illuminated the extensive heterogeneity present in brain cancers, further complicating treatment strategies. In the case of glioblastoma, therapeutic advancement is impeded by several critical factors, including pronounced interpatient variability, intratumoral heterogeneity, and the formidable challenge of delivering drugs across the blood-brain barrier [54, 55].

The standard treatment protocol for glioblastoma multiforme (GBM) includes maximal surgical resection followed by concurrent chemoradiotherapy with temozolomide (TMZ) [20, 56]. Despite this aggressive multimodal approach, outcomes remain poor. According to the 2020 CBTRUS Statistical Report, the five-year survival rate for GBM patients in the United States is only 7.2% [7, 57]. Even with complete surgical resection and adjuvant therapies, nearly all GBM tumors recur locally, underscoring the disease's resistance to conventional treatment modalities [58]. Several factors contribute to the complexity of GBM management, including the inability to achieve complete tumor resection, pronounced genetic heterogeneity, the protective nature of the blood-brain barrier (BBB), and the presence of an immunosuppressive tumor microenvironment. These challenges collectively hinder therapeutic efficacy and highlight the urgent need for innovative treatment strategies.

Intertumor and Intratumor Heterogeneity

The development of targeted therapies for glioblastoma multiforme (GBM) remains a formidable challenge, primarily due to pronounced intertumor and intratumor heterogeneity [58]. To better understand this complexity, The Cancer Genome Atlas (TCGA) has classified GBMs into four distinct molecular subtypes based on genetic and epigenetic profiles: mesenchymal, classical, proneural, and neural [59]. Mesenchymal GBMs are characterized by mutations in the neurofibromin 1 (NF1) tumor suppressor gene, along with frequent alterations in PTEN and TP53. Classical GBMs exhibit high proliferative activity and are typically defined by EGFR amplification in the absence of TP53 mutations. Proneural GBMs are commonly associated with TP53 mutations and frequently harbor alterations in IDH1 and PDGFRA [60]. Neural GBMs are distinguished by the expression of multiple genes that are also found in normal, non-neoplastic neurons. This molecular classification underscores the biological diversity of GBM and highlights the need for subtype-specific therapeutic approaches to improve treatment efficacy and patient outcomes.

Blood-Brain Barrier

One of the most formidable challenges in the treatment of brain tumors, particularly glioblastoma multiforme (GBM), is the effective delivery of chemotherapeutic agents across the blood-brain barrier (BBB). The BBB is a highly selective physiological barrier that separates the systemic circulation from the central nervous system (CNS) parenchyma. It is primarily composed of tightly connected endothelial cells lining the cerebral vasculature, which regulate the entry of substances into the brain [61, 62].

Within tumor regions, however, BBB integrity is often heterogeneous. Some areas exhibit increased vascular permeability, while others maintain intact vessels and vascular shunts, resulting in inconsistent drug penetration. Even when chemotherapeutic agents reach the tumor microenvironment, achieving therapeutic intracellular concentrations remains difficult due to the activation of efflux pumps by glioblastoma cells, which actively expel drugs from the cytoplasm [63]. Moreover, a substantial portion of GBM tissue retains a locoregionally intact BBB, compounded by elevated efflux pump activity. This dual barrier, structural and functional, significantly impairs drug accumulation within tumor cells and limits the efficacy of systemic therapies [64].

Immunosuppressive Microenvironment

The treatment of GBM is further complicated by its microenvironment. Some studies describe GBMs with an immunosuppressive microenvironment as "cold tumors," characterized by a lack of pre-existing T-cell infiltration, which makes them resistant to immune checkpoint inhibitors [65]. These tumors typically exhibit a deficiency of tumor antigens, poor antigen presentation, and a high presence of immunosuppressive cells [65]. As a result, immune checkpoint inhibitors have demonstrated only modest effectiveness in these cases [66]. In contrast, "hot tumors" are densely infiltrated with T cells, rendering them more immunogenic. Strategies that enhance anti-tumor immunity may help convert "cold" tumors into "hot" ones, potentially improving treatment outcomes [67].

SIGNALING PATHWAYS IN BRAIN TUMOR

Primary brain tumors, particularly glioblastoma, exhibit marked resistance to conventional therapies and are associated with significant morbidity. Recent large-scale genome sequencing efforts have yielded valuable insights into the genetic abnormalities and cellular signaling pathways that drive brain tumor pathogenesis. However, the molecular mechanisms underlying these tumors are highly diverse and complex, making it challenging to translate these findings into improved clinical outcomes. Nonetheless, ongoing research aimed at identifying and targeting molecular vulnerabilities continues to advance [68, 69].

Glioblastoma is among the most extensively studied cancers, largely due to its inclusion in The Cancer Genome Atlas (TCGA) initiative. TCGA aims to provide comprehensive molecular characterizations of selected tumors by cataloging genomic, epigenomic, transcriptomic, and proteomic features, alongside clinical data from affected patients [70]. Tumors selected for TCGA analysis were chosen based on their poor prognosis and substantial public health burden, with glioblastoma being one of the first three cancers prioritized for investigation [71].

TCGA has significantly enhanced our understanding of glioma biology, highlighting critical molecular pathways and identifying distinct glioblastoma subtypes with prognostic relevance: **proneural**, **neural**, **classical**, and **mesenchymal**. Among these, the proneural subtype is more frequently observed in younger individuals and is commonly associated with Grade II and III gliomas, which may progress to glioblastoma [72, 73].

PI3K/AKT/mTOR Pathway

Central nervous system tumors encompass a diverse range of neoplasms arising from various cell lineages. In adults, malignant gliomas are the most common type, while medulloblastoma (MB) is prevalent in pediatric populations. The development of GBM is frequently associated with molecular alterations in the PI3K/Akt/mTOR and epidermal growth factor receptor (EGFR) pathways. It is estimated that around 88% of genetic alterations are common, including the overexpression of EGFR, activating mutations in PI3CA (p110) or PIK3R1 (P85), and the loss of PTEN expression [74].

PI3K/AKT/mTOR Pathway in Brain Tumor

Patients with glioblastoma multiforme (GBM) who exhibit activation of the phosphatidylinositol 3-kinase (PI3K)/Akt/mammalian target of rapamycin (mTOR) signaling pathway tend to have a poorer prognosis compared to those without pathway activation. As a result, therapeutic strategies targeting the epidermal growth factor receptor (EGFR) and PI3K/Akt/mTOR pathways have been proposed to improve treatment outcomes [75, 76]. Numerous preclinical and clinical trials are currently evaluating inhibitors of these pathways, either as monotherapies or in combination with standard-of-care treatments. Investigating the potential of these inhibitors to restore treatment sensitivity is a particularly promising area of research [77, 78]. The pathogenesis of GBM is further associated with altered or elevated expression of several key regulatory proteins, including transforming growth factor β (TGF-β), vascular endothelial growth factor (VEGF), EGFR, cyclin-dependent kinase inhibitor 2A (CDKN2A), nuclear factor-κB (NF-κB), and components of the PI3K/Akt/mTOR (PAM) signaling axis [32]. This intracellular signaling platform plays a central role in regulating stem cell division, cellular growth, and metabolic activity [33]. The PAM pathway governs a wide array of cellular processes including cell cycle progression, metabolism, migration, and apoptosis by integrating extracellular signals through receptor tyrosine kinases (RTKs) and G protein-coupled receptors. PI3K catalyzes the phosphorylation of the 3'-hydroxyl group of phosphatidylinositol lipids, generating second messengers that recruit cytoplasmic signaling proteins to the plasma membrane. These include TEC family tyrosine kinases, AKT, and various members of the AGC protein kinase family, as well as multiple regulators of small GTPase activity [79, 80].

Alteration in PI3K/AKT/mTOR Pathway

The PI3K/Akt/mTOR signaling pathway, activated by the epidermal growth factor receptor (EGFR), is one of the most pivotal molecular cascades in tumor cells and plays a central role in the initiation and progression of gliomas. EGFR, a receptor tyrosine kinase (RTK), is essential for regulating various cellular processes, including adhesion, differentiation, apoptosis, migration, and proliferation [76, 79].

EGFR consists of three primary domains: an extracellular ligand-binding domain, a transmembrane domain, and an intracellular tyrosine kinase domain. Upon binding to ligands such as transforming growth factor alpha (TGFα) and epidermal growth factor (EGF), EGFR undergoes homodimerization or heterodimerization, leading to autophosphorylation of its intracellular tyrosine kinase domain [50, 81]. This phosphorylation event activates multiple downstream signaling pathways, most notably the PI3K/Akt/mTOR axis [79, 81].

Phosphatidylinositol 3-kinase (PI3K) is a family of lipid kinases that regulate cellular metabolism, survival, migration, differentiation, and proliferation [82, 83]. Class I PI3Ks, which are particularly relevant in tumorigenesis, comprise a regulatory subunit (p85) and a catalytic subunit (p110), with isoforms including α, β, and γ [79, 84]. Upon RTK activation, phosphorylated tyrosine residues interact with p85, which in turn activates p110. The activated p110 subunit phosphorylates phosphatidylinositol-4,5-bisphosphate (PIP2) to generate phosphatidylinositol-3,4,5-trisphosphate (PIP3), facilitating the recruitment of Akt to the inner plasma membrane [85, 86].

Akt is subsequently phosphorylated at two key sites—Thr308 and Ser473—activating its serine/threonine kinase function [87, 88]. One of its major downstream targets is mTORC1, which regulates ribosome biogenesis and protein synthesis. mTOR functions both as an upstream regulator and a downstream effector within the PI3K pathway [89]. Upon activation, Akt inhibits the tuberous sclerosis complex (TSC1/2), thereby promoting mTORC1 signaling. This leads to the phosphorylation of ribosomal protein S6 kinase (pS6K), eukaryotic initiation factor 4E (eIF4E), and eukaryotic initiation factor 4E-binding protein 1 (4EBP1), all of which are involved in protein translation and cell growth.

In parallel, mTORC2 contributes to cytoskeletal organization, cell survival, metabolism, and proliferation by phosphorylating Akt at Ser473 [90, 91]. Another critical regulator of this pathway is PTEN (phosphatase and tensin homolog deleted on chromosome ten), a tumor suppressor that antagonizes PI3K signaling. Clinical studies have demonstrated that mutations or loss of function in either EGFR or PTEN can result in persistent activation of the PI3K/Akt/mTOR pathway, thereby promoting tumorigenesis and contributing to resistance against cancer therapies.

pRb Pathway

The retinoblastoma protein (pRB) functions as a critical tumor suppressor by binding to and inhibiting E2F family transcription factors, thereby playing a pivotal role in regulating cell cycle progression. This regulation is primarily mediated by cyclin-dependent kinase (CDK) complexes, and the pRB protein is encoded by the RB1 gene [83]. Extensive research has explored the roles of proto-oncogenes and tumor suppressor genes in the pathogenesis of various human malignancies. Among these, the RB1 gene is particularly significant, as it encodes the Rb protein, an essential regulator of cellular proliferation [92]. The RB pathway serves as a key checkpoint in the cell cycle, preventing uncontrolled cell division and maintaining genomic integrity.

In normal cells, the Rb protein remains in an unphosphorylated state and binds to E2F transcription factors, inhibiting their activity and blocking progression into the S phase. This mechanism is crucial for maintaining control at the G1/S checkpoint and ensuring orderly cell cycle progression [83, 93]. Upon stimulation by mitogenic signals, cyclin-CDK complexes phosphorylate the Rb protein, leading to the release of E2F. This release activates the transcription of genes required for DNA synthesis and cell cycle advancement, thereby promoting cellular proliferation. Disruptions in the RB pathway such as mutations, deletions, or functional loss of the Rb protein can result in unchecked cell cycle progression and contribute to tumorigenesis. Such abnormalities are commonly observed in a wide range of cancers, including glioblastoma, underscoring the pathway's importance in maintaining cellular homeostasis.

pRb Pathway in Brain Tumor

The retinoblastoma (RB1) gene, located on chromosome 13q14.1–q14.2, plays a critical role in the malignant progression of astrocytomas. Mutations or deletions in this chromosomal region are observed in over 20% of high-grade gliomas, including glioblastoma multiforme (GBM) [94]. Studies have demonstrated that GBM typically expresses reduced levels of the pRB protein, and the transition from low-grade to intermediate-grade gliomas is frequently associated with deletion of the 13q region.

During the G1 phase of the cell cycle, cyclin D forms complexes with cyclin-dependent kinases CDK4 and CDK6, which phosphorylate pRB, leading to its functional inactivation. This phosphorylation event causes the release of E2F transcription factors from pRB, thereby enabling the transcription of genes required for entry into the S phase. In GBM, the CDK inhibitor CDKN2B, which normally inhibits CDK4 and CDK6 activity, is frequently inactivated. Loss of CDKN2B function contributes to unchecked CDK activity and subsequent pRB inactivation, facilitating uncontrolled cell cycle progression [77, 95].

Moreover, GBM often exhibits amplification of CDK4 and CDK6, further emphasizing their roles in astrocytic tumorigenesis and disease progression. Beyond its role in cell cycle regulation, pRB also functions as an adaptor and transcriptional cofactor, modulating the expression of genes involved in differentiation, proliferation, apoptosis, chromatin remodeling, and cellular senescence [96, 97].

The E2F family of transcription factors, particularly E2F1 through E2F3, is essential for cell proliferation and the G1-to-S phase transition. These factors regulate genes involved in DNA replication and metabolism. The inhibitory interaction between pRB and E2F suppresses the transcription of these genes, a process that requires the recruitment of histone deacetylases (HDACs), DNA methyltransferases (DNMTs), and ATP-dependent chromatin remodeling complexes to the promoters of target genes [98, 99]. Notably, E2F1–E2F3 also function as transcriptional activators and have been implicated in oncogenic processes across multiple malignancies, including gliomas, liposarcomas, and breast and ovarian cancers [98, 100].

Regulation of pRB Pathway

The modulation of phosphorylation status is critical for regulating the activity of the retinoblastoma protein (pRb) throughout the cell cycle, particularly during the G1 phase. In quiescent cells, pRb remains in a hypo-phosphorylated state, enabling it to bind E2F transcription factors and other regulatory targets, thereby suppressing cell proliferation. Upon receiving mitogenic signals, cyclin-dependent kinases (CDKs) sequentially phosphorylate pRb, disrupting its interaction with E2Fs [98, 99]. This phosphorylation event releases pRb from E2F-bound gene promoters, allowing trans-activators to initiate the transcription of genes required for S phase entry. One notable E2F target gene activated following pRb phosphorylation is Skp2, which promotes cell cycle progression through both E2F- and Skp2-dependent pathways. Although pRb contains sixteen potential CDK phosphorylation sites, not all have been confirmed to be phosphorylated *in vivo* [101, 102].

The regulation of pRb activity also involves phosphatases and cyclin-dependent kinase inhibitors (CDKIs), which mediate its dephosphorylation and reactivation. CDKIs respond to anti-proliferative signals such as cellular senescence, differentiation, and DNA damage, and they play a pivotal role in positively regulating pRb by inhibiting active cyclin-CDK complexes [83, 102]. Consequently, CDKI-mediated inhibition of cyclin-CDKs leads to cell cycle arrest *via* a pRb-dependent mechanism. In glioblastoma multiforme (GBM), where pRb function is frequently disrupted, targeting CDK4/6 with selective inhibitors has emerged as a promising therapeutic strategy. The CDK4/CDK6/Cyclin D complex is known to inactivate the pRb pathway, and its inhibition may restore pRb function and suppress tumor progression [77]. Both the pRb and p53 pathways are commonly altered in GBM, although mutation frequencies vary across studies. Therapeutic approaches aimed at modulating these signaling networks hold considerable potential for advancing GBM treatment and improving patient outcomes [103].

Alteration in pRB Pathways in Brain Tumor

Alterations in the pRB signaling pathway have been recognized as major contributors to cancer progression and malignant transformation. Inactivation of negative regulators such as cyclin-dependent kinase inhibitors (CDKIs) and hyperactivation of positive regulators, including cyclins and cyclin-dependent kinases (CDKs), are commonly observed across nearly all human malignancies [104]. Specifically, dysregulation of the pRB pathway has been reported in approximately 78% of glioblastoma cases, as well as in other cancers such as lung, breast, urothelial, and prostate carcinomas [77].

The RB1 tumor suppressor gene is implicated in a wide range of intracranial tumors, including gliomas, meningiomas, pituitary adenomas, and vestibular schwannomas. Emerging evidence also highlights the role of epigenetic mechanisms, particularly promoter hypermethylation in the downregulation or loss of pRB expression in glioblastomas and pituitary tumors [105]. These findings suggest that epigenetic modifications may play a critical role in influencing pRB activity and contribute to tumorigenesis.

A deeper understanding of the molecular and epigenetic alterations affecting the pRB pathway could offer valuable insights into novel therapeutic targets and strategies for treating glioblastoma and other malignancies. Targeting these regulatory mechanisms may enhance the efficacy of current treatments and support the development of precision oncology approaches.

NF-κB Pathway

The activation of the nuclear factor kappa-light-chain-enhancer of activated B cells (NF-κB) signaling pathway plays a pivotal role in the malignant progression of glioblastoma (GBM), contributing to poor patient prognoses [106]. The NF-κB transcription factor family regulates a wide array of cellular signaling pathways in response to inflammatory stimuli and other stressors. Both the canonical and non-canonical NF-κB pathways are involved in modulating the behavior of activated cells, influencing processes such as survival, proliferation, and immune response [107].

In the canonical NF-κB pathway, the degradation of inhibitor of κB (IκB) proteins leads to the release of NF-κB dimers, which translocate to the nucleus and bind to κB sites in the promoters of target genes. This activation regulates the transcription of genes involved in inflammation, cell survival, and immune modulation. NF-κB is ubiquitously expressed in neurons, where it contributes to neuronal information processing and provides neuroprotection against various insults [108, 109]. By controlling the expression of proinflammatory enzymes, chemokines, cytokines, and adhesion molecules, NF-κB influences both neuronal survival and the inflammatory milieu of the central nervous system (CNS).

Beyond neurons, NF-κB transcription factors are abundantly expressed in glial cells and cerebral vasculature, underscoring their role in coordinating CNS inflammation. Chronic inflammation within the brain can promote tumorigenesis by altering genetic and epigenetic landscapes of both tumor cells and the surrounding microenvironment. This dual role of inflammation facilitating immune evasion and promoting genetic instability has been recognized as contributing to two hallmarks of cancer: avoiding immune destruction and promoting tumor growth.

NF-κB functions as a master regulator at the intersection of inflammation and tumorigenesis, orchestrating cellular responses that support tumor development [110, 111]. Elevated NF-κB activity, along with increased levels of proinflammatory cytokines in GBM tissue, fosters a pro-tumorigenic microenvironment that accelerates tumor growth and progression [106]. Elucidating the role of NF-κB in GBM pathophysiology may offer valuable insights into therapeutic strategies targeting inflammatory signaling pathways to improve clinical outcomes.

NF-κB Pathway in Brain Tumor

Glioblastoma (GBM) frequently exhibits constitutive and aberrant activation of the nuclear factor kappa-light-chain-enhancer of activated B cells (NF-κB) signaling pathway. This dysregulation is driven by multiple mechanisms and is closely associated with hallmark cancer traits, including enhanced cell survival, increased invasiveness, and resistance to conventional therapies [111, 113]. As one of the most aggressive and prevalent malignant brain tumors in adults, GBM presents substantial therapeutic challenges, with high rates of recurrence and poor responsiveness to standard treatment modalities [112].

NF-κB activation in GBM is mediated through both protein kinase B (AKT)-dependent and -independent pathways, often involving aberrant signaling from platelet-derived growth factor receptors (PDGFR) and epidermal growth factor receptors (EGFR). These receptors are frequently overexpressed or mutated in GBM and play critical roles in promoting tumor proliferation, invasion, and survival [114 - 116]. Oncogenic signaling from EGFR and PDGFR has been shown to amplify NF-κB activity, thereby enhancing the tumor-promoting functions of this pathway.

Recent studies have highlighted the overactivation of NF-κB in GBM, with increased expression and mutations in the EGFR gene serving as key drivers of this phenomenon [115 - 117]. Alterations in EGFR signaling are among the most significant activators of NF-κB, contributing to the malignant phenotype of glioblastoma cells. In addition to EGFR dysregulation, several other genetic abnormalities are commonly observed in GBM, including the loss of INK4A, loss of PTEN, increased MDM2 expression, and EGFR gene amplification or mutation [118, 119]. These genetic alterations further complicate GBM pathophysiology and underscore the importance of targeting NF-κB and its upstream regulators as part of a comprehensive therapeutic strategy. Continued investigation into the molecular networks driving NF-κB activation may yield novel therapeutic approaches aimed at overcoming treatment resistance and improving clinical outcomes for patients with GBM.

p53 Pathway

The p53 protein, commonly referred to as tumor protein, is an essential explorer of the human genome. It causes apoptosis and the cell cycle to stop in response to internal and external cell stresses. Therefore, TP53 plays a crucial role as a tumor suppressor gene, stopping the cell cycle if genome replication or stability is compromised [112, 121]. There is no easy and straightforward response concerning when and how p53 acts. That being said, it is evident that p53 responds to environmental disturbances in a very flexible and diverse manner, utilizing this reaction to either preserve cellular homeostasis or determine cell death [120]. By regulating transcriptional levels and protein-protein interactions, p53 is a connecting element for numerous cellular signaling pathways, balancing and modulating various biological processes [122].

p53 Pathway in Brain Tumor

Glioblastoma multiforme (GBM) is the most aggressive form of brain cancer, associated with a dismal prognosis and significant therapeutic resistance. GBM is classified into two major subtypes: primary GBM, which arises *de novo*, and secondary GBM, which evolves from lower-grade gliomas. These subtypes exhibit distinct mutation frequencies in the TP53 tumor suppressor gene, with mutations present in approximately 30% of primary GBMs and up to 65% of secondary GBMs [121, 123]. Notably, abnormalities in p53 signaling are observed in approximately 87% of sporadic gliomas, making it the most frequently altered pathway in these malignancies [124, 125]. In primary GBMs, mutations are distributed broadly across the genome, whereas in secondary GBMs and low-grade astrocytic gliomas, mutations tend to cluster at specific codons, particularly 248 and 272 [126, 127]. The p53 pathway regulates over 2,500 genes involved in carcinogenesis and interacts with other signaling networks through complex feedback loops involving histone acetyltransferases, methyltransferases, ubiquitin ligases, and protein kinases [128, 129].

Key regulators of p53 in brain tumors include negative regulators such as MDM2 and MDM4, and positive regulators like CDKN2A and its alternative reading frame product (ARF). MDM2 inhibits p53 by binding directly to it and promoting its degradation *via* ubiquitin ligase activity. Overexpression of MDM2 is a common mechanism for suppressing p53 function, thereby facilitating uncontrolled cell proliferation [130, 131].

Functionally, p53 acts as a tumor suppressor by regulating genes involved in stem cell differentiation, cell cycle arrest, apoptosis, and senescence. It is activated in response to cellular stressors such as hypoxia, DNA damage, oncogene activation, and genotoxic insults [120, 122, 132]. In GBM, TP53 mutations are associated with more invasive, proliferative, and stem-like tumor phenotypes, as well as reduced sensitivity to DNA-damaging agents like cisplatin [128, 133]. Homozygous deletion of CDKN2A/ARF occurs in approximately 60% of GBM cases and is more prevalent in younger patients [133].

Multiple mechanisms contribute to p53 inactivation in GBM. These include disruptions in the ARF/MDM2/MDM4/p53 axis, loss of DNA damage response components such as ATM and CHEK2, and mutations in tumor suppressors like CHD5 [120, 134]. Transcriptional suppression of p53 also plays a role, for example, Parkin, a transcriptional target of p53, has been implicated in glioma development. Increased Parkin expression in murine models infected with p53-bearing viruses suggests a functional interaction between these proteins [128, 134]. Metabolic regulation further influences p53 activity. Mutations in IDH1 (R132H/R132Q) suppress p53 expression by stabilizing hypoxia-inducible factor-2α (HIF-2α), which induces miR-380-5p, a microRNA that negatively regulates p53. A negative correlation between p53 protein levels and IDH1 R132H has been observed in human glioma samples, indicating a bidirectional relationship between metabolic alterations and p53 status [120, 128].

Recent findings also implicate the transcription factor NFIA in GBM progression through the transcriptional repression of p53, p21, and PAI1. The overexpression of Bcl2L12 in primary GBMs inhibits p53 transactivation, thereby preventing p53-mediated apoptosis and senescence [135, 136]. Additionally, macrophage migration inhibitory factor (MIF) has been identified as a physical inhibitor of p53, highly expressed in brain tumor-initiating cells (BTICs), and shown to promote tumor growth in xenograft models. In conclusion, the impairment of oxidative metabolism, such as complex I inhibition or reduced mitochondrial DNA copy number can lead to p53 inactivation and facilitate the transformation of neural stem cells, which are believed to be the origin of high-grade gliomas [135, 137]. This intricate interplay between p53 signaling, metabolic regulation, and tumor biology underscores the complexity of GBM and highlights promising therapeutic targets for intervention (Fig. 1).

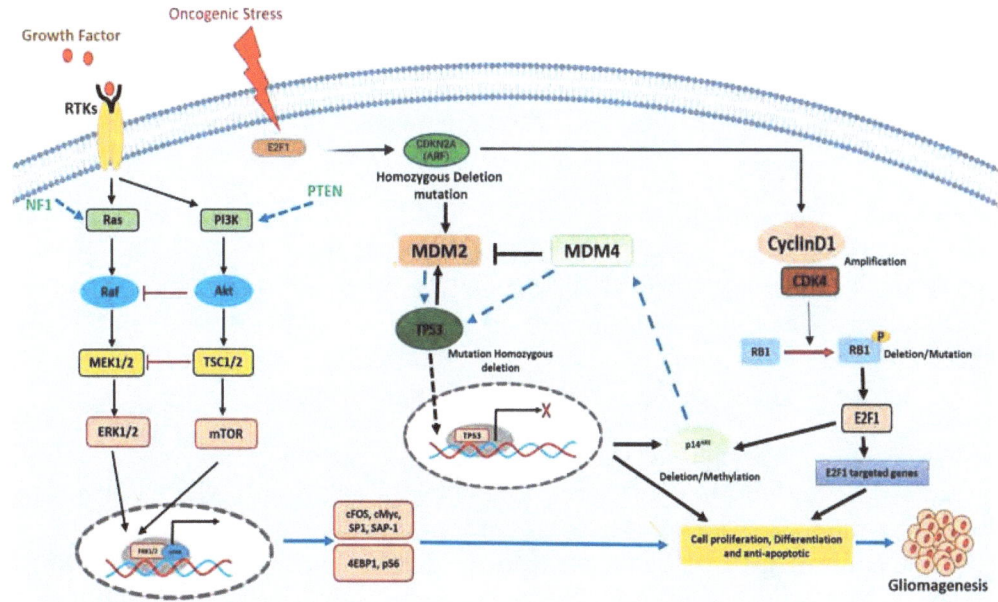

Fig. (1). Schematic representation of pathways that are altered in brain tumor.

RAS-MAPK Pathway

Three closely related human RAS genes H-Ras, N-Ras, and K-Ras can mutate into oncogenes, particularly in the context of Rat Sarcoma (RAS) virus infection. These genes belong to the G protein family and are pivotal in cell signaling. The activation state of RAS is determined by the binding of guanosine triphosphate (GTP) or guanosine diphosphate (GDP). When GTP binds to RAS, it activates the protein, while GDP binding results in its inactivation [138, 139].

Once activated, RAS plays a critical role in regulating various cellular processes, including cell division, proliferation, apoptosis, and signal transduction. It exerts its effects primarily through the activation of downstream signaling pathways, notably the mitogen-activated protein kinase (MAPK) pathway. Activated RAS directly binds to RAF kinase, which subsequently initiates a cascade of phosphorylation events leading to MAPK pathway activation [138, 140]. Additionally, RAS is involved in modulating the phosphoinositide 3-kinase (PI3K) pathway, further contributing to its role in regulating cellular functions and influencing carcinogenesis [138].

RAS-MAPK Pathway in Brain Tumor

Recent research in transcriptomics and sequencing has illuminated the role of the RAS-MAPK pathway in the initiation and progression of neuroblastoma (NB).

The ALK (anaplastic lymphoma kinase) gene gained prominence in 2008 after it was found that most familial cases of NB harbored mutations in this gene [141]. Notably, ALK mutations were identified as the most frequently altered somatic gene in primary NB, particularly in high-risk patients.

Current studies indicate that approximately 6–10% of sporadic primary NB tumors and 12–14% of high-risk cases exhibit ALK mutations [142, 143]. A comparative analysis of ALK mutation rates across various tumor types revealed that NB has a higher frequency of these mutations than brain, colorectal, lung, and blood malignancies [144]. This suggests that ALK mutations are more pivotal in NB development than in many other common cancers, with only breast and skin cancers showing higher frequencies of ALK mutations [145].

Furthermore, aberrations in the RAS/MAPK signaling pathway have been implicated in multiple malignancies, including glioblastoma (GBM) and high-grade astrocytomas, where RAS overexpression has been observed [138]. The overexpression of growth factor receptors, such as EGFR and PDGFR, which regulate RAS, indicates a positive correlation between the pathophysiology of gliomas and the expression of these receptors [146]. This aberrant activation of the RAS/MAPK pathway presents a potential therapeutic target for glioma treatments [32, 138].

Activation of RAS/MAPK Pathways in Brain Tumor

Activating the PI3K/Akt/mTOR and RAS/MAPK pathways *via* RTK signaling controls cellular functions like growth, survival, differentiation, and migration in normal cells where receptor activity is regulated [147]. Nevertheless, RTKs are aberrantly activated in cancer, which leads to the oncogenic phenotype through one of the following mechanisms: (1) encouraging cancer cells to overproduce growth factors; (2) overexpressing and amplifying the RTK itself, enabling hypersensitivity to low ligand concentrations; (3) acquiring mutations in their kinase or ligand-binding domains; (4) fusing kinase domains with motifs of unrelated proteins; or (5) chromosomal translocation, resulting in a chimeric product with increased kinase activity [148].

The RAS-MAPK pathway activation is associated with poor prognosis in primary neuroblastoma (NB) tumors. High activity levels in this pathway are linked to mutations in known activation-causing genes, such as SOS1, PTPN11, and ALK. Recent studies have identified recurring abnormalities in genes like DMD, PHOX2B, and CIC across various activated cancers, showing that different mutations can have distinct effects on this pathway both *in vitro* and *in vivo*. For instance, the deletion of CIC *in vivo* leads to a significant increase in hallmark gene expression and tumor growth rate, occurring independently of phosphorylated ERK, suggesting that CIC functions outside the RAS-MAPK pathway [149, 150].

The human RAS family consists of three closely related genes: H-Ras, N-Ras, and K-Ras. RAS proteins, as members of the G protein family, toggle between active (GTP-bound) and inactive (GDP-bound) states. RAF kinase plays a crucial role in regulating the MAPK pathway and other downstream signaling pathways, further activated through direct binding to active RAS [151]. Overall, RAS is pivotal in governing cell division, proliferation, signal transduction, apoptosis, and carcinogenesis, while also influencing other pathways, including the PI3K pathway (Fig. **1**).

STAT3 and ZIP4 Pathway

The Signal Transducer and Activator of Transcription (STAT) family comprises seven structurally related and highly conserved cytoplasmic proteins STAT1 through STAT7 characterized by the presence of Src homology 2 (SH2) domains and their critical roles as transcription factors. Among the STAT family, STAT3 has been extensively studied in the context of brain tumors, particularly glioblastoma multiforme (GBM) [152]. Elevated levels of STAT3 have been observed in various malignancies, including GBM, where its activation by epidermal growth factor (EGF) highlights its involvement in tumor biology and progression [153]. Beyond its oncogenic associations, STAT3 is also essential for the differentiation of astrocytes and neural stem cells, indicating its broader role in neural development.

Interestingly, the function of STAT3 in GBM appears to be context-dependent. While traditionally linked to tumor-promoting activities, recent studies suggest that STAT3 may exhibit both oncogenic and tumor-suppressive properties, contingent upon the genetic and molecular characteristics of the tumor microenvironment [154, 155]. This duality underscores the complexity of STAT3 signaling in glioblastoma, highlighting the need for nuanced therapeutic approaches targeting this pathway [156].

STAT3 and ZIP4 Pathway in Brain Tumor

The Signal Transducer and Activator of Transcription (STAT) family comprises seven structurally related and highly conserved cytoplasmic proteins STAT1 through STAT7 characterized by the presence of Src homology 2 (SH2) domains and their critical roles as transcription factors. Among the STAT family, STAT3 has been extensively studied in the context of brain tumors, particularly glioblastoma multiforme (GBM) [152]. Elevated levels of STAT3 have been observed in various malignancies, including GBM, where its activation by epidermal growth factor (EGF) highlights its involvement in tumor biology and progression [153]. Beyond its oncogenic associations, STAT3 is also essential for the differentiation of astrocytes and neural stem cells, indicating its broader role in neural development. Interestingly, the function of STAT3 in GBM appears to be context-dependent. While traditionally linked to tumor-promoting activities, recent studies suggest that STAT3 may exhibit both oncogenic and tumor-suppressive properties, contingent upon the genetic and molecular characteristics of the tumor microenvironment [154, 155]. This duality underscores the complexity of STAT3 signaling in glioblastoma, highlighting the need for nuanced therapeutic approaches targeting this pathway [156].

WNT Pathway

WNT (Wingless/Integrated) signaling plays a fundamental role in numerous cellular processes during the development and maintenance of the central nervous system (CNS) [157, 158]. This pathway is involved in critical biological functions such as embryonic stem cell formation, tissue regeneration, cell differentiation, and immune cell regulation. Recent studies have emphasized the significance of WNT signaling in modulating immune responses and its potential role in preventing various diseases [159].

In the context of neural development, WNT signaling is essential for the regulation of neural stem cells (NSCs) in key regions, including the fetal ventricular zone, the postnatal subventricular zone, and the hippocampus [160]. The pathway functions through signal transduction mechanisms that convey extracellular signals from cell surface receptors to intracellular targets in the cytoplasm and nucleus [161].

Canonical WNT/β-catenin pathway – This pathway involves β-catenin, T-cell factor (TCF), and lymphocyte enhancer-binding factor (LEF), and is primarily responsible for regulating NSC self-renewal, proliferation, and fate determination. Non-canonical pathways – These include the planar cell polarity (PCP) pathway and the WNT/Ca^{2+} pathway, which are independent of β-catenin and are involved in processes such as cell migration, polarity, and cytoskeletal organization [162, 163]. The canonical WNT/β-catenin pathway is tightly regulated, particularly in terms of β-catenin levels and subcellular localization, as β-catenin serves as a key transcriptional co-activator mediating WNT-responsive gene expression. Proper modulation of this pathway is crucial for maintaining NSC function and CNS integrity.

WNT Pathway in Brain Tumor

The WNT/β-catenin signaling pathway has been shown to play a critical role in neuroblastoma (NB) proliferation. Elevated WNT/β-catenin activity correlates with increased MYCN expression, particularly in NB cell lines that initially lack MYCN amplification. This upregulation is accompanied by the increased expression of canonical WNT ligands WNT1, WNT6, WNT7A, and WNT10B as well as elevated levels of FZD2 and LRP5, indicating their significant involvement in NB pathogenesis [159, 164].

Silencing WNT1 *via* RNA interference markedly reduces cell viability in the SH-SY5Y NB cell line, underscoring its functional importance. Additionally, relapsed NB tumors exhibit elevated levels of LGR5, a stem cell marker associated with an aggressive disease. LGR5 expression has been linked to MYCN amplification, tumor aggressiveness, and recurrence in NB [165, 166]. R-spondins enhance WNT/β-catenin signaling by binding to LGR receptors, further amplifying oncogenic signaling. Knockdown of LGR5 in NB cell lines such as SK-N-BE, SK-N-AS, and SH-SY5Y induces apoptotic cell death, independent of β-catenin signaling [167].

Moreover, LGR5 knockdown leads to reduced phosphorylation of MEK/ERK in the MAPK pathway and negatively impacts the AKT/mTOR signaling axis [168 - 170]. Recent studies have identified RAS/RAF mutations upstream regulators of MAPK signaling as common events acquired during NB relapse, contributing to a more aggressive phenotype. WNT3A/RSPO2 signaling has been shown to alter the transcriptomic landscape of NB cell lines, affecting not only WNT-related molecules but also BMP4, Cyclin D1, and phosphorylation of the RB protein, all of which are critical for NB differentiation and progression [171, 172].

In glioblastoma (GBM), noncanonical WNT signaling pathways, specifically WNT/Ca^{2+} and planar cell polarity (PCP), are predominantly activated. These pathways are closely associated with cytoskeletal reorganization, a process essential for epithelial-to-mesenchymal transition (EMT) and cell migration [173, 174]. In embryonic models such as Xenopus and zebrafish, noncanonical WNT ligands WNT5A and WNT11 stimulate cell migration through the activation of these pathways [117, 175, 176].

The expression of WNT inhibitory factor 1 (WIF1), a secreted antagonist of WNT signaling, further supports the utilization of noncanonical pathways in GBM. WIF1 can inhibit both the WNT/Ca^{2+} and canonical WNT/β-catenin pathways. Additionally, activation of WNT/β-catenin signaling increases c-MYC expression, which promotes EMT in mammary epithelial cells [177, 178]. In mouse embryonic stem cells, noncanonical WNT signaling induces EMT through NFAT activation [177].

Further evidence from pancreatic ductal adenocarcinoma demonstrates that NFATc1 regulates EMT transcription factors Snail and Zeb1, reinforcing the role of both canonical and noncanonical WNT pathways in partial EMT phenotypes. EMT represents just one of many developmental signaling mechanisms exploited by glioma stem cells (GSCs) to drive tumor progression [175, 179].

CONCLUSION

Despite extensive research into the molecular biology and dysregulated signaling networks of brain tumors, including glioblastoma, these malignancies remain among the most difficult to treat and are associated with poor prognoses. Recent advances in understanding the molecular signaling pathways involved in brain tumor pathogenesis have expanded the potential for targeted therapeutic interventions. This chapter has explored key signaling pathways and therapeutic strategies relevant to brain tumor development. Currently available treatments may offer only modest improvements in survival, and curative outcomes remain rare. Consequently, there is an urgent need for continued research to elucidate the mechanisms underlying brain tumor progression and to identify novel therapeutic targets within emerging signaling pathways. The growing body of knowledge surrounding molecular markers and tumorigenic pathways holds promise for the development of more effective and personalized treatment strategies.

Several critical signaling pathways, including the pRB, p53, NF-κB, RAS/MAPK, STAT3, ZIP3, and WNT pathways, are implicated in the development of various brain tumor types. These pathways play essential roles in regulating cell cycle progression, apoptosis, and cellular differentiation. Their involvement in tumorigenesis not only enhances our understanding of disease mechanisms but also provides valuable molecular targets for therapeutic intervention.

REFERENCES

[1]   Tang W, Fan W, Lau J, Deng L, Shen Z, Chen X. Emerging blood–brain-barrier-crossing nanotechnology for brain cancer theranostics. Chem Soc Rev 2019; 48(11): 2967-3014.
        [http://dx.doi.org/10.1039/C8CS00805A] [PMID: 31089607]

[2]   Louis DN, Perry A, Wesseling P, *et al.* The 2021 WHO classification of tumors of the central nervous system: a summary. Neuro-oncol 2021; 23(8): 1231-51.
        [http://dx.doi.org/10.1093/neuonc/noab106] [PMID: 34185076]

[3]   Kalluri AL, Shah PP, Lim M. The tumor immune microenvironment in primary CNS neoplasms: a review of current knowledge and therapeutic approaches. Int J Mol Sci 2023; 24(3): 2020.
        [http://dx.doi.org/10.3390/ijms24032020] [PMID: 36768342]

[4]   Ilic I, Ilic M. International patterns and trends in the brain cancer incidence and mortality: An observational study based on the global burden of disease. Heliyon 2023; 9(7): e18222.

[http://dx.doi.org/10.1016/j.heliyon.2023.e18222] [PMID: 37519769]

[5] Rindi G, Klimstra DS, Abedi-Ardekani B, *et al.* A common classification framework for neuroendocrine neoplasms: an International Agency for Research on Cancer (IARC) and World Health Organization (WHO) expert consensus proposal. Mod Pathol 2018; 31(12): 1770-86.
[http://dx.doi.org/10.1038/s41379-018-0110-y] [PMID: 30140036]

[6] Wei R, Zhou J, Bui B, Liu X. Glioma actively orchestrate a self-advantageous extracellular matrix to promote recurrence and progression. BMC Cancer. 2024;24:974.
[http://dx.doi.org/10.1186/s12885-024-12751-3]

[7] Ostrom QT, Cioffi G, Waite K, Kruchko C, Barnholtz-Sloan JS. CBTRUS statistical report: primary brain and other central nervous system tumors diagnosed in the United States in 2014–2018. Neuro-oncol 2021; 23(12) (Suppl. 3): iii1-iii105.
[http://dx.doi.org/10.1093/neuonc/noab200] [PMID: 34608945]

[8] Zarco N, Norton E, Quiñones-Hinojosa A, Guerrero-Cázares H. Overlapping migratory mechanisms between neural progenitor cells and brain tumor stem cells. Cell Mol Life Sci 2019; 76(18): 3553-70.
[http://dx.doi.org/10.1007/s00018-019-03149-7] [PMID: 31101934]

[9] Burnett BA, Womeldorff MR, Jensen R. Meningioma: Signaling pathways and tumor growth Handbook of Clinical Neurology 169. Elsevier 2020; pp. 137-50.
[http://dx.doi.org/10.1016/B978-0-12-804280-9.00009-3]

[10] Gupta S, Bi WL, Dunn IF. Medical management of meningioma in the era of precision medicine. Neurosurg Focus 2018; 44(4): E3.
[http://dx.doi.org/10.3171/2018.1.FOCUS17754] [PMID: 29606052]

[11] Pinker B, Barciszewska AM. mTOR signaling and potential therapeutic targeting in meningioma. Int J Mol Sci 2022; 23(4): 1978.
[http://dx.doi.org/10.3390/ijms23041978] [PMID: 35216092]

[12] Whitehead CE, Ziemke EK, Frankowski-McGregor CL, Mumby RA, Chung J, Li J. A first-in-class selective inhibitor of EGFR and PI3K offers a single-molecule approach to targeting adaptive resistance. Nature Cancer 2024; pp. 1-17.

[13] Roda D, Veiga P, Melo JB, Carreira IM, Ribeiro IP. Principles in the Management of Glioblastoma. Genes (Basel) 2024; 15(4): 501.
[http://dx.doi.org/10.3390/genes15040501] [PMID: 38674436]

[14] McNeill RS, Canoutas DA, Stuhlmiller TJ, *et al.* Combination therapy with potent PI3K and MAPK inhibitors overcomes adaptive kinome resistance to single agents in preclinical models of glioblastoma. Neuro-oncol 2017; 19(11): 1469-80.
[http://dx.doi.org/10.1093/neuonc/nox044] [PMID: 28379424]

[15] Ikliptikawati DK, Hirai N, Makiyama K, *et al.* Nuclear transport surveillance of p53 by nuclear pores in glioblastoma. Cell Rep 2023; 42(8): 112882.
[http://dx.doi.org/10.1016/j.celrep.2023.112882] [PMID: 37552992]

[16] Grommes C, DeAngelis LM. Primary CNS Lymphoma. J Clin Oncol 2017; 35(21): 2410-8.
[http://dx.doi.org/10.1200/JCO.2017.72.7602] [PMID: 28640701]

[17] Le Rhun E, Preusser M, Roth P, *et al.* Molecular targeted therapy of glioblastoma. Cancer Treat Rev 2019; 80: 101896.
[http://dx.doi.org/10.1016/j.ctrv.2019.101896] [PMID: 31541850]

[18] Harrison RA, de Groot JF. Cell signaling pathways in brain tumors. Top Magn Reson Imaging 2017; 26(1): 15-26.
[http://dx.doi.org/10.1097/RMR.0000000000000112] [PMID: 28079711]

[19] Schaff LR, Mellinghoff IK. Glioblastoma and other primary brain malignancies in adults: a review. JAMA 2023; 329(7): 574-87.
[http://dx.doi.org/10.1001/jama.2023.0023] [PMID: 36809318]

[20] Wu W, Klockow JL, Zhang M, *et al.* Glioblastoma multiforme (GBM): An overview of current therapies and mechanisms of resistance. Pharmacol Res 2021; 171: 105780.
[http://dx.doi.org/10.1016/j.phrs.2021.105780] [PMID: 34302977]

[21] Seker-Polat F, Pinarbasi Degirmenci N, Solaroglu I, Bagci-Onder T. Tumor cell infiltration into the brain in glioblastoma: from mechanisms to clinical perspectives. Cancers (Basel) 2022; 14(2): 443.
[http://dx.doi.org/10.3390/cancers14020443] [PMID: 35053605]

[22] Ullah N, Javed A, Alhazmi A, Hasnain SM, Tahir A, Ashraf R. TumorDetNet: A unified deep learning model for brain tumor detection and classification. PLoS One 2023; 18(9): e0291200.
[http://dx.doi.org/10.1371/journal.pone.0291200] [PMID: 37756305]

[23] Lin X, DeAngelis LM. Treatment of brain metastases. J Clin Oncol 2015; 33(30): 3475-84.
[http://dx.doi.org/10.1200/JCO.2015.60.9503] [PMID: 26282648]

[24] Nabors B, Portnow J, Hattangadi-Gluth J, Horbinski C. NCCN CNS tumor guidelines update for 2023. Oxford University Press US 2023; pp. 2114-6.

[25] Ogasawara C, Philbrick BD, Adamson DC. Meningioma: a review of epidemiology, pathology, diagnosis, treatment, and future directions. Biomedicines 2021; 9(3): 319.
[http://dx.doi.org/10.3390/biomedicines9030319] [PMID: 33801089]

[26] Fahlström A, Dwivedi S, Drummond K. Multiple meningiomas: Epidemiology, management, and outcomes. Neurooncol Adv 2023; 5 (Suppl. 1): i35-48.
[http://dx.doi.org/10.1093/noajnl/vdac108] [PMID: 37287575]

[27] Singh D, Tonjam R, Chaudhary T, *et al.* A short appraisal on gold nanoparticles: recent advances and applications. Curr Nanomed. 2021;11(3):168–76.
[http://dx.doi.org/10.2174/2468187312666211220122455]

[28] Pandey P, Khan F, Upadhyay TK, Seungjoon M, Park MN, Kim B. New insights about the PDGF/PDGFR signaling pathway as a promising target to develop cancer therapeutic strategies. Biomed Pharmacother 2023; 161: 114491.
[http://dx.doi.org/10.1016/j.biopha.2023.114491] [PMID: 37002577]

[29] Tyagi T, Jain K, Gu SX, Qiu M, Gu VW, Melchinger H. A guide to molecular and functional investigations of platelets to bridge basic and clinical sciences. Nature cardiovascular research. 2022;1(3):223-37.
[http://dx.doi.org/10.1038/s44161-022-00021-z]

[30] Orofiamma LA, Vural D, Antonescu CN. Control of cell metabolism by the epidermal growth factor receptor. Biochim Biophys Acta Mol Cell Res 2022; 1869(12): 119359.
[http://dx.doi.org/10.1016/j.bbamcr.2022.119359] [PMID: 36089077]

[31] Wee P, Wang Z. Epidermal growth factor receptor cell proliferation signaling pathways. Cancers (Basel) 2017; 9(5): 52.
[http://dx.doi.org/10.3390/cancers9050052] [PMID: 28513565]

[32] Khabibov M, Garifullin A, Boumber Y, *et al.* Signaling pathways and therapeutic approaches in glioblastoma multiforme (Review). Int J Oncol 2022; 60(6): 69.
[http://dx.doi.org/10.3892/ijo.2022.5359] [PMID: 35445737]

[33] Iimori K, Kurita A, Yazumi S. Metastasis of pancreatic cancer to cerebellum resembles meningitis. Clin Gastroenterol Hepatol 2021; 19(8): A23.
[http://dx.doi.org/10.1016/j.cgh.2020.04.082] [PMID: 32389883]

[34] Dal Bello S, Martinuzzi D, Tereshko Y, *et al.* The present and future of optic pathway glioma therapy. Cells 2023; 12(19): 2380.
[http://dx.doi.org/10.3390/cells12192380] [PMID: 37830595]

[35] Nunno VD, Aprile M, Gatto L, *et al.* Novel insights toward diagnosis and treatment of glioneuronal and neuronal tumors in young adults. CNS Oncol 2024; 13(1): 2357532.

[http://dx.doi.org/10.1080/20450907.2024.2357532] [PMID: 38873961]

[36]   Bale TA, Rosenblum MK. The 2021 WHO classification of tumors of the central nervous system: An update on pediatric low-grade gliomas and glioneuronal tumors. Brain Pathol 2022; 32(4): e13060.
[http://dx.doi.org/10.1111/bpa.13060] [PMID: 35218102]

[37]   Byun YH, Park CK. Classification and diagnosis of adult glioma: a scoping review. Brain Neurorehabil 2022; 15(3): e23.
[http://dx.doi.org/10.12786/bn.2022.15.e23] [PMID: 36742083]

[38]   Dragoo DD, Taher A, Wong VK, *et al.* PTEN hamartoma tumor syndrome/Cowden syndrome: genomics, oncogenesis, and imaging review for associated lesions and malignancy. Cancers (Basel) 2021; 13(13): 3120.
[http://dx.doi.org/10.3390/cancers13133120] [PMID: 34206559]

[39]   Fyllingen EH, Bø LE, Reinertsen I, *et al.* Survival of glioblastoma in relation to tumor location: a statistical tumor atlas of a population-based cohort. Acta Neurochir (Wien) 2021; 163(7): 1895-905.
[http://dx.doi.org/10.1007/s00701-021-04802-6] [PMID: 33742279]

[40]   Poyuran R, Mahadevan A, Mhatre R, *et al.* Neuropathological spectrum of drug resistant epilepsy: 15-years-experience from a tertiary care centre. J Clin Neurosci 2021; 91: 226-36.
[http://dx.doi.org/10.1016/j.jocn.2021.07.014] [PMID: 34373032]

[41]   Métais A, Appay R, Pagès M, *et al.* Low-grade epilepsy-associated neuroepithelial tumours with a prominent oligodendroglioma-like component: The diagnostic challenges. Neuropathol Appl Neurobiol 2022; 48(2): e12769.
[http://dx.doi.org/10.1111/nan.12769] [PMID: 34551121]

[42]   Singh D, Tiwari P, Nagdev S. Particulate vaccine dispersions emerge as a novel carrier for deep pulmonary immunization. Curr Nanomedicine. 2023;13(2):71–4.
[http://dx.doi.org/10.2174/2468187313666230714124009]

[43]   Giulioni M, Marucci G, Martinoni M, *et al.* Epilepsy associated tumors: Review article. World J Clin Cases 2014; 2(11): 623-41.
[http://dx.doi.org/10.12998/wjcc.v2.i11.623] [PMID: 25405186]

[44]   Huang D, Wang Y, Thompson JW, *et al.* Cancer-cell-derived GABA promotes β-catenin-mediated tumour growth and immunosuppression. Nat Cell Biol 2022; 24(2): 230-41.
[http://dx.doi.org/10.1038/s41556-021-00820-9] [PMID: 35145222]

[45]   Chieffo DPR, Lino F, Ferrarese D, Belella D, Della Pepa GM, Doglietto F. Brain tumor at diagnosis: from cognition and behavior to quality of life. Diagnostics (Basel) 2023; 13(3): 541.
[http://dx.doi.org/10.3390/diagnostics13030541] [PMID: 36766646]

[46]   Palmieri A, Valentinis L, Zanchin G. Update on headache and brain tumors. Cephalalgia 2021; 41(4): 431-7.
[http://dx.doi.org/10.1177/0333102420974351] [PMID: 33249916]

[47]   Hadidchi S, Surento W, Lerner A, *et al.* Headache and brain tumor. Neuroimaging Clin N Am 2019; 29(2): 291-300.
[http://dx.doi.org/10.1016/j.nic.2019.01.008] [PMID: 30926118]

[48]   Taylor LP. Mechanism of brain tumor headache. Headache 2014; 54(4): 772-5.
[http://dx.doi.org/10.1111/head.12317] [PMID: 24628259]

[49]   Cuneo A, Murinova N, Eds. Headache management in individuals with brain tumor Seminars in neurology. Thieme Medical Publishers, Inc. 2024.

[50]   Zhang M, Yu H, Cao G, *et al.* Enhanced focal cortical dysplasia detection in pediatric frontal lobe epilepsy with asymmetric radiomic and morphological features. Front Neurosci 2023; 17: 1289897.
[http://dx.doi.org/10.3389/fnins.2023.1289897] [PMID: 38033536]

[51]   Dziedzic TA, Bala A, Balasa A, Olejnik A, Marchel A. Cortical and white matter anatomy relevant for

the lateral and superior approaches to resect intraaxial lesions within the frontal lobe. Sci Rep 2022; 12(1): 21402.
[http://dx.doi.org/10.1038/s41598-022-25375-z] [PMID: 36496517]

[52] Jacquemin V, Antoine M, Dom G, Detours V, Maenhaut C, Dumont JE. Dynamic cancer cell heterogeneity: Diagnostic and therapeutic implications. Cancers (Basel) 2022; 14(2): 280.
[http://dx.doi.org/10.3390/cancers14020280] [PMID: 35053446]

[53] Couturier CP, Nadaf J, Li Z, *et al.* Glioblastoma scRNA-seq shows treatment-induced, immune-dependent increase in mesenchymal cancer cells and structural variants in distal neural stem cells. Neuro-oncol 2022; 24(9): 1494-508.
[http://dx.doi.org/10.1093/neuonc/noac085] [PMID: 35416251]

[54] Jose A, Kulkarni P, Thilakan J, *et al.* Integration of pan-omics technologies and three-dimensional *in vitro* tumor models: An approach toward drug discovery and precision medicine. Mol Cancer 2024; 23(1): 50.
[http://dx.doi.org/10.1186/s12943-023-01916-6] [PMID: 38461268]

[55] Park JH, de Lomana ALG, Marzese DM, *et al.* A systems approach to brain tumor treatment. Cancers (Basel) 2021; 13(13): 3152.
[http://dx.doi.org/10.3390/cancers13133152] [PMID: 34202449]

[56] Bou-Gharios J, Noël G, Burckel H. Preclinical and clinical advances to overcome hypoxia in glioblastoma multiforme. Cell Death Dis 2024; 15(7): 503.
[http://dx.doi.org/10.1038/s41419-024-06904-2] [PMID: 39003252]

[57] Ringel F, Pape H, Sabel M, *et al.* Clinical benefit from resection of recurrent glioblastomas: results of a multicenter study including 503 patients with recurrent glioblastomas undergoing surgical resection. Neuro-oncol 2016; 18(1): 96-104.
[http://dx.doi.org/10.1093/neuonc/nov145] [PMID: 26243790]

[58] Ramón y Cajal S, Sesé M, Capdevila C, *et al.* Clinical implications of intratumor heterogeneity: challenges and opportunities. J Mol Med (Berl) 2020; 98(2): 161-77.
[http://dx.doi.org/10.1007/s00109-020-01874-2] [PMID: 31970428]

[59] Behnan J, Finocchiaro G, Hanna G. The landscape of the mesenchymal signature in brain tumours. Brain 2019; 142(4): 847-66.
[http://dx.doi.org/10.1093/brain/awz044] [PMID: 30946477]

[60] Sturm D, Bender S, Jones DTW, *et al.* Paediatric and adult glioblastoma: multiform (epi)genomic culprits emerge. Nat Rev Cancer 2014; 14(2): 92-107.
[http://dx.doi.org/10.1038/nrc3655] [PMID: 24457416]

[61] Mo F, Pellerino A, Soffietti R, Rudà R. Blood–brain barrier in brain tumors: biology and clinical relevance. Int J Mol Sci 2021; 22(23): 12654.
[http://dx.doi.org/10.3390/ijms222312654] [PMID: 34884457]

[62] Daneman R, Prat A. The blood-brain barrier. Cold Spring Harb Perspect Biol 2015; 7(1): a020412.
[http://dx.doi.org/10.1101/cshperspect.a020412] [PMID: 25561720]

[63] Mollaei M, Hassan ZM, Khorshidi F, Langroudi L. Chemotherapeutic drugs: Cell death- and resistance-related signaling pathways. Are they really as smart as the tumor cells? Transl Oncol 2021; 14(5): 101056.
[http://dx.doi.org/10.1016/j.tranon.2021.101056] [PMID: 33684837]

[64] Zhao Y, Yue P, Peng Y, *et al.* Recent advances in drug delivery systems for targeting brain tumors. Drug Deliv 2023; 30(1): 1-18.
[http://dx.doi.org/10.1080/10717544.2022.2154409] [PMID: 36597214]

[65] Lin H, Liu C, Hu A, Zhang D, Yang H, Mao Y. Understanding the immunosuppressive microenvironment of glioma: mechanistic insights and clinical perspectives. J Hematol Oncol 2024; 17(1): 31.

[http://dx.doi.org/10.1186/s13045-024-01544-7] [PMID: 38720342]

[66]   Liu YT, Sun ZJ. Turning cold tumors into hot tumors by improving T-cell infiltration. Theranostics 2021; 11(11): 5365-86.
[http://dx.doi.org/10.7150/thno.58390] [PMID: 33859752]

[67]   Ni J, Zhang Z, Ge M, Chen J, Zhuo W. Immune-based combination therapy to convert immunologically cold tumors into hot tumors: an update and new insights. Acta Pharmacol Sin 2023; 44(2): 288-307.
[http://dx.doi.org/10.1038/s41401-022-00953-z] [PMID: 35927312]

[68]   Angom RS, Nakka NMR, Bhattacharya S. Advances in glioblastoma therapy: an update on current approaches. Brain Sci 2023; 13(11): 1536.
[http://dx.doi.org/10.3390/brainsci13111536] [PMID: 38002496]

[69]   Vengoji R, Macha MA, Batra SK, Shonka NA. Natural products: a hope for glioblastoma patients. Oncotarget 2018; 9(31): 22194-219.
[http://dx.doi.org/10.18632/oncotarget.25175] [PMID: 29774132]

[70]   Tiong KL, Sintupisut N, Lin MC, *et al.* An integrated analysis of the cancer genome atlas data discovers a hierarchical association structure across thirty three cancer types. PLOS Digit Health 2022; 1(12): e0000151.
[http://dx.doi.org/10.1371/journal.pdig.0000151] [PMID: 36812605]

[71]   Hanif F, Muzaffar K, Perveen K, Malhi SM, Simjee ShU. Glioblastoma multiforme: a review of its epidemiology and pathogenesis through clinical presentation and treatment. Asian Pacific journal of cancer prevention. APJCP 2017; 18(1): 3-9.
[PMID: 28239999]

[72]   Zakharova G, Efimov V, Raevskiy M, *et al.* Reclassification of TCGA diffuse glioma profiles linked to transcriptomic, epigenetic, genomic and clinical data, according to the 2021 WHO CNS tumor classification. Int J Mol Sci 2022; 24(1): 157.
[http://dx.doi.org/10.3390/ijms24010157] [PMID: 36613601]

[73]   Dunn GP, Rinne ML, Wykosky J, *et al.* Emerging insights into the molecular and cellular basis of glioblastoma. Genes Dev 2012; 26(8): 756-84.
[http://dx.doi.org/10.1101/gad.187922.112] [PMID: 22508724]

[74]   Barzegar Behrooz A, Talaie Z, Jusheghani F, Łos MJ, Klonisch T, Ghavami S. Wnt and PI3K/Akt/mTOR survival pathways as therapeutic targets in glioblastoma. Int J Mol Sci 2022; 23(3): 1353.
[http://dx.doi.org/10.3390/ijms23031353] [PMID: 35163279]

[75]   Hashemi M, Etemad S, Rezaei S, *et al.* Progress in targeting PTEN/PI3K/Akt axis in glioblastoma therapy: Revisiting molecular interactions. Biomed Pharmacother 2023; 158: 114204.
[http://dx.doi.org/10.1016/j.biopha.2022.114204] [PMID: 36916430]

[76]   Arya R, Jain S, Paliwal S, *et al.* BACE1 inhibitors: A promising therapeutic approach for the management of Alzheimer's disease. Asian Pacific Journal of Tropical Biomedicine. 2024; 14(9):p 369-381.
[http://dx.doi.org/10.4103/apjtb.apjtb_192_24]

[77]   Goel S, Bergholz JS, Zhao JJ. Targeting CDK4 and CDK6 in cancer. Nat Rev Cancer 2022; 22(6): 356-72.
[http://dx.doi.org/10.1038/s41568-022-00456-3] [PMID: 35304604]

[78]   Eckerdt FD, Bell JB, Gonzalez C, *et al.* Combined PI3Kα-mTOR targeting of glioma stem cells. Sci Rep 2020; 10(1): 21873.
[http://dx.doi.org/10.1038/s41598-020-78788-z] [PMID: 33318517]

[79]   He Y, Sun MM, Zhang GG, *et al.* Targeting PI3K/Akt signal transduction for cancer therapy. Signal Transduct Target Ther 2021; 6(1): 425.

[http://dx.doi.org/10.1038/s41392-021-00828-5] [PMID: 34916492]

[80]   Rauch J, Volinsky N, Romano D, Kolch W. The secret life of kinases: functions beyond catalysis. Cell Commun Signal 2011; 9(1): 23.
[http://dx.doi.org/10.1186/1478-811X-9-23] [PMID: 22035226]

[81]   Bai X, Sun P, Wang X, *et al.* Structure and dynamics of the EGFR/HER2 heterodimer. Cell Discov 2023; 9(1): 18.
[http://dx.doi.org/10.1038/s41421-023-00523-5] [PMID: 36781849]

[82]   Razali NN, Raja Ali RA, Muhammad Nawawi KN, Yahaya A, Mohd Rathi ND, Mokhtar NM. Roles of phosphatidylinositol-3-kinases signaling pathway in inflammation-related cancer: Impact of rs10889677 variant and buparlisib in colitis-associated cancer. World J Gastroenterol 2023; 29(40): 5543-56.
[http://dx.doi.org/10.3748/wjg.v29.i40.5543] [PMID: 37970476]

[83]   Engeland K. Cell cycle regulation: p53-p21-RB signaling. Cell Death Differ 2022; 29(5): 946-60.
[http://dx.doi.org/10.1038/s41418-022-00988-z] [PMID: 35361964]

[84]   Rathinaswamy MK, Burke JE. Class I phosphoinositide 3-kinase (PI3K) regulatory subunits and their roles in signaling and disease. Adv Biol Regul 2020; 75: 100657.
[http://dx.doi.org/10.1016/j.jbior.2019.100657] [PMID: 31611073]

[85]   Kim CW, Lee JM, Park SW. Divergent roles of the regulatory subunits of class IA PI3K. Front Endocrinol (Lausanne) 2024; 14: 1152579.
[http://dx.doi.org/10.3389/fendo.2023.1152579] [PMID: 38317714]

[86]   Riehle RD, Cornea S, Degterev A. Role of phosphatidylinositol 3, 4, 5-trisphosphate in cell signaling. Lipid-mediated Protein Signaling 2013; pp. 105-39.
[http://dx.doi.org/10.1007/978-94-007-6331-9_7]

[87]   Blaustein M, Piegari E, Martínez Calejman C, *et al.* Akt is S-palmitoylated: a new layer of regulation for Akt. Front Cell Dev Biol 2021; 9: 626404.
[http://dx.doi.org/10.3389/fcell.2021.626404] [PMID: 33659252]

[88]   Sarbassov DD, Guertin DA, Ali SM, Sabatini DM. Phosphorylation and regulation of Akt/PKB by the rictor-mTOR complex. Science 2005; 307(5712): 1098-101.
[http://dx.doi.org/10.1126/science.1106148] [PMID: 15718470]

[89]   Guertin DA, Sabatini DM. An expanding role for mTOR in cancer. Trends Mol Med 2005; 11(8): 353-61.
[http://dx.doi.org/10.1016/j.molmed.2005.06.007] [PMID: 16002336]

[90]   Dunkerly-Eyring BL, Pan S, Pinilla-Vera M, *et al.* Single serine on TSC2 exerts biased control over mTORC1 activation mediated by ERK1/2 but not Akt. Life Sci Alliance 2022; 5(6): e202101169.
[http://dx.doi.org/10.26508/lsa.202101169] [PMID: 35288456]

[91]   Zhou Y, Patel M, Singh R, Monje M. Neuron–glioma synapses and electrical coupling: emerging targets in glioma therapy. Nat Rev Cancer. 2024;24(3):145–59.
[http://dx.doi.org/10.1038/s41571-024-00891-2]

[92]   Datta N, Chakraborty S, Basu M, Ghosh MK. Tumor suppressors having oncogenic functions: the double agents. Cells 2020; 10(1): 46.
[http://dx.doi.org/10.3390/cells10010046] [PMID: 33396222]

[93]   Konagaya Y, Rosenthal D, Ratnayeke N, Fan Y, Meyer T. An intermediate Rb–E2F activity state safeguards proliferation commitment. Nature 2024; 631(8020): 424-31.
[http://dx.doi.org/10.1038/s41586-024-07554-2] [PMID: 38926571]

[94]   Giambra M, Messuti E, Di Cristofori A, *et al.* Characterizing the genomic profile in high-grade gliomas: from tumor core to peritumoral brain zone, passing through glioma-derived tumorspheres. Biology (Basel) 2021; 10(11): 1157.
[http://dx.doi.org/10.3390/biology10111157] [PMID: 34827152]

[95]   Wang Y, Li Y, Liu D, *et al.* A potential anti-glioblastoma compound LH20 induces apoptosis and arrest of human glioblastoma cells *via* CDK4/6 inhibition. Molecules 2023; 28(13): 5047.
[http://dx.doi.org/10.3390/molecules28135047] [PMID: 37446710]

[96]   Mouery BL, Baker EM, Mei L, *et al.* APC/C prevents a noncanonical order of cyclin/CDK activity to maintain CDK4/6 inhibitor–induced arrest. Proc Natl Acad Sci USA 2024; 121(30): e2319574121.
[http://dx.doi.org/10.1073/pnas.2319574121] [PMID: 39024113]

[97]   Crozier L, Foy R, Mouery BL, *et al.* CDK4/6 inhibitors induce replication stress to cause long-term cell cycle withdrawal. EMBO J 2022; 41(6): e108599.
[http://dx.doi.org/10.15252/embj.2021108599] [PMID: 35037284]

[98]   Lee C, Kim JK. Chromatin regulators in retinoblastoma: Biological roles and therapeutic applications. J Cell Physiol 2021; 236(4): 2318-32.
[http://dx.doi.org/10.1002/jcp.30022] [PMID: 32840881]

[99]   Guzman F, Fazeli Y, Khuu M, Salcido K, Singh S, Benavente CA. Retinoblastoma tumor suppressor protein roles in epigenetic regulation. Cancers (Basel) 2020; 12(10): 2807.
[http://dx.doi.org/10.3390/cancers12102807] [PMID: 33003565]

[100]   Vélez-Cruz R, Johnson D. The retinoblastoma (RB) tumor suppressor: pushing back against genome instability on multiple fronts. Int J Mol Sci 2017; 18(8): 1776.
[http://dx.doi.org/10.3390/ijms18081776] [PMID: 28812991]

[101]   Janostiak R, Torres-Sanchez A, Posas F, de Nadal E. Understanding retinoblastoma post-translational regulation for the design of targeted cancer therapies. Cancers (Basel) 2022; 14(5): 1265.
[http://dx.doi.org/10.3390/cancers14051265] [PMID: 35267571]

[102]   Zhou L, Ng DSC, Yam JC, *et al.* Post-translational modifications on the retinoblastoma protein. J Biomed Sci 2022; 29(1): 33.
[http://dx.doi.org/10.1186/s12929-022-00818-x] [PMID: 35650644]

[103]   Yuan K, Wang X, Dong H, Min W, Hao H, Yang P. Selective inhibition of CDK4/6: A safe and effective strategy for developing anticancer drugs. Acta Pharm Sin B 2021; 11(1): 30-54.
[http://dx.doi.org/10.1016/j.apsb.2020.05.001] [PMID: 33532179]

[104]   Riess C, Irmscher N, Salewski I, *et al.* Cyclin-dependent kinase inhibitors in head and neck cancer and glioblastoma—backbone or add-on in immune-oncology? Cancer Metastasis Rev 2021; 40(1): 153-71.
[http://dx.doi.org/10.1007/s10555-020-09940-4] [PMID: 33161487]

[105]   Juric V, Murphy B. Cyclin-dependent kinase inhibitors in brain cancer: current state and future directions. Cancer Drug Resist 2020; 3(1): 48-62.
[http://dx.doi.org/10.20517/cdr.2019.105] [PMID: 35582046]

[106]   Shi P, Xu J, Cui H. The recent research progress of NF-κB signaling on the proliferation, migration, invasion, immune escape and drug resistance of glioblastoma. Int J Mol Sci 2023; 24(12): 10337.
[http://dx.doi.org/10.3390/ijms241210337] [PMID: 37373484]

[107]   Prescott JA, Mitchell JP, Cook SJ. Inhibitory feedback control of NF-κB signalling in health and disease. Biochem J 2021; 478(13): 2619-64.
[http://dx.doi.org/10.1042/BCJ20210139] [PMID: 34269817]

[108]   Guo Q, Jin Y, Chen X, *et al.* NF-κB in biology and targeted therapy: new insights and translational implications. Signal Transduct Target Ther 2024; 9(1): 53.
[http://dx.doi.org/10.1038/s41392-024-01757-9] [PMID: 38433280]

[109]   Liu T, Zhang L, Joo D, Sun S-C. NF-κB signaling in inflammation. Signal Transduct Target Ther 2017; 2(1): 1-9.

[110]   Zhao H, Wu L, Yan G, *et al.* Inflammation and tumor progression: signaling pathways and targeted intervention. Signal Transduct Target Ther 2021; 6(1): 263.
[http://dx.doi.org/10.1038/s41392-021-00658-5] [PMID: 34248142]

[111] Greten FR, Grivennikov SI. Inflammation and cancer: triggers, mechanisms, and consequences. Immunity 2019; 51(1): 27-41.
[http://dx.doi.org/10.1016/j.immuni.2019.06.025] [PMID: 31315034]

[112] Uddin MS, Kabir MT, Mamun AA, *et al.* Natural small molecules targeting NF-κB signaling in glioblastoma. Front Pharmacol 2021; 12: 703761.
[http://dx.doi.org/10.3389/fphar.2021.703761] [PMID: 34512336]

[113] Soubannier V, Stifani S. NF-κB signalling in glioblastoma. Biomedicines 2017; 5(2): 29.
[http://dx.doi.org/10.3390/biomedicines5020029] [PMID: 28598356]

[114] Chen Y, Gupta R, Al-Muftah M, Zhao X. Glioblastoma therapy in 2025: integrating immunotherapy, nanomedicine, and AI-guided precision oncology. Nat Rev Clin Oncol. 2025;22(7):451–68.
[http://dx.doi.org/10.1038/s41571-025-00987-8] [PMID: 38712813]

[115] Dewdney B, Jenkins MR, Best SA, *et al.* From signalling pathways to targeted therapies: unravelling glioblastoma's secrets and harnessing two decades of progress. Signal Transduct Target Ther 2023; 8(1): 400.
[http://dx.doi.org/10.1038/s41392-023-01637-8] [PMID: 37857607]

[116] Shih AH, Holland EC. Platelet-derived growth factor (PDGF) and glial tumorigenesis. Cancer Lett 2006; 232(2): 139-47.
[http://dx.doi.org/10.1016/j.canlet.2005.02.002] [PMID: 16139423]

[117] Qin A, Musket A, Musich PR, Schweitzer JB, Xie Q. Receptor tyrosine kinases as druggable targets in glioblastoma: Do signaling pathways matter? Neurooncol Adv 2021; 3(1): vdab133.
[http://dx.doi.org/10.1093/noajnl/vdab133] [PMID: 34806012]

[118] Mekala JR, Adusumilli K, Chamarthy S, Angirekula HSR. Novel sights on therapeutic, prognostic, and diagnostics aspects of non-coding RNAs in glioblastoma multiforme. Metab Brain Dis 2023; 38(6): 1801-29.
[http://dx.doi.org/10.1007/s11011-023-01234-2] [PMID: 37249862]

[119] Wang Y, El-Sayed M, Tang L, Monje M. Nanotechnology-enabled immunotherapy for glioblastoma: crossing barriers and reprogramming the microenvironment. Nat Nanotechnol. 2025;20(4):312–28.
[http://dx.doi.org/10.1038/s41565-025-01456-9]

[120] Wang H, Guo M, Wei H, Chen Y. Targeting p53 pathways: mechanisms, structures and advances in therapy. Signal Transduct Target Ther 2023; 8(1): 92.
[http://dx.doi.org/10.1038/s41392-023-01347-1] [PMID: 36859359]

[121] Marei HE, Althani A, Afifi N, *et al.* p53 signaling in cancer progression and therapy. Cancer Cell Int 2021; 21(1): 703.
[http://dx.doi.org/10.1186/s12935-021-02396-8] [PMID: 34952583]

[122] Abuetabh Y, Wu HH, Chai C, *et al.* DNA damage response revisited: the p53 family and its regulators provide endless cancer therapy opportunities. Exp Mol Med 2022; 54(10): 1658-69.
[http://dx.doi.org/10.1038/s12276-022-00863-4] [PMID: 36207426]

[123] Luo J, Junaid M, Hamid N, Duan JJ, Yang X, Pei DS. Current understanding of gliomagenesis: from model to mechanism. Int J Med Sci 2022; 19(14): 2071-9.
[http://dx.doi.org/10.7150/ijms.77287] [PMID: 36483593]

[124] Gousias K, Theocharous T, Simon M. Mechanisms of cell cycle arrest and apoptosis in glioblastoma. Biomedicines 2022; 10(3): 564.
[http://dx.doi.org/10.3390/biomedicines10030564] [PMID: 35327366]

[125] Zhang Y, Dube C, Gibert M Jr, *et al.* The p53 pathway in glioblastoma. Cancers (Basel) 2018; 10(9): 297.
[http://dx.doi.org/10.3390/cancers10090297] [PMID: 30200436]

[126] Verdugo E, Puerto I, Medina MÁ. An update on the molecular biology of glioblastoma, with clinical

implications and progress in its treatment. Cancer Commun (Lond) 2022; 42(11): 1083-111.
[http://dx.doi.org/10.1002/cac2.12361] [PMID: 36129048]

[127] Kim HJ, Park JW, Lee JH. Genetic architectures and cell-of-origin in glioblastoma. Front Oncol 2021;
10: 615400.
[http://dx.doi.org/10.3389/fonc.2020.615400] [PMID: 33552990]

[128] Borrero LJH, El-Deiry WS. Tumor suppressor p53: Biology, signaling pathways, and therapeutic
targeting. Biochimica et Biophysica Acta (BBA)-. Rev Can 2021; 1876(1): 188556.

[129] Capuozzo M, Santorsola M, Bocchetti M, *et al.* p53: from fundamental biology to clinical applications
in cancer. Biology (Basel) 2022; 11(9): 1325.
[http://dx.doi.org/10.3390/biology11091325] [PMID: 36138802]

[130] Chinnam M, Xu C, Lama R, *et al.* MDM2 E3 ligase activity is essential for p53 regulation and cell
cycle integrity. PLoS Genet 2022; 18(5): e1010171.
[http://dx.doi.org/10.1371/journal.pgen.1010171] [PMID: 35588102]

[131] Klein AM, de Queiroz RM, Venkatesh D, Prives C. The roles and regulation of MDM2 and MDMX: it
is not just about p53. Genes Dev 2021; 35(9-10): 575-601.
[http://dx.doi.org/10.1101/gad.347872.120] [PMID: 33888565]

[132] Toledo F, Wahl GM. MDM2 and MDM4: p53 regulators as targets in anticancer therapy. Int J
Biochem Cell Biol 2007; 39(7-8): 1476-82.
[http://dx.doi.org/10.1016/j.biocel.2007.03.022] [PMID: 17499002]

[133] Chen X, Zhang T, Su W, *et al.* Mutant p53 in cancer: from molecular mechanism to therapeutic
modulation. Cell Death Dis 2022; 13(11): 974.
[http://dx.doi.org/10.1038/s41419-022-05408-1] [PMID: 36400749]

[134] Xiong Y, Zhang Y, Xiong S, Williams-Villalobo AE. A glance of p53 functions in brain development,
neural stem cells, and brain cancer. Biology (Basel) 2020; 9(9): 285.
[http://dx.doi.org/10.3390/biology9090285] [PMID: 32932978]

[135] Fischer M, Sammons MA. Determinants of p53 DNA binding, gene regulation, and cell fate decisions.
Cell Death Differ 2024; 31(7): 836-43.
[http://dx.doi.org/10.1038/s41418-024-01326-1] [PMID: 38951700]

[136] Aubrey BJ, Kelly GL, Janic A, Herold MJ, Strasser A. How does p53 induce apoptosis and how does
this relate to p53-mediated tumour suppression? Cell Death Differ 2018; 25(1): 104-13.
[http://dx.doi.org/10.1038/cdd.2017.169] [PMID: 29149101]

[137] Fingerle-Rowson G, Petrenko O, Metz CN, *et al.* The p53-dependent effects of macrophage migration
inhibitory factor revealed by gene targeting. Proc Natl Acad Sci USA 2003; 100(16): 9354-9.
[http://dx.doi.org/10.1073/pnas.1533295100] [PMID: 12878730]

[138] Bahar ME, Kim HJ, Kim DR. Targeting the RAS/RAF/MAPK pathway for cancer therapy: from
mechanism to clinical studies. Signal Transduct Target Ther 2023; 8(1): 455.
[http://dx.doi.org/10.1038/s41392-023-01705-z] [PMID: 38105263]

[139] Lu J, Getz G, Miska EA, Alvarez-Saavedra E, Lamb J, Peck D. MicroRNA expression profiles
classify human cancers. Nature 2005; 435(7043): 834-8.

[140] Braicu C, Buse M, Busuioc C, *et al.* A comprehensive review on MAPK: a promising therapeutic
target in cancer. Cancers (Basel) 2019; 11(10): 1618.
[http://dx.doi.org/10.3390/cancers11101618] [PMID: 31652660]

[141] Ou K, Liu X, Li W, Yang Y, Ying J, Yang L. ALK rearrangement–positive pancreatic cancer with
brain metastasis has remarkable response to ALK inhibitors: A case report. Front Oncol 2021; 11:
724815.
[http://dx.doi.org/10.3389/fonc.2021.724815] [PMID: 34568053]

[142] Rosswog C, Fassunke J, Ernst A, *et al.* Genomic ALK alterations in primary and relapsed

neuroblastoma. Br J Cancer 2023; 128(8): 1559-71.
[http://dx.doi.org/10.1038/s41416-023-02208-y] [PMID: 36807339]

[143] Mathien S, Tesnière C, Meloche S. Regulation of mitogen-activated protein kinase signaling pathways by the ubiquitin-proteasome system and its pharmacological potential. Pharmacol Rev 2021; 73(4): 1434-67.
[http://dx.doi.org/10.1124/pharmrev.120.000170] [PMID: 34732541]

[144] Bresler SC, Weiser DA, Huwe PJ, *et al.* ALK mutations confer differential oncogenic activation and sensitivity to ALK inhibition therapy in neuroblastoma. Cancer Cell 2014; 26(5): 682-94.
[http://dx.doi.org/10.1016/j.ccell.2014.09.019] [PMID: 25517749]

[145] Souza VGP, de Araújo RP, Santesso MR, *et al.* Advances in the molecular landscape of lung cancer brain metastasis. Cancers (Basel) 2023; 15(3): 722.
[http://dx.doi.org/10.3390/cancers15030722] [PMID: 36765679]

[146] Mao H, LeBrun DG, Yang J, Zhu VF, Li M. Deregulated signaling pathways in glioblastoma multiforme: molecular mechanisms and therapeutic targets. Cancer Invest 2012; 30(1): 48-56.
[http://dx.doi.org/10.3109/07357907.2011.630050] [PMID: 22236189]

[147] Glaviano A, Foo ASC, Lam HY, *et al.* PI3K/AKT/mTOR signaling transduction pathway and targeted therapies in cancer. Mol Cancer 2023; 22(1): 138.
[http://dx.doi.org/10.1186/s12943-023-01827-6] [PMID: 37596643]

[148] Joshi G, Singh PK, Negi A, Rana A, Singh S, Kumar R. Growth factors mediated cell signalling in prostate cancer progression: Implications in discovery of anti-prostate cancer agents. Chem Biol Interact 2015; 240: 120-33.
[http://dx.doi.org/10.1016/j.cbi.2015.08.009] [PMID: 26297992]

[149] Eleveld TF, Schild L, Koster J, *et al.* RAS–MAPK pathway-driven tumor progression is associated with loss of CIC and other genomic aberrations in neuroblastoma. Cancer Res 2018; 78(21): 6297-307.
[http://dx.doi.org/10.1158/0008-5472.CAN-18-1045] [PMID: 30115695]

[150] Billhaq DH, Lee S. The role of the guanosine nucleotide-binding protein in the corpus luteum. Animals (Basel) 2021; 11(6): 1524.
[http://dx.doi.org/10.3390/ani11061524] [PMID: 34073800]

[151] Jiang J, Jiang L, Maldonato BJ, *et al.* Translational and therapeutic evaluation of RAS-GTP inhibition by RMC-6236 in RAS-driven cancers. Cancer Discov 2024; 14(6): 994-1017.
[http://dx.doi.org/10.1158/2159-8290.CD-24-0027] [PMID: 38593348]

[152] Abal M, Planaguma J, Gil-Moreno A, *et al.* Molecular pathology of endometrial carcinoma: transcriptional signature in endometrioid tumors. Histol Histopathol 2006; 21(2): 197-204.
[PMID: 16329044]

[153] Verhoeven Y, Tilborghs S, Jacobs J, De Waele J, Quatannens D, Deben C, Eds. The potential and controversy of targeting STAT family members in cancer Seminars in cancer biology. Elsevier 2020.

[154] Wu M, Song D, Li H, *et al.* Negative regulators of STAT3 signaling pathway in cancers. Cancer Manag Res 2019; 11: 4957-69.
[http://dx.doi.org/10.2147/CMAR.S206175] [PMID: 31213912]

[155] Tan MSY, Sandanaraj E, Chong YK, *et al.* A STAT3-based gene signature stratifies glioma patients for targeted therapy. Nat Commun 2019; 10(1): 3601.
[http://dx.doi.org/10.1038/s41467-019-11614-x] [PMID: 31399589]

[156] Fu W, Hou X, Dong L, Hou W. Roles of STAT3 in the pathogenesis and treatment of glioblastoma. Front Cell Dev Biol 2023; 11: 1098482.
[http://dx.doi.org/10.3389/fcell.2023.1098482] [PMID: 36923251]

[157] Zhang Y, Bharadwaj U, Logsdon CD, Chen C, Yao Q, Li M. ZIP4 regulates pancreatic cancer cell growth by activating IL-6/STAT3 pathway through zinc finger transcription factor CREB. Clin Cancer Res 2010; 16(5): 1423-30.

[http://dx.doi.org/10.1158/1078-0432.CCR-09-2405] [PMID: 20160059]

[158] Laribee RN, Boucher AB, Madireddy S, Pfeffer LM. The STAT3-regulated autophagy pathway in glioblastoma. Pharmaceuticals (Basel) 2023; 16(5): 671.
[http://dx.doi.org/10.3390/ph16050671] [PMID: 37242454]

[159] Barker N. Adult intestinal stem cells: critical drivers of epithelial homeostasis and regeneration. Nat Rev Mol Cell Biol 2014; 15(1): 19-33.
[http://dx.doi.org/10.1038/nrm3721] [PMID: 24326621]

[160] Kalani MYS, Cheshier SH, Cord BJ, *et al.* Wnt-mediated self-renewal of neural stem/progenitor cells. Proc Natl Acad Sci USA 2008; 105(44): 16970-5.
[http://dx.doi.org/10.1073/pnas.0808616105] [PMID: 18957545]

[161] Habas R, Dawid IB. Dishevelled and Wnt signaling: is the nucleus the final frontier? J Biol 2005; 4(1): 2.
[http://dx.doi.org/10.1186/jbiol22] [PMID: 15720723]

[162] Willert K, Nusse R. Wnt Proteins. Cold Spring Harb Perspect Biol 2012; 4(9): a007864.
[http://dx.doi.org/10.1101/cshperspect.a007864] [PMID: 22952392]

[163] Staal FJT, Luis TC, Tiemessen MM. WNT signalling in the immune system: WNT is spreading its wings. Nat Rev Immunol 2008; 8(8): 581-93.
[http://dx.doi.org/10.1038/nri2360] [PMID: 18617885]

[164] Martinez-Font E, Pérez-Capó M, Ramos R, *et al.* Impact of wnt/β-catenin inhibition on cell proliferation through cdc25a downregulation in soft tissue sarcomas. Cancers (Basel) 2020; 12(9): 2556.
[http://dx.doi.org/10.3390/cancers12092556] [PMID: 32911761]

[165] Yu M, Qin K, Fan J, *et al.* The evolving roles of Wnt signaling in stem cell proliferation and differentiation, the development of human diseases, and therapeutic opportunities. Genes Dis 2024; 11(3): 101026.
[http://dx.doi.org/10.1016/j.gendis.2023.04.042] [PMID: 38292186]

[166] MacDonald BT, Tamai K, He X. Wnt/β-catenin signaling: components, mechanisms, and diseases. Dev Cell 2009; 17(1): 9-26.
[http://dx.doi.org/10.1016/j.devcel.2009.06.016] [PMID: 19619488]

[167] Becker J, Wilting J. WNT Signaling in Neuroblastoma. Cancers (Basel) 2019; 11(7): 1013.
[http://dx.doi.org/10.3390/cancers11071013] [PMID: 31331081]

[168] Li T, Chan RWS, Lee CL, *et al.* WNT5A interacts with FZD5 and LRP5 to regulate proliferation and self-renewal of endometrial mesenchymal stem-like cells. Front Cell Dev Biol 2022; 10: 837827.
[http://dx.doi.org/10.3389/fcell.2022.837827] [PMID: 35295855]

[169] Zhan T, Ambrosi G, Wandmacher AM, *et al.* MEK inhibitors activate Wnt signalling and induce stem cell plasticity in colorectal cancer. Nat Commun 2019; 10(1): 2197.
[http://dx.doi.org/10.1038/s41467-019-09898-0] [PMID: 31097693]

[170] Safa A, Abak A, Shoorei H, Taheri M, Ghafouri-Fard S. MicroRNAs as regulators of ERK/MAPK pathway: A comprehensive review. Biomed Pharmacother 2020; 132: 110853.
[http://dx.doi.org/10.1016/j.biopha.2020.110853] [PMID: 33068932]

[171] Mlakar V, Morel E, Mlakar SJ, Ansari M, Gumy-Pause F. A review of the biological and clinical implications of RAS-MAPK pathway alterations in neuroblastoma. J Exp Clin Cancer Res 2021; 40(1): 189.
[http://dx.doi.org/10.1186/s13046-021-01967-x] [PMID: 34103089]

[172] McCubrey JA, Steelman LS, Chappell WH, *et al.* Mutations and deregulation of Ras/Raf/MEK/ERK and PI3K/PTEN/Akt/mTOR cascades which alter therapy response. Oncotarget 2012; 3(9): 954-87.
[http://dx.doi.org/10.18632/oncotarget.652] [PMID: 23006971]

[173] Nurmagambetova A, Mustyatsa V, Saidova A, Vorobjev I. Morphological and cytoskeleton changes in cells after EMT. Sci Rep 2023; 13(1): 22164.
[http://dx.doi.org/10.1038/s41598-023-48279-y] [PMID: 38092761]

[174] Datta A, Deng S, Gopal V, *et al.* Cytoskeletal dynamics in epithelial-mesenchymal transition: insights into therapeutic targets for cancer metastasis. Cancers (Basel) 2021; 13(8): 1882.
[http://dx.doi.org/10.3390/cancers13081882] [PMID: 33919917]

[175] Manfreda L, Rampazzo E, Persano L. Wnt signaling in brain tumors: A challenging therapeutic target. Biology (Basel) 2023; 12(5): 729.
[http://dx.doi.org/10.3390/biology12050729] [PMID: 37237541]

[176] Zuccarini M, Giuliani P, Ziberi S, *et al.* The role of Wnt signal in glioblastoma development and progression: a possible new pharmacological target for the therapy of this tumor. Genes (Basel) 2018; 9(2): 105.
[http://dx.doi.org/10.3390/genes9020105] [PMID: 29462960]

[177] Song P, Gao Z, Bao Y, *et al.* Wnt/β-catenin signaling pathway in carcinogenesis and cancer therapy. J Hematol Oncol 2024; 17(1): 46.
[http://dx.doi.org/10.1186/s13045-024-01563-4] [PMID: 38886806]

[178] Zhao H, Ming T, Tang S, *et al.* Wnt signaling in colorectal cancer: pathogenic role and therapeutic target. Mol Cancer 2022; 21(1): 144.
[http://dx.doi.org/10.1186/s12943-022-01616-7] [PMID: 35836256]

[179] Palamaris K, Felekouras E, Sakellariou S. Epithelial to mesenchymal transition: key regulator of pancreatic ductal adenocarcinoma progression and chemoresistance. Cancers (Basel) 2021; 13(21): 5532.
[http://dx.doi.org/10.3390/cancers13215532] [PMID: 34771695]

[180] Barber SM, Sadrameli SS, Lee JJ, *et al.* Chordoma—current understanding and modern treatment paradigms. J Clin Med 2021; 10(5): 1054.
[http://dx.doi.org/10.3390/jcm10051054] [PMID: 33806339]

[181] Sun X, Hornicek F, Schwab JH. Chordoma: an update on the pathophysiology and molecular mechanisms. Curr Rev Musculoskelet Med 2015; 8(4): 344-52.
[http://dx.doi.org/10.1007/s12178-015-9311-x] [PMID: 26493697]

[182] Lara-Velazquez M, Mehkri Y, Panther E, *et al.* Current advances in the management of adult craniopharyngiomas. Curr Oncol 2022; 29(3): 1645-71.
[http://dx.doi.org/10.3390/curroncol29030138] [PMID: 35323338]

[183] Massimi L, Palombi D, Musarra A, *et al.* Adamantinomatous craniopharyngioma: evolution in the management. Childs Nerv Syst 2023; 39(10): 2613-32.
[http://dx.doi.org/10.1007/s00381-023-06143-4] [PMID: 37728836]

[184] Alarifi N, Del Bigio MR, Beiko J. Adult gangliocytoma arising within the lateral ventricle: A case report and review of the literature. Surg Neurol Int 2022; 13: 11.
[http://dx.doi.org/10.25259/SNI_814_2021] [PMID: 35127211]

[185] Odia Y. Gangliocytomas and gangliogliomas: review of clinical, pathologic and genetic features. Clin Oncol (Las Vegas) 2016; 1.

[186] Wang JZ, Landry AP, Raleigh DR, *et al.* Meningioma: International Consortium on Meningiomas consensus review on scientific advances and treatment paradigms for clinicians, researchers, and patients. Neuro-oncol 2024; 26(10): 1742-80.
[http://dx.doi.org/10.1093/neuonc/noae082]

[187] Inetas-Yengin G, Bayrak OF. Related mechanisms, current treatments, and new perspectives in meningioma. Genes Chromosomes Cancer 2024; 63(5): e23248.
[http://dx.doi.org/10.1002/gcc.23248] [PMID: 38801095]

[188] Peng W, Wu P, Yuan M, *et al*. Potential molecular mechanisms of recurrent and progressive meningiomas: A review of the latest literature. Front Oncol 2022; 12: 850463.
[http://dx.doi.org/10.3389/fonc.2022.850463] [PMID: 35712491]

[189] Shao Z, Liu L, Zheng Y, *et al*. Molecular mechanism and approach in progression of meningioma. Front Oncol 2020; 10: 538845.
[http://dx.doi.org/10.3389/fonc.2020.538845] [PMID: 33042832]

[190] Chung PED, Gendoo DMA, Ghanbari-Azarnier R, *et al*. Modeling germline mutations in pineoblastoma uncovers lysosome disruption-based therapy. Nat Commun 2020; 11(1): 1825.
[http://dx.doi.org/10.1038/s41467-020-15585-2] [PMID: 32286280]

[191] Schulz M, Afshar-Bakshloo M, Koch A, *et al*. Management of pineal region tumors in a pediatric case series. Neurosurg Rev 2021; 44(3): 1417-27.
[http://dx.doi.org/10.1007/s10143-020-01323-1] [PMID: 32504201]

[192] Derwich A, Sykutera M, Brominska B, Rubiś B, Ruchała M, Sawicka-Gutaj N. The role of activation of PI3K/AKT/mTOR and RAF/MEK/ERK pathways in aggressive pituitary adenomas—new potential therapeutic approach—a systematic review. Int J Mol Sci 2023; 24(13): 10952.
[http://dx.doi.org/10.3390/ijms241310952] [PMID: 37446128]

[193] McDonald WC. Pituitary adenoma classification: Tools to improve the current system. Free Neuropathology 2024; p. 5.

[194] Li Y, Ren X, Gao W, *et al*. The biological behavior and clinical outcome of pituitary adenoma are affected by the microenvironment. CNS Neurosci Ther 2024; 30(5): e14729.
[http://dx.doi.org/10.1111/cns.14729] [PMID: 38738958]

[195] Hilton DA, Hanemann CO. Schwannomas and their pathogenesis. Brain Pathol 2014; 24(3): 205-20.
[http://dx.doi.org/10.1111/bpa.12125] [PMID: 24450866]

[196] Nahar Metu CL, Sutihar SK, Sohel M, *et al*. Unraveling the signaling mechanism behind astrocytoma and possible therapeutics strategies: A comprehensive review. Cancer Rep 2023; 6(10): e1889.
[http://dx.doi.org/10.1002/cnr2.1889] [PMID: 37675821]

[197] Molofsky AV, Deneen B. Astrocyte development: a guide for the perplexed. Glia 2015; 63(8): 1320-9.
[http://dx.doi.org/10.1002/glia.22836] [PMID: 25963996]

[198] Rogers HA, Mayne C, Chapman RJ, Kilday JP, Coyle B, Grundy RG. PI3K pathway activation provides a novel therapeutic target for pediatric ependymoma and is an independent marker of progression-free survival. Clin Cancer Res 2013; 19(23): 6450-60.
[http://dx.doi.org/10.1158/1078-0432.CCR-13-0222] [PMID: 24077346]

[199] Lan Z, Li X, Zhang X. Glioblastoma: An update in pathology, molecular mechanisms and biomarkers. Int J Mol Sci 2024; 25(5): 3040.
[http://dx.doi.org/10.3390/ijms25053040] [PMID: 38474286]

[200] Dutra LF. ZEB1 como marcador de sobrevida subgrupo-específico em meduloblastoma e sua interação com RNAs não codificantes. 2023.

[201] Borowska A, Jóźwiak J. Medulloblastoma: molecular pathways and histopathological classification. Arch Med Sci 2016; 3(3): 659-66.
[http://dx.doi.org/10.5114/aoms.2016.59939] [PMID: 27279861]

[202] Bou Zerdan M, Assi HI. Oligodendroglioma: a review of management and pathways. Front Mol Neurosci 2021; 14: 722396.
[http://dx.doi.org/10.3389/fnmol.2021.722396] [PMID: 34675774]

[203] Engelhard HH, Stelea A, Cochran EJ. Oligodendroglioma: pathology and molecular biology. Surg Neurol 2002; 58(2): 111-7.
[http://dx.doi.org/10.1016/S0090-3019(02)00751-6] [PMID: 12453646]

CHAPTER 7

Investigating the Impact of Immunotherapy on Brain Tumors

Pranjal Gujarathi[1], Deepa Mandlik[2] and Meghraj Suryawanshi[3,4,*]

[1] *Department of Pharmacology, Vidhyadeep Institute of Pharmacy, Vidhyadeep University, Surat, Gujarat-394110, India*

[2] *Department of Pharmacology, Bharati Vidyapeeth (Deemed to be University), Poona College of Pharmacy, Pune-411038, Maharashtra, India*

[3] *Department of Pharmaceutics, Sandip Institute of Pharmaceutical Sciences (SIPS), Affiliated to Savitribai Phule Pune University (SPPU, Pune), Nashik, Maharashtra-422213, India*

[4] *AllWell Nutritech LLP Dharangaon, Maharashtra-425105, India*

Abstract: Immunotherapy has become a viable treatment option for brain tumors, particularly gliomas and brain malignancies that have metastasized. This review examines the clinical outcomes of recent clinical studies and the mechanisms by which immunotherapy improves anti-tumor immune responses. Cancer vaccines, chimeric antigen receptor (CAR) T-cell therapy and immune checkpoint inhibitors are important tactics. Immune checkpoint inhibitors strengthen the natural defenses against cancer by blocking proteins that hinder the immune system from attacking cancer cells. Through the modulation of an individual's T cells to target particular cancer antigens, CAR T-cell treatment provides a customized course of treatment. The primary intent of cancer vaccination is to prepare the host immune system for identifying and combating tumor cells. Notwithstanding these developments, problems still exist. For example, the blood-brain barrier limits the amount of medicinal entity that may reach the brain and the tumor's immunosuppressive milieu impairs the activity of immune cells. Combination therapies, which combine immunotherapy with conventional treatments like radiation and chemotherapy, or employ numerous immunotherapeutic drugs, show promise for overcoming these challenges. Approaches to personalized therapy that are adapted to each patient's unique immunologic and genetic profile are also being investigated to increase effectiveness and patient survival. Further research will be needed to optimize these treatments and overcome their current limitations. Immunotherapy possesses the ability to dramatically reinforce outcomes for individuals with brain tumors, which are notoriously difficult to treat, by addressing the particular difficulties that these malignancies present. It has shown promise in ameliorating brain malignancies, particularly glioblastoma (GBM), but identifying biomarkers to predict treatment outcomes remains a significant challenge. Prospective biomarkers, adoptive cell transfer treatment, and novel drug delivery strategies are all being studied in

* **Corresponding author Meghraj Suryawanshi:** Department of Pharmaceutics, Sandip Institute of Pharmaceutical Sciences (SIPS), Affiliated To Savitribai Phule Pune University (SPPU, Pune), Nashik, Maharashtra-422213, India; Tel: +918668430089; E-mail: suryawanshimeghraj917@gmail.com

Prashant Tiwari, Pankaj Kumar Singh & Sunil Kumar Kadiri (Eds.)
All rights reserved-© 2025 Bentham Science Publishers

current and upcoming clinical trials. There is optimism for improved GBM outcomes with the introduction of novel drugs, such as immune checkpoint inhibitors, chimeric antigen receptor (CAR) T-cell therapy, oncolytic virotherapy, and vaccination. However, the fruitful utilization of immunotherapies for brain cancers requires the improvement of biopsy collecting methods as well as the development of more practical animal models.

Keywords: Brain tumors, BBB, CAR-T therapy, Immunotherapy.

INTRODUCTION

Although nervous system and brain malignancies account for a very minor percentage of yearly cancer occurrence (1.4%), they account for nearly twice as many cancer deaths (2.7%) [1]. Most malignant brain Tumors are neuroepithelial Tumors, with glioblastoma accounting for 46% of cases. Of these, only 5% of patients remain alive for more than five years [2]. The glaring lack of treatments that can both get past anatomical obstacles and spare sensitive neuronal tissue is the cause of this low survival rate. According to Nabors *et al.* (2015), the healthiest possible surgical excision of brain tumors is the existing strategies of treatment, along with modulated actinotherapy and chemotherapeutic agents [3].

However, because the tumor is in an area of the brain that is sensitive or inoperable, surgical removal is not always feasible [4, 5]. Furthermore, it is difficult to eradicate every cancer cell because the majority of brain tumors have aggressive fronts that extend past the original tumor core and into healthy tissue. Recurrence following surgery is therefore frequent [6]. Additionally, systemic chemotherapy is less effective in treating brain tumors because it cannot fully eradicate resident tumor-renewing populations and is unable to penetrate some brain tumor tissues due to blood-brain barrier's (BBB) transport-restrictive properties, drug adaptive resistance, elevated effective dose necessities, neural tissue sensitivity, and exacerbation of side effects [7, 8]. The influence of ongoing advancements in drug administration and discovery, which have significantly enhanced results for systemic cancers, has been very minimal when it comes to brain Tumors. Although immunotherapies are not now part of the standard of treatment, they may be able to lessen some of the difficulties associated with treating brain Tumors. Although it has proven challenging to target inside the brain, immune system modulations may allow one to take advantage of normal entry points, and memory of the immune system to stop reappearance later a standard therapy regimen may improve the prognosis for brain cancer. One remarkable finding supporting the promise of brain neoplasm immunotherapy is the possibility of marginally improved outcomes for brain Tumors that metastasize and spread into non-neuronal immunity a very uncommon occurrence [9]. Moreover, Volovitz *et al.* demonstrated in immune-compromised animals that an anti-cancer impact in the brain may be obtained simply by transplanting severe brain cancer into the peripheral [10]. This implies that immune-driven treatments for brain cancer hold considerable promise. Besides the difficulty of designing immunotherapies, the brain presents anatomical and physiological challenges for the effective application of such a therapy in the case of brain neoplasm.

TYPES OF BRAIN TUMORS

Primary and secondary Tumors are the two main classes into which a wide group of neoplasms that make up Central Nervous System (CNS) Tumors can be roughly separated. Primary Tumors, such as gliomas, start from progenitor or mature cells found in the CNS, while secondary neoplasms are found in different regions in the body and proliferate hematogenously to the CNS.

Primary Tumors of CNS

Meningiomas

The most prevalent primary brain tumor, meningiomas, are typically slow-growing, benign neoplasms that originate from the arachnoidal cells of the middle layer of the meninges [11, 12]. Meningiomas are comparatively prevalent, making up as much as 30% of CNS Tumors, and they typically affect middle-aged and older persons. Based on histology, meningiomas are categorised by World Health Organisation (WHO) in classes I through III. Benign meningiomas, or grade I meningiomas, are the most frequent type and have a good prognosis. However, according to Durand *et al.* (2009), meningiomas classified as WHO class II and III are highly severe and have a 78% and 44% 5-year survival rate, respectively [13].

Glioblastoma

According to Louis *et al.* (2016), gliomas, which comprise astrocytomas, oligodendrogliomas, ependymomas, and some uncommon histologies, are the most frequent malignant CNS Tumors [11]. The frequent and dangerous type of astrocytoma is glioblastoma, a grade IV tumor. With an occurrence of 3.2/100,000 and a median age at the time of diagnosis of 64 years, GBM accountable for for 15% of all primary and 45% of malignant primary brain Tumors [11].

Secondary Tumors of CNS (Metastases)

The most prevalent type of brain cancer is metastatic brain tumors, which originate outside the central nervous system (CNS) and reach the brain either through hematogenous spread or direct invasion from adjacent tissues, such as head and neck malignancies. Brain metastases occur in approximately 10% of pediatric cancer patients and nearly 30% of adults with systemic malignancies.

In adults, the most common primary sources of brain metastases include melanoma (skin cancer), breast cancer, and lung cancer. Additionally, prostate, breast, and lung cancers frequently metastasize to the spinal epidural space. Neuroimaging studies typically reveal contrast-enhancing lesions surrounded by vasogenic edema, indicative of metastatic involvement. While cerebellar metastases are observed in roughly 15% of cases, the cerebral hemispheres are affected in the majority of patients over 75% particularly at the junction of the cerebral cortex and white matter, a region prone to metastatic seeding due to its rich vascular supply. Metastases often present as multiple discrete lesions rather than a single isolated mass, reflecting the systemic nature of their spread [14].

Immunotherapy Approaches for Brain Tumors

Immune Checkpoint Inhibition

By combating immunological suppression, immune response therapy can boost the body's natural defenses on its own or in conjunction with other therapies to increase their efficacy. It works by obstructing pathways that impede cytotoxic T-cells' normal action [15]. Tumour cells express proteins, such as cytotoxic T-lymphocyte-associated protein 4 (CTLA4) and programmed cell death 1 (PD1), that interact with co-inhibitory receptors on T-cells [16]. As a brain tumor treatment, antibodies made to target these receptors are presently being researched, along with methods to hinder their suppressive ligands from emerging (such as PDL1 on dendritic cells (DCs), macrophages, and tumor cells) [17]. Since PD1+ TILs show higher clonality than PD1 populations, PD1 expression had previously been well-known as a clinical diagnostic for tumor-infiltrating lymphocyte (TIL) activation and (dys) function in Glioblastoma multiforme (GBM) [18]. In the Phase I trial for GBM, nivolumab, an antibody against the PD1 receptor, was assessed for safety in conjunction with concurrent temozolomide and radiation therapy. The trial reported a minimal number of side events [19]. In GBM, a strategy that targets CTLA4 has also been tested. 40% of patients responded partially to the anti-CTLA4 monoclonal antibody imipomumab when used in conjunction with concurrent adjuvants and bevacizumab. According to preliminary 1-year data, 44% of the 16 patients survived, which is higher than the predicted 3-7% 1-year survival rate for GBM [20]. The BBB continues to be a crucial barrier that prevents systemic medication delivery, and checkpoint inhibitor administration to the brain is impeded in the same way as any other systemic drug delivery. It is uncertain if immunosuppression rife in the tumor bed would be sufficiently suppressed even with an adequate inhibitor present at the position of the brain malignancy, and even if enough potent killer T-lymphocytes are unable to be inhibited, along with maintaining its antitumor action at therapeutic levels. Furthermore, patients receiving these treatments are more vulnerable to the evolution of inflammation and autoimmune diseases since the blockage of checkpoint pathways is not antigen-specific, which is disastrous for a healthy neural environment [21] (Fig. 1).

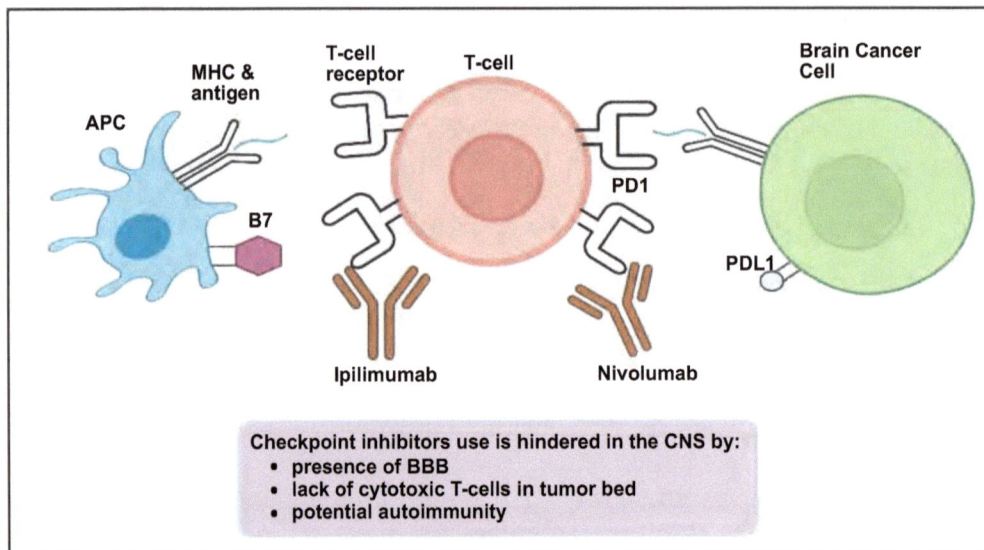

Fig. (1). Checkpoint inhibitor therapy for brain tumour.

Adoptive Cell Therapy (ACT)

Other TILs and T-cells can be directly triggered in ACT as an alternative to depending on the activation of DC [22]. According to Jean Bart and Swartz (2015), the most basic version of ACT involves expanding isolated TILs from tumor biopsies, assuming that these TILs are unique to the tumor, before injecting the patient and finally coming back to the tumor. These customized methods rely on the individual's immunity and knowledge of the tissue's lymph system. TILs that are frequently used in antigen-presenting therapy (ACT) are natural killer T lymphocytes and chimeric antigen receptor (CAR) T-cells [23] (Fig. **2**). Mature circulating lymphocytes stimulated *in-vitro,* accompanied by interferon-γ, interleukin-2, and CD3 monoclonal antibodies, are the source of cytokine-induced killer cells [24]. According to Fesnak *et al.* (2016), CAR T-cells are designed to link stimulation of an extracellular domain from within that is exclusive to one or more costimulatory tumor-associated antigens [23]. Human leukocyte antigen (HLA)-unrestrictive antigen identification is a feature that makes both natural killer T lymphocytes as well as CAR T-cells useful for use in ACT. These cells have also been used in GBM clinical studies. A 5.6-month increase in average viability was observed in Phase III randomized research on GBM when compared to the standard of treatment. While not statistically significant, this improvement was observed when autologous natural killer T lymphocytes were injected intravenously along with contemporaneous Temozolomide [24]. According to preliminary findings from Phase I studies, CAR T-cells designed to attack GBM tumor antigen (IL-13Rα2) were safely delivered intracranially and recognized in tumor cyst fluid for at least seven days. A portion of these sufferer showed a reduction in the expression of the IL-13Rα2 antigen, and one patient had an N79% regression of the recurring tumor mass [25]. T-cell treatment has obstacles similar to cellular vaccination therapies, despite its great promise. These include manufacturing and technical difficulties such as maintaining a cytotoxic phenotype and expanding *ex vivo*. If immune cells identify antigens unique to the cancer's source, immunotherapies may unintentionally target healthy cells. Immune cells outside the CNS may not be able to identify some CNS cell types, a problem that is made worse by brain tumors. Furthermore, a major barrier to the intravenous delivery of autologous cells is the blood-brain barrier.

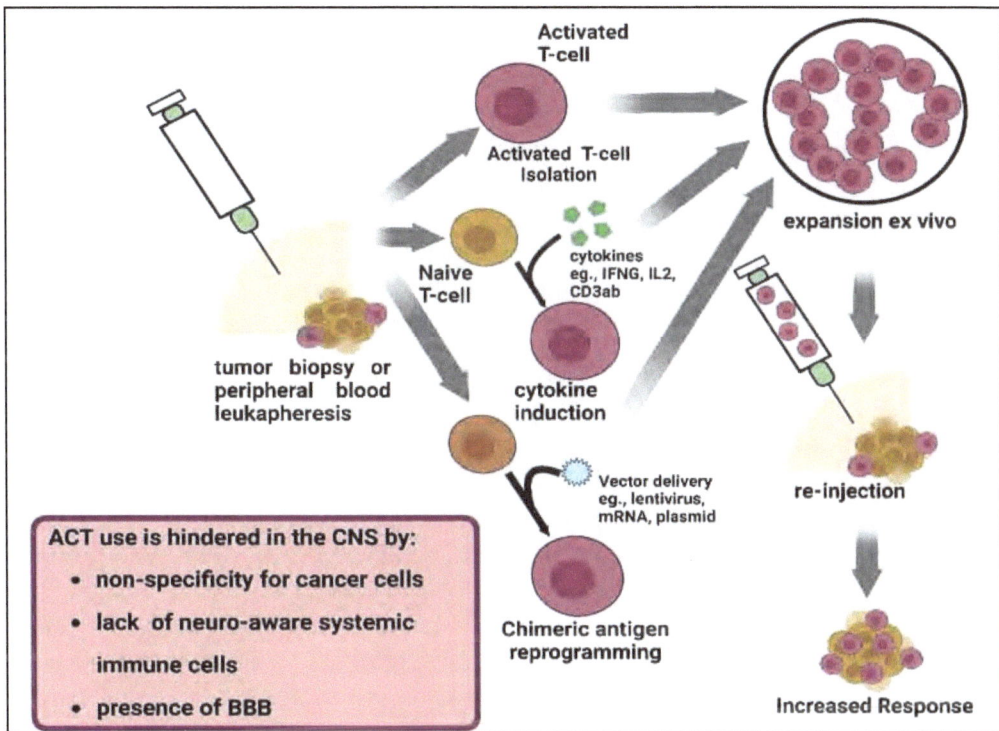

Fig. (2). Adaptive cell therapy for brain tumour.

Cancer Vaccines

Conventional vaccines against viral infections, such as influenza, activate dendritic cells (DCs) by combining a danger signal, typically an adjuvant, with an attenuated or live virus. Upon uptake of the viral antigen, DCs process and present antigenic peptides on major histocompatibility complex (MHC) molecules. These activated DCs then migrate to lymph nodes *via* lymphatic channels, where they present the processed antigens to T-lymphocytes. This interaction stimulates the proliferation of T-cells into transient effector cells and long-lived memory cells, enabling rapid immune responses upon future antigen exposure and establishing immunological memory.

Cancer vaccines, particularly those targeting glioblastoma (GBM), are primarily therapeutic in nature, although a small subset, such as those for hepatocellular carcinoma and cervical cancer, serve preventive roles [26]. Therapeutic cancer vaccines face considerable challenges due to the complex coevolution of tumor characteristics and host immune responses [27]. Among the most extensively studied cancer vaccine platforms are neoantigen-based vaccines, dendritic cell vaccines, whole tumor cell vaccines, and peptide-based vaccines [28].

Innovative vaccine strategies incorporating diverse adjuvants, combination therapies, and advanced delivery systems have demonstrated therapeutic potential in brain tumors, often with minimal adverse effects [29]. However, the immune-privileged nature of the brain, characterized by limited resident immune cells, can hinder robust immunological responses, making the inclusion of adjuvants particularly critical. Consequently, vaccine designs are being optimized to overcome the immunosuppressive microenvironment of brain tumors. Brain cancer vaccines are broadly categorized into two primary subtypes: peptide vaccines and antigen-presenting cell (APC) vaccines, each offering distinct mechanisms of action and therapeutic potential (Fig. **3**).

Peptide Vaccines

Peptide vaccines function by delivering tumor-associated antigen (TAA)-specific peptides that elicit a targeted T-lymphocyte response at the tumor site. These peptides are internalized by antigen-presenting cells (APCs), often in conjunction with adjuvants and carrier molecules to enhance immunogenicity. Once processed, the peptides are presented on the plasma membrane *via* major histocompatibility complex (MHC) molecules referred to as human leukocyte antigens (HLA) in humans, specifically MHC class I molecules [30]. APCs then migrate through the lymphatic system to prime T-cells, enabling them to recognize and attack tumor cells expressing the corresponding antigens [31]. Despite significant efforts to identify glioma-specific antigens for therapeutic targeting, antigen expression varies widely across tumors, complicating the characterization of a consistent antigenic repertoire [32, 33]. One notable example is rindopepimut, a peptide vaccine designed to target the EGFRvIII mutation, which is overexpressed in a subset of glioblastoma cells. Although early-phase trials showed promise, the Phase III clinical trial revealed no significant improvement in overall survival [34]. Moreover, tumors from relapsed patients often lacked EGFRvIII expression, indicating antigen loss or heterogeneity [35]. This outcome underscores a critical limitation of peptide vaccines in brain tumors: their specificity may restrict efficacy to only a subset of tumor cells. The heterogeneous and evolving nature of brain neoplasms allows EGFRvIII-negative cells to escape immune targeting and continue proliferating, even when EGFRvIII-positive cells are eliminated [36]. Such antigenic variability and tumor plasticity present significant challenges for peptide-based immunotherapies, as tumor cells can dynamically alter their antigenic profile to evade immune surveillance [37, 38].

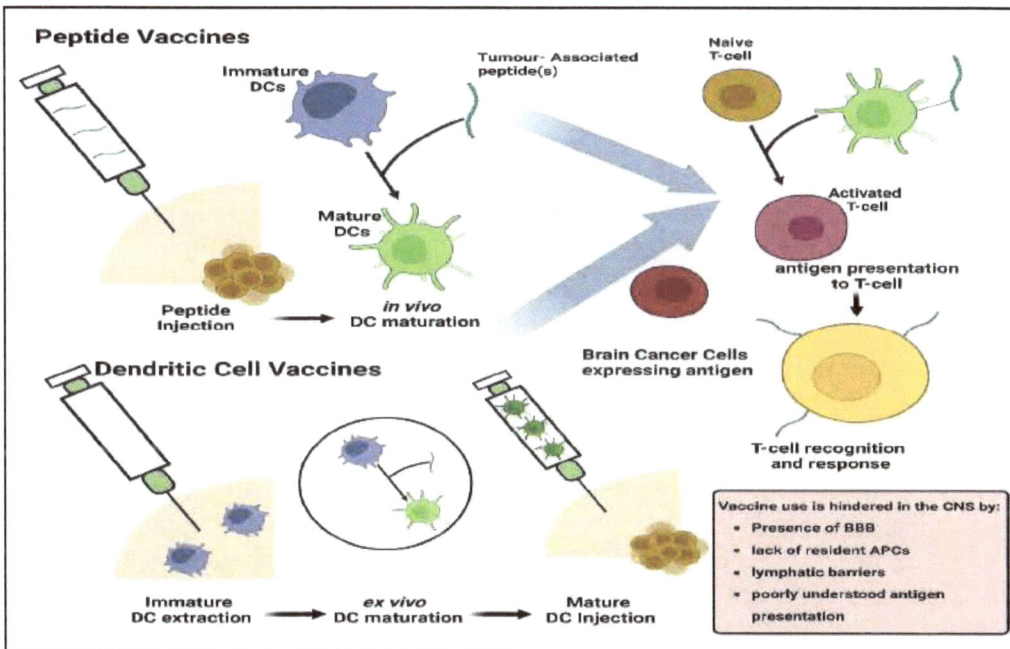

Fig. (3). Vaccine-based immunotherapy for brain tumors.

Dendritic Cell Vaccines

Ex vivo stimulation of dendritic cells (DCs) represents a promising strategy for cancer vaccination, particularly in the context of brain tumors. Unlike peptide vaccines, which rely on *in vivo* antigen presentation, DC-based immunotherapy utilizes autologous dendritic cells derived from peripheral blood monocytes. These immature DCs are primed *ex vivo* with tumor-associated antigens (TAAs), allowing for controlled maturation and antigen loading prior to reinfusion into the patient [39].

Immature DCs are capable of internalizing and processing antigens. Upon exposure to inflammatory signals, they undergo maturation and acquire the ability to present antigens in an MHC-restricted manner, essential for effective T-cell activation [40]. Once reinjected, mature DCs migrate to secondary lymphoid organs, where they stimulate CD8$^+$ cytotoxic T-lymphocytes (CTLs). These activated T-cells recognize tumor cells *via* unique antigens presented on MHC class I molecules and initiate targeted lysis of neoplastic cells. One key advantage of DC vaccines over peptide vaccines is their ability to present a broader antigenic repertoire. By exposing DCs to tumor cell lysates *ex vivo*, a more comprehensive pool of antigens is processed and presented, potentially enhancing immune activation [41]. A variety of DC vaccine formulations have undergone clinical evaluation for the treatment of brain tumors [40, 42]. For example, ICT-107, an autologous DC vaccine pulsed with six glioma-associated antigens, has been studied for the treatment of recurrent glioblastoma.

While DC vaccines address several limitations of peptide-based approaches, particularly in antigen selection and presentation, optimizing their efficacy remains a major challenge. The therapeutic impact is often constrained by the short-lived activity of injected DCs and the limited proportion that successfully migrates to lymph nodes. Td-preconditioning, as demonstrated by Mitchell *et al.*, offers a promising strategy to enhance DC trafficking by leveraging innate immune cell recruitment mechanisms [43].

Nevertheless, brain tumors pose unique challenges due to the restricted access to CNS lymphatics, limited antigen presentation, and incomplete understanding of the cytotoxic T-cell response within the brain. Although antigen presentation mechanisms are well characterized in neuroimmunology, the effectiveness of DC/T-cell interactions central to DC vaccine success remains under debate. Even localized reinjection of activated DCs into the tumor microenvironment depends heavily on the host's natural immune induction, adding further complexity to therapeutic outcomes.

Monoclonal Antibodies

Monoclonal antibodies (mAbs) represent a form of passive immunotherapy, which does not necessarily rely on the host immune system for efficacy. These antibodies can be administered in various formats: naked antibodies, which directly disrupt target signaling pathways; antibody-drug conjugates (ADCs), which deliver cytotoxic agents specifically to tumor cells; and immune-activating antibodies, which are designed to stimulate the host immune response against tumor cells [44].

Antibodies are typically selected to target highly expressed or aberrantly expressed surface receptors on tumor cells, particularly those implicated in tumorigenesis. In glioblastoma (GBM), the most frequently targeted receptor is the epidermal growth factor receptor (EGFR), especially its mutant form EGFRvIII, which is commonly overexpressed in GBM cells. One notable ADC is ABT-414, which targets both EGFR and EGFRvIII. Upon binding to the tumor cell, ABT-414 releases a potent anti-microtubule toxin, inducing cell death. In ongoing Phase I clinical trials, newly diagnosed GBM patients treated with ABT-414 demonstrated a median progression-free survival of 6.1 months, while the 6-month progression-free survival rate for recurrent GBM was reported at 25.3% [45, 46]. Another monoclonal antibody, nimotuzumab, an anti-EGFR inhibitor, has been used in combination with irinotecan for treating pediatric high-grade gliomas. However, this combination has yielded only modest improvements in overall survival [47].

The limited efficacy of monoclonal antibody therapies in brain tumors is largely attributed to several key challenges. Restricted penetration across the blood-brain barrier (BBB), which limits drug delivery to tumor sites. Tumor heterogeneity and antigenic variability, which reduce antibody-binding effectiveness. Insufficient immune effector cells, such as macrophages and natural killer (NK) cells, which are necessary for antibody-dependent cellular cytotoxicity (ADCC) and phagocytosis. Moreover, antibody-based strategies for selectively targeting brain tumor cells remain underexplored. The immunosuppressive microenvironment and lack of robust innate immune cell recruitment further hinder the therapeutic potential of monoclonal antibodies in GBM.

INVESTIGATING THE IMPACT OF IMMUNOTHERAPY ON BRAIN TUMORS

Challenges in Immunotherapy for Brain Tumors

Monoclonal antibodies (mAbs) represent a form of passive immunotherapy, which does not necessarily rely on the host immune system for efficacy. These antibodies can be administered in various formats: naked antibodies, which directly disrupt target signaling pathways; antibody-drug conjugates (ADCs), which deliver cytotoxic agents specifically to tumor cells; and immune-activating antibodies, which are designed to stimulate the host immune response against tumor cells [44].

Antibodies are typically selected to target highly expressed or aberrantly expressed surface receptors on tumor cells, particularly those implicated in tumorigenesis. In glioblastoma (GBM), the most frequently targeted receptor is the epidermal growth factor receptor (EGFR), especially its mutant form EGFRvIII, which is commonly overexpressed in GBM cells. One notable ADC is ABT-414, which targets both EGFR and EGFRvIII. Upon binding to the tumor cell, ABT-414 releases a potent anti-microtubule toxin, inducing cell death. In ongoing Phase I clinical trials, newly diagnosed GBM patients treated with ABT-414 demonstrated a median progression-free survival of 6.1 months, while the 6-month progression-free survival rate for recurrent GBM was reported at 25.3% [45, 46]. Another monoclonal antibody, nimotuzumab, an anti-EGFR inhibitor, has been used in combination with irinotecan for treating pediatric high-grade gliomas. However, this combination has yielded only modest improvements in overall survival [47].

The limited efficacy of monoclonal antibody therapies in brain tumors is largely attributed to several key challenges. Restricted penetration across the blood-brain barrier (BBB) limits drug delivery to tumor sites and tumor heterogeneity and antigenic variability reduce antibody-binding effectiveness. Insufficient immune effector cells, such as macrophages and natural killer (NK) cells, are necessary for antibody-dependent cellular cytotoxicity (ADCC) and phagocytosis. Moreover, antibody-based strategies for selectively targeting brain tumor cells remain underexplored. The immunosuppressive microenvironment and lack of robust innate immune cell recruitment further hinder the therapeutic potential of monoclonal antibodies in GBM.

The Blood-Brain Barrier (BBB)

The central nervous system (CNS) is protected by three primary physiological barriers that restrict the passage of external elements into brain tissue. Blood-Brain Barrier (BBB) is a highly selective boundary between cerebral microvasculature and brain parenchyma, whereas Blood-Meningeal Barrier is located between the blood vessels of the meninges and the meningeal tissue. Also, choroid plexus barrier (Blood–Cerebrospinal Fluid Barrier) prevents blood-borne substances from entering the cerebrospinal fluid (CSF) within the brain's ventricles [48].

Among these, the BBB is the most selective, followed by the blood-meningeal barrier, while the blood–CSF barrier is comparatively less restrictive [49]. Under normal physiological conditions, blood-borne cells cannot cross the BBB due to its unique structural composition [50, 51]. The BBB is formed by endothelial cells lining cerebral microvessels, which are tightly connected by tight junctions that severely limit vesicular transport. Consequently, only small, hydrophobic molecules with a molecular weight below approximately 400 Da can passively diffuse across the BBB [52, 53]. Active efflux transport systems, such as the ATP-binding cassette (ABC) transporters, actively expel foreign substances. These features collectively pose significant challenges for delivering therapeutic agents to the CNS, limiting the efficacy of many anticancer drugs in treating brain tumors [54]. Although many primary brain tumors exhibit aberrant or leaky vasculature due to rapid angiogenesis, this increased permeability is often heterogeneous and inconsistent across the tumor mass. Emerging strategies, such as the use of nanoparticles, have shown promise in traversing the BBB *via* active transcellular transport mechanisms across endothelial cells [55, 56].

Tumor Microenvironment (TEM)

Brain tumor-mediated immunosuppression arises from a complex and coordinated interplay among diverse cellular populations, cytokines, and extracellular matrix components within the tumor microenvironment (TME) [57]. The TME plays a pivotal role in limiting the infiltration and activation of tumor-infiltrating lymphocytes (TILs) and other immune effector cells, thereby impairing anti-tumor immune responses. This immunosuppressive milieu is driven by the secretion of inhibitory cytokines such as interleukins, transforming growth factor-beta (TGF-β), and prostaglandin E2 (PGE2), which collectively suppress cytotoxic immune cell activity within the TME [58]. T-cell proliferation and effector functions are inhibited, while FoxP3$^+$ regulatory T cells (Tregs) are preferentially activated. Additionally, tumor-associated macrophages (TAMs) and microglia secrete immunosuppressive cytokines that promote tumor growth and induce programmed cell death in cytotoxic T-cells through multiple signaling pathways [59]. This immunosuppression is further reinforced by tumor-associated molecules such as CD70 and gangliosides [60].

A key molecular interaction contributing to immune suppression involves the S100B protein and the receptor for advanced glycation end products (RAGE). This interaction activates the signal transducer and activator of transcription 3 (STAT3) pathway in TAMs and microglia, leading to reduced production of immunostimulatory cytokines and enhanced tumor-supportive functions [61]. STAT3 is a central mediator of tumor–microenvironment communication, operating through receptors for interleukin-6 (IL-6) class cytokines [62]. Multiple lines of evidence implicate STAT3 in promoting tumor progression within the TME. Elevated STAT3 expression in hypoxic tumor regions enhances the secretion of vascular endothelial growth factor (VEGF) and hypoxia-inducible factor 1 (HIF-1), facilitating angiogenesis and malignancy [63, 64]. STAT3 also plays a critical role in maintaining glioma stem cells, suppressing both innate and adaptive immune responses, and inducing immune tolerance *via* Treg activation [65 - 67].

In hypoxic conditions, CNS macrophages differentiate into TAMs and undergo M2 polarization, a process heavily influenced by STAT3 signaling [68]. Within the TME, a variety of cytokines and chemokines act as potent immunosuppressive agents. Interleukin-10 (IL-10), predominantly secreted by macrophages and tumor cells, is widely expressed across neoplasms and inhibits the production of interferon (IFN) and tumor necrosis factor (TNF) [69, 70]. IL-10 also downregulates human leukocyte antigen (HLA) class II molecules, inducing a non-responsive state in TILs. Furthermore, programmed cell death ligand 1 (PD-L1) expression is elevated on microglia in proximity to tumor cells, contributing to T-cell apoptosis and immune evasion [71]. These findings underscore the profoundly immunosuppressive nature of the brain tumor microenvironment and highlight the need to target TME components to enhance innate immune responses and improve the efficacy of immunotherapeutic strategies.

Intracellular Trafficking

Traditionally, brain tumors have been considered immunologically "cold" because of the lack of significant tumor-infiltrating lymphocytes [72]. While lymphocytes may infiltrate tumors, they tend to become exhausted and ineffective. Immune surveillance critically identifies and clears pathogens from the body [73]. The BBB tightly regulates the ingress of immune cells into the central nervous system. Entry of leukocytes from the blood involves adhesion molecules [73]. The expression of this entity varies depending on the disease but tends to increase in neuroinflammatory conditions [74]. Ten brain tumors show variability in both the quantity and location of immune cells within tumor tissue. A recent study from the Heimberger laboratory examined immune cell populations across different regions within glioblastoma tumors. They used RNA sequencing data from the Ivy Glioblastoma Atlas findings. Researchers observed lower expression of T cell markers and B cell markers at the tumor edge and within tumor tissue, including necrotic zones. However, immune cell markers were considerably high in the vascular portion of the glioblastomas [75].

Cancer-Stem Cell (TSC) Compartment

Tumor-initiating cells (TICs), also referred to as cancer stem cells (CSCs), comprise a heterogeneous population of multipotent, undifferentiated cells capable of self-renewal and high tumorigenicity. These cells play a pivotal role in the progression of glioblastoma multiforme (GBM) and immune evasion. TICs influence T-lymphocyte activity through the STAT3 signaling pathway, a key mechanism by which tumors exert immunosuppressive effects [76 - 79]. Certain tumor cells also secrete macrophage inhibitory cytokine-1 (MIC-1), which has been associated with poor clinical outcomes [80].

TICs contribute to the recruitment and polarization of monocytes into immunosuppressive M2 macrophages within the tumor microenvironment (TME), further enhancing immune evasion [81]. Although a universally accepted marker profile for TICs remains elusive, several surface markers and molecular mediators have been identified, including CD133, CD90, CD44, L1CAM, A2B5, and GPD1 [82]. These markers, along with functional attributes such as *in vivo* tumorigenic potential and pluripotency, are used to characterize TICs. Continued research into these defining features is essential for advancing our understanding of stem cell biology and tumor evolution [83]. The TME in GBM is a highly heterogeneous and dynamic system, comprising diverse cellular populations and complex signaling networks. Studies have identified three primary niches for glioblastoma stem cells. Perivascular niche characterized by leaky, irregular blood vessels; stem cells in this niche regulate vascular integrity and promote angiogenesis [84 - 86].

Perinecrotic/hypoxic niche defined by regions of low oxygen and tissue necrosis; hypoxia-inducible factors (HIFs) play a central role in maintaining stem cell function [87]. Immune niche involves interactions with immune cells that contribute to immune suppression and tumor persistence. Among hypoxia-related regulators, HIF-2α has emerged as a critical factor in sustaining GBM stem cell activity. While HIF-1α is induced under extreme hypoxia (~1% oxygen) and is expressed in both stem and non-stem tumor cells, HIF-2α responds to moderate hypoxia and is exclusively expressed in cancer stem cells, not in normal immune cells [88, 89]. This unique expression profile positions HIF-2α as a promising therapeutic target, potentially enabling interventions that disrupt glioblastoma at its cellular origin.

The Interplay Between Tumor and Neuronal Cells

While neuronal-glioma interactions are increasingly understood to be crucial features of tumor biology and the development of carcinoma, the precise impacts of these interactions on disease advancement and the TME are just beginning to be explored. Stanford University has conducted meaningful research on glioma's electrical and synaptic incorporation inside neural networks. Previous studies have demonstrated the synaptic interrelation of neurons and normal oligodendrocyte precursor cells (OPCs), whereby signaling at this intersection of neurons can affect the multiplication and viability of the OPCs [90]. An analysis of single-cell gene expression data from pre-treatment autopsy samples of adult and pediatric severe brain cancers revealed enrichment of synaptic genes within specific malignant cell populations. This suggests that distinct cancer cell types may serve unique roles within the TME [91]. Researchers used electron microscopy to find the microstructure of the neuronal intersection and electrophysiological measurements. They found evidence of excitatory postsynaptic currents in glioma cells, providing functional support for neuron-t--glioma intersection. These synapses could be blocked by tetrodotoxin [92]. It is believed that bidirectional interactions exist between neurons and tumors. Tumors were found to enhance neuronal activity, and neuronal activity in turn may promote tumor growth. This perspective is particularly high weightage given the high prevalence of seizures associated with intracranial neoplasm. Research has validated that tumors are capable of inducing neuronal hyperactivity and triggering seizure activity [93, 94]. Some studies have used magnetoencephalography to identify areas of high (HFC) and low (LFC) functional connection in the brain tumors of temporal lobe glioblastoma patients. Researchers found that the TSP1 protein moderates tumor-induced neuronal connectivity changes and helps cancer network integration. This has a negative impact on patient behavior and survival. Additional studies are still needed to completely understand how brain tumors interact functionally with neurons and their clinical effects [95].

Iatrogenic Effects

Corticosteroids remain the standard treatment for managing immune-related adverse events (irAEs) associated with cancer immunotherapies, particularly immune checkpoint inhibitors. Their immunosuppressive effects were first demonstrated in 1976, with findings especially relevant to intracranial tumors. In a seminal study by Fauci *et al.*, patients administered dexamethasone, hydrocortisone, or prednisone showed a marked but transient reduction in circulating lymphocyte and monocyte levels, with nadirs occurring 4–6 hours post-administration. Cell counts typically returned to baseline within 24 hours, occasionally rebounding above normal levels [96].

In brain tumor immunotherapy, corticosteroids are often used alongside chemotherapy, most commonly temozolomide (TMZ). However, TMZ-induced lymphopenia raises concerns when combined with immunotherapy, as it may impair immune responses [97]. Animal models of glioblastoma multiforme (GBM) have shown that dose-intensified TMZ enhances CAR T-cell proliferation and persistence compared to standard TMZ dosing or CAR T-cell therapy alone [98]. Research by Karachi *et al.* revealed that standard TMZ dosing significantly reduced $CD4^+$ and $CD8^+$ T-cell counts, leading to T-cell exhaustion and diminished efficacy of anti-PD-1 therapy. In contrast, lower or metronomic TMZ dosing preserved cytotoxic T-cell function and improved tumor targeting [99].

Interestingly, preliminary studies have observed reduced PD-L1 expression on GBM cells following TMZ treatment, suggesting that chemotherapy may modulate the tumor's immunosuppressive profile [100]. Emerging cancer immunotherapies offer potential advantages over traditional chemotherapy, particularly in terms of target specificity, immune activation, and tolerability. These therapies aim to stimulate the host's immune system to recognize and eliminate tumor cells, often with fewer systemic side effects. Common adverse events include mild symptoms such as gastrointestinal discomfort, oral mucositis, and cytopenias. However, serious irAEs may occur, including abnormal angiogenesis, severe inflammation, and autoimmune disorders, necessitating careful risk assessment and management [101].

A key concern is cytokine release syndrome (CRS), triggered by excessive cytokine production following immune cell activation and damage. CRS has been associated with neurological complications, including tremors, cerebellar dysfunction, seizures, and in rare cases, fatal cerebral edema. Immune-related adverse reactions have been reported in approximately 85% of patients treated with ipilimumab and 70% of those receiving PD-1 axis inhibitors. Notable irAEs include colitis, thyroiditis, and hypophysitis, with some cases requiring intensive care [102 - 106]. Clinical trials have shown increased rates of pituitary inflammation with CTLA-4 inhibition and thyroid dysfunction with PD-(L)1 inhibition. Additionally, rare cases of type 1 diabetes and primary adrenal insufficiency have been documented. As immunotherapies become more widely adopted, ongoing research is essential to understand better and mitigate immune-related toxicities, ensuring patient safety and treatment efficacy [107].

Elucidation of Tumor Imaging

The successful application of immunotherapy in treating brain neoplasms presents a paradox: distinguishing accurate treatment response from tumor progression can be challenging. In cancers such as melanoma, where immune checkpoint inhibitors (ICIs) have demonstrated efficacy, serial tumor biopsies often reveal extensive immune cell infiltration and histologic signs of inflammation. However, the central nervous system (CNS) lacks a robust lymphatic drainage system, unlike most peripheral organs. This anatomical limitation can result in treatment-induced immune responses manifesting as edema, increased perfusion, enhanced contrast uptake, and mass effect radiographic features that may mimic tumor progression. This phenomenon, known as immune pseudoprogression, has been sparsely reported in the context of brain tumors treated with ICIs, but it remains a critical diagnostic challenge [108]. Misinterpreting immune-related changes as disease progression may lead to premature discontinuation of effective therapies or unnecessary interventions.

Emerging evidence suggests that patients receiving ICIs or other targeted immunotherapies may experience radiation necrosis at a higher rate following stereotactic radiosurgery (SRS) compared to those treated with chemotherapy or no systemic therapy [109, 110]. These findings underscore the importance of carefully sequencing immunotherapy and radiosurgery to balance therapeutic efficacy with safety. However, further well-designed clinical studies are needed to validate these observations and guide optimal treatment protocols. To improve response assessment, researchers are exploring radiomic features such as sphericity, heterogeneity, and border sharpness as reproducible indicators of tumor immune states. These imaging biomarkers may help predict responses to ICIs and differentiate between actual progression and pseudoprogression [111, 112]. Advancing such predictive tools is essential for refining immunotherapy strategies and enhancing clinical decision-making in neuro-oncology.

RECENT ADVANCES IN IMMUNOTHERAPY

Targeting Novel Checkpoints in Immunotherapy

CTLA-4 and PD-1 blockade are currently the primary areas of focus in basic findings and clinical work in the field of immune checkpoint pathways. However, continued exploration of additional immune checkpoint pathways could enable the development of combination treatment strategies with the potential to improve responses and broaden the applicability of immune checkpoint inhibition to benefit more patients with brain tumors [113 - 115]. Researchers have identified additional checkpoints for tumor therapy, including CD47 and CD73. Unlike PD-L1, which signals the adaptive immune system not to detect tumor cells, the CD47 protein signals the innate immune system not to engulf tumor cells. CD47 binds to its receptor SIRPα on phagocytic cells, inhibiting the ability of macrophages to phagocytose or engulf tumor cells [116, 117]. Several preclinical studies have examined safely and effectively targeting CD47 as an immune checkpoint molecule for brain tumor therapy. Blocking CD47 using an anti-CD47 antibody stimulated macrophage phagocytosis of GBM and thereby decreased tumor burden *in vivo*. The interaction between the protein CD47 and its receptor SIRPα on immune cells has been shown to inhibit phagocytosis of tumor cells. Recently, scientists have explored targeting the CD47-SIRPα interaction as a potential therapeutic approach for cancer. Preclinical animal studies have evaluated the safety and effectiveness of blocking CD47 in treating glioblastoma. Results from these early-stage investigations have been encouraging [114, 116, 117].

While immune checkpoint (IC) blockade therapy has shown success in treating late-stage cancers, a substantial proportion of patients remain unresponsive to these interventions. Evidence suggests that multiple, distinct immunosuppressive mechanisms may operate concurrently within the tumor microenvironment (TME), contributing to therapeutic resistance. One such mechanism involves the enzymatic conversion of pro-inflammatory extracellular ATP into the immunosuppressive molecule adenosine (ADA). This pathway significantly dampens anti-tumor immune responses. The conversion process begins with CD39, which degrades ATP to AMP, and CD38, which converts NAD^+ to ADP-ribose. Both pathways converge at CD73 (also known as 5'-nucleotidase), which catalyzes the final step degrading AMP into adenosine. Studies have demonstrated that targeting CD73 on both host immune cells and tumor cells can reduce adenosine-mediated immunosuppression and inhibit tumor progression in solid tumor models. These findings highlight CD73 as a promising therapeutic target for overcoming resistance to immunotherapy [114]. Currently, novel drug candidates targeting the adenosine signaling pathway are entering clinical trials for patients with advanced cancers. These agents are being evaluated both as monotherapies and in combination with conventional treatments and immunotherapies, aiming to disrupt non-redundant resistance mechanisms and improve patient outcomes [118].

Combination Approach

Glioblastoma multiforme (GBM) remains one of the most challenging cancers to treat, despite extensive research into its cellular signaling pathways, genetic alterations, and tumor microenvironment (TME) characteristics. Treatment efficacy is hindered by several factors, including limited drug distribution, tumor heterogeneity, and the activation of drug resistance mechanisms, which often lead to tumor recurrence and poor patient outcomes [119, 120]. Given the limitations of single-agent therapies, researchers and clinicians are increasingly exploring combination treatment strategies. Among these, immune checkpoint inhibitor (ICI) combinations have shown promise, particularly due to their success in other malignancies such as melanoma. Several combination regimens are currently under investigation for GBM, aiming to enhance therapeutic efficacy. However, these approaches are still in early stages, and their clinical impact remains to be fully determined.

While combination therapies offer new hope, they also present significant challenges. Patients may be exposed to highly toxic drug combinations, which can result in antagonistic side effects or intolerable toxicity. Even when side effects are manageable, it is often difficult to disentangle the individual contributions of each therapeutic agent to the overall efficacy of the regimen. Recent innovations in multi-modal delivery systems have shown potential to overcome these barriers. For example, Li *et al.* utilized neutrophils to deliver nanosensitizers, enhancing the effects of ultrasound therapy, chemotherapy, and immunotherapy in GBM.

This cellular-level integration of therapeutic modalities may improve treatment outcomes for this aggressive cancer. Ongoing research will focus on validating the safety and efficacy of this approach in preclinical models and clinical trials [121].

In another study, Wang *et al.* employed perfluorocarbon (PFC) liquid-filled silica microshells, activated by focused ultrasound, to induce localized tissue damage and promote immunogenic cell death (ICD). When combined with PD-1 blockade, this strategy transformed the TME from an immunosuppressive "cold" state to a proinflammatory "hot" state, significantly increasing immune cell infiltration. Specifically, combination therapy increased CD45$^+$ cells by over 20-fold, CD8$^+$ cytotoxic T cells by over 100-fold, and IFN-γ expression by over 200-fold compared to monotherapy [122].

Kadiyala *et al.* developed a chemo-immunotherapy platform using CpG, a Toll-like receptor-9 (TLR-9) agonist, and docetaxel (DTX). These agents were encapsulated in high-density lipoprotein nanodiscs (DTX-sHDL-CpG), which offer extended plasma circulation and targeted delivery. This approach aims to synergize immune activation with cytotoxic chemotherapy, enhancing therapeutic precision and minimizing systemic toxicity [123].

Immunogenic Cell Death Aided Immune Cell Activation

Inducing immunogenic cell death (ICD) has emerged as a compelling strategy in cancer immunotherapy, offering the dual benefit of tumor eradication and immune system activation. ICD promotes the release and presentation of stress-related molecules and tumor-associated antigens that are typically sequestered within malignant cells, thereby enhancing immune recognition and targeting of cancer cells [124, 125]. Cellular injury, such as that caused by ischemia-reperfusion, leads to the release of damage-associated molecular patterns (DAMPs), which serve as potent immunostimulatory signals. These DAMPs activate antigen-presenting cells (APCs), enabling them to deliver both antigenic and adjuvant signals to T-cells. This mechanism represents a promising avenue for stimulating durable anti-tumor immunity.

Several therapeutic modalities are capable of inducing ICD, including chemotherapy, radiotherapy and hyperthermia. These approaches aim to create an inflammatory tumor microenvironment (TME) that facilitates immune cell infiltration, promotes tumor clearance, and establishes long-term immunological memory. Continued research to optimize ICD-inducing pathways may significantly improve clinical outcomes across various cancer types [124, 126].

In a recent study, Zhang *et al.* developed bradykinin (BK) aggregation-induced emission nanoparticles (BK@AIE NPs), which selectively permeate the blood-brain barrier (BBB) and exhibit strong absorption in the near-infrared (NIR) region. Upon photothermal therapy, these nanoparticles induced ICD, leading to the release of tumor antigens and activation of the host immune system. Following immune activation, key cytokines involved in immune regulation such as interleukin-2 (IL-2), IL-10, IL-12, IL-1β, interferon-gamma (IFN-γ), and tumor necrosis factor-alpha (TNF-α) were significantly upregulated in serum. This cytokine surge contributed to a robust anti-tumor immune response, highlighting the therapeutic potential of BK@AIE NPs in brain tumor immunotherapy [127, 128].

Individualized Vaccine Therapy

Cancer immunotherapy has demonstrated considerable promise, particularly in late-stage malignancies that elicit a robust immune response. These therapies harness the body's natural defense mechanisms to recognize and eliminate cancer cells, offering hope to patients with advanced, treatment-resistant disease. While outcomes vary, immunotherapy has led to durable remissions in a subset of patients, underscoring its transformative potential. Ongoing research aims to expand these benefits to a broader patient population. However, many cancers remain unresponsive to immunotherapy, including gliomas, which are characterized by: A low somatic mutation burden and Sparse T-cell infiltration. A profoundly immunosuppressive tumor microenvironment (TME) [129, 130]. Although the significance of the TME in shaping immunotherapy outcomes is well recognized, comprehensive analyses of its characteristics across cancer types have been limited. A deeper understanding of TME heterogeneity and its influence on treatment resistance is critical for advancing immunotherapeutic strategies.

Recent efforts have focused on elucidating key mechanisms underlying immunotherapy responses, including. The composition and suppressive nature of the TME, Antigen presentation pathways and The density and phenotype of tumor-infiltrating immune cells. These investigations have led to refined TME classifications, enabling stratification of tumors based on immune activity. One widely accepted framework categorizes tumors into three groups. Inflamed tumors Characterized by abundant tumor-infiltrating lymphocytes (TILs), high interferon-gamma (IFN-γ) production, and elevated PD-L1 expression, Non-inflamed tumors – Display minimal lymphocyte infiltration, low PD-L1 expression, and reduced MHC-I antigen presentation.

Excluded tumors – Exhibit immune cell infiltration restricted to the peritumoral region, with T-cells unable to penetrate the tumor core [131]. In a landmark study, Torsson *et al.* utilized data from The Cancer Genome Atlas (TCGA) to conduct a comprehensive immunogenomic analysis across 33 cancer types. This investigation identified six distinct immune subtypes. Immunologically quiet, Inflammatory, Wound healing, IFN-γ dominant, Lymphocyte deficient and TGF-β dominant.

These subtypes were defined based on variables such as neoantigen burden, macrophage and lymphocyte signatures, Th1/Th2 cell ratios, immunomodulatory gene expression, and clinical prognosis [132]. Such stratification offers valuable insights into the immune landscape of tumors and may guide the development of personalized immunotherapy approaches.

Clinical Trials and Outcomes

Summary of Key Clinical Trials

Immunotherapy, particularly with ICI, personalized cancer vaccines, oncolytic viruses, and CAR T-cell therapy, shows promise in the treatment of glioblastoma [133]. Active immunotherapy using dendritic cells, specifically with autologous DCs loaded with autologous tumor cells, has the potential to increase median overall survival and prolong tumor progression-free survival with minimal complications [134]. However, a meta-analysis found no statistically significant differences in 1-year overall survival or progression-free survival when immunotherapy was combined with standard chemotherapy and radiotherapy compared to chemotherapy and radiotherapy alone [135]. Despite this, another systematic review and meta-analysis found that dendritic cell-based vaccines were associated with significantly longer overall survival and 2-year survival compared to conventional therapy [136]. Table **1** displays information about key clinical trials.

Table 1. Summary of key clinical trials.

Sr. No.	Class of Immunotherapy	Study Phase	Number of Patients	Outcomes	References
1	CAR T-cell	Phase-1	03	The clinical trial data did not reveal any dose-limiting toxicities. Laboratory findings provided clinical evidence of localized immune activation within the central nervous system.	[137]

(Table 1) cont.....

Sr. No.	Class of Immunotherapy	Study Phase	Number of Patients	Outcomes	References
2	Autologous Vaccines	Prospective	09	Preparation for ADCV is attainable and safe. DIPG-specific immune action was identified in primary blood mononuclear cells and cerebral spinal fluid.	[138]
3	Peptide Vaccines	Prospective	12	The results found that 9 out of 10 patients had an effective immune reaction to at least one of the GAA treatments. Devoid of dose-limiting toxicities. The median progression-free survival was 4.1 months. The 6-month progression-free survival rate was 33% and the 6-month overall survival rate was 73%. Preliminary findings provided evidence of both immunologic and clinical responses to the treatment. The median overall survival was 12.9 months.	[139]
4		Prospective	14	A Phase 1 study was conducted to evaluate the safety and effectiveness of a novel immunotherapy agent in 12 patients with advanced cancers. All 12 patients developed a positive immune response to at least one tumor-associated antigen after treatment. No dose-limiting toxicities were reported. Median progression-free survival was 9.0 months, with 85% of patients progression-free at 6 months and 42% at 12 months. Preliminary evidence suggests the agent elicited both immunologic and clinical responses.	[140]
5	Oncolytic Virus	Phase 1	12	4/11 patients remained alive 18 months after treatment. No viral shedding was detected in any patients. Immune responses were observed in all 11 patients. Tissue samples collected from 4 patients after treatment showed substantial immune cell infiltration, indicating the treatment may have converted immunologically "cold" tumors to "hot" tumors more susceptible to immune-mediated destruction. The median overall survival for the 11 patients was 12.2 months.	[141]

(Table 1) cont.....

Sr. No.	Class of Immunotherapy	Study Phase	Number of Patients	Outcomes	References
6			08	Three patients remained alive for more than twenty-four months, including one with anaplastic astrocytoma. The median survival was 8.9 months for patients treated at dose level one. One patient with an ependymoma had a median survival of twenty-five months when treated at dose level two. Glioma-modified cytotoxin immunotoxin appears to be safely administered with radiotherapy, with or without temozolomide, in pediatric patients.	[142]
7		Phase 1	29	The MTD was determined to be 2.6 mg/m2. Immune correlates were observed through a serum response. Plans were made to initiate a subsequent Phase 2 clinical trial.	[143]
8	Immunomodulatory	Phase 2	52	Two high-grade glioma patients achieved either an objective response or long-term stable disease. There was no objective response or long-term stable disease observed in the medulloblastoma or diffuse intrinsic pontine glioma patients. Treatment with POM monotherapy did not meet the primary endpoint of success. The median overall survival was: 5.1 months for high-grade glioma patients, 3.8 months for diffuse intrinsic pontine glioma patients, 12.0 months for ependymoma patients, and 11.6 months for medulloblastoma patients.	[144]

(Table 1) cont.....

Sr. No.	Class of Immunotherapy	Study Phase	Number of Patients	Outcomes	References
9	Other	Phase 1, dose escalation Study	09	Four out of five patients with progressive disease showed radiographic improvement. At higher doses, natural killer cells increased in the cerebrospinal fluid during treatment. Direct infusion of a patient's natural killer cells into the ventricles of the brain was determined to be safe and viable.	[145]
10		Retrospective	41	A study investigated the feasibility of multimodal immunotherapy in patients. No significant toxicities were observed. Treatment was associated with a shift towards a Type 1 immune response profile. Median overall survival was 9.1 months.	[146]

Future Directions

Immunotherapy has demonstrated potential in the management of brain neoplasm, particularly glioblastoma (GBM), but identifying biomarkers for treatment response remains a challenge [147]. Ongoing research and upcoming trials are exploring innovative drug delivery systems, potential biomarkers, and the use of adoptively transferred cell therapies [148]. The development of new therapies, such as immune checkpoint blockade, chimeric antigen receptor T (CAR T) cell therapy, oncolytic virotherapy, and vaccine therapy, offers hope for improving GBM outcomes [149]. However, the effective use of immunotherapies to treat brain tumors requires more representative animal models and improved biopsy collection methods [148].

CONCLUSION

Gliomas and metastatic brain malignancies are among the central nervous system tumors for which immunotherapy has shown promising therapeutic potential. This review highlights the clinical outcomes of recent trials and explores the mechanisms by which immunotherapeutic strategies enhance anti-tumor immune responses. Key approaches include immune checkpoint inhibitors (ICIs), which block inhibitory proteins that prevent immune cells from attacking cancer cells, thereby amplifying the host immune response. Chimeric antigen receptor (CAR) T-cell therapy is a personalized treatment that involves genetically modifying a patient's T cells to recognize and target specific tumor antigens.

Cancer vaccines are designed to activate the immune system. These vaccines stimulate recognition and destruction of tumor cells by presenting tumor-associated antigens. Despite the promise of these therapies, several unique challenges complicate their application in brain tumors. The blood-brain barrier (BBB) restricts the delivery of therapeutic agents to the CNS, while the immunosuppressive tumor microenvironment (TME) hinders immune cell infiltration and function. Nevertheless, advances in combination therapies and personalized medicine offer new avenues to overcome these barriers. Combination strategies, which integrate immunotherapy with conventional treatments such as radiation and chemotherapy, or combine multiple immunotherapeutic agents, have shown potential to enhance efficacy and mitigate resistance. Concurrently, personalized medicine approaches tailored to each patient's genetic and immunologic profile are being actively explored to optimize treatment outcomes. In conclusion, while immunotherapy for brain tumors continues to face significant obstacles, ongoing research and innovation are paving the way for more effective and individualized treatment options. By addressing the specific challenges posed by brain tumors, immunotherapy holds the potential to significantly improve survival rates and quality of life for affected patients. Continued investigation is essential to refine these therapies and unlock their full clinical potential, offering renewed hope for the treatment of these formidable cancers.

LIST OF ABBREVATIONS

BBB   Blood-brain barrier CART Chimeric antigen receptor T cells

CNS   Central nervous system CIK Cytokine-induced killer

CTLA4 Cytotoxic T-lymphocyte-associated protein 4 APCs Antigen-presenting cells

PD1   Programmed cell death 1 GBM Glioblastoma multiforme

TIL   Tumor-infiltrating lymphocyte

WHO   World Health Organisation MHC Major Histocompatibility Complex

AUTHOR'S CONTRIBUTION

Pranjal Gujarathi was responsible for the conceptualization, methodology, and data analysis. Deepa Mandlik contributed by collecting data and drafting the manuscript. Meghraj Suryawanshi conducted the literature review and was involved in editing and revising the manuscript.

REFERENCES

[1]   SEER cancer statistics review. National Cancer Institute. 2016:1975–2013.

[2]   Ostrom QT, Gittleman H, Farah P, Ondracek A, Chen Y, Wolinsky Y. CBTRUS statistical report: Primary brain and central nervous system tumors diagnosed in the United States in 2006-2010. Neuro

Oncol 2013; Nov 15 Suppl 2 (Suppl 2):ii1–56.

[3]   Nabors LB, Portnow J, Ammirati M, *et al.* Central nervous system cancers, version 1.2015. J Natl Compr Canc Netw 2015; 13(10): 1191-202.
[http://dx.doi.org/10.6004/jnccn.2015.0148] [PMID: 26483059]

[4]   Wiesner SM, Freese A, Ohlfest JR. Emerging concepts in glioma biology: implications for clinical protocols and rational treatment strategies. Neurosurg Focus E3 2005;19(4):1–6.
[http://dx.doi.org/10.3171/foc.2005.19.4.4]

[5]   Ricard D, Idbaih A, Ducray F, Lahutte M, Hoang-Xuan K, Delattre JY. Primary brain tumours in adults. Lancet 2012; 379(9830): 1984-96.
[http://dx.doi.org/10.1016/S0140-6736(11)61346-9] [PMID: 22510398]

[6]   Bellail AC, Hunter SB, Brat DJ, Tan C, Van Meir EG. Microregional extracellular matrix heterogeneity in brain modulates glioma cell invasion. Int J Biochem Cell Biol 2004; 36(6): 1046-69.
[http://dx.doi.org/10.1016/j.biocel.2004.01.013] [PMID: 15094120]

[7]   Bellail AC, Hunter SB, Brat DJ, Tan C, Van Meir EG. Microregional extracellular matrix heterogeneity in brain modulates glioma cell invasion. Int J Biochem Cell Biol 2004; 36(6): 1046-69.
[http://dx.doi.org/10.1016/j.biocel.2004.01.013] [PMID: 15094120]

[8]   Blakeley J. Drug delivery to brain tumors. Curr Neurol Neurosci Rep 2008; 8(3): 235-41.
[http://dx.doi.org/10.1007/s11910-008-0036-8] [PMID: 18541119]

[9]   Pietschmann S, von Bueren AO, Kerber MJ, Baumert BG, Kortmann RD, Müller K. An individual patient data meta-analysis on characteristics, treatments and outcomes of glioblastoma/ gliosarcoma patients with metastases outside of the central nervous system. PLoS One 2015; 10(4): e0121592.
[http://dx.doi.org/10.1371/journal.pone.0121592] [PMID: 25860797]

[10]   Volovitz I, Marmor Y, Azulay M, *et al.* Split immunity: immune inhibition of rat gliomas by subcutaneous exposure to unmodified live tumor cells. J Immunol 2011; 187(10): 5452-62.
[http://dx.doi.org/10.4049/jimmunol.1003946] [PMID: 21998458]

[11]   Ostrom QT, Gittleman H, Fulop J, Liu M, Blanda R, Kromer C. CBTRUS statistical report: primary brain and central nervous system tumors diagnosed in the United States in 2008-2012. Neuro Oncol 2015;17 Suppl 4(suppl 4) :iv1–62.

[12]   Louis DN, Perry A, Reifenberger G, *et al.* The 2016 world health organization classification of tumors of the central nervous system: A summary. Acta Neuropathol 2016; 131(6): 803-20.
[http://dx.doi.org/10.1007/s00401-016-1545-1] [PMID: 27157931]

[13]   Durand A, Labrousse F, Jouvet A, *et al.* WHO grade II and III meningiomas: a study of prognostic factors. J Neurooncol 2009; 95(3): 367-75.
[http://dx.doi.org/10.1007/s11060-009-9934-0] [PMID: 19562258]

[14]   Ramkissoon S. Surgical pathology of neoplasms of the central nervous system Pathobiology of human disease. Elsevier 2014; pp. 3592-606.

[15]   Zamarin D, Postow MA. Immune checkpoint modulation: Rational design of combination strategies. Pharmacol Ther 2015; 150: 23-32.
[http://dx.doi.org/10.1016/j.pharmthera.2015.01.003] [PMID: 25583297]

[16]   Chen L, Flies DB. Molecular mechanisms of T cell co-stimulation and co-inhibition. Nat Rev Immunol 2013; 13(4): 227-42.
[http://dx.doi.org/10.1038/nri3405] [PMID: 23470321]

[17]   Powles T, Eder JP, Fine GD, *et al.* MPDL3280A (anti-PD-L1) treatment leads to clinical activity in metastatic bladder cancer. Nature 2014; 515(7528): 558-62.
[http://dx.doi.org/10.1038/nature13904] [PMID: 25428503]

[18]   Davidson T, Hsu M, Sedighim S, *et al.* ATIM-26. PD-1 expression by tumor infiltrating lymphocytes in glioblastoma is a marker of both activation and exhaustion. Neuro-oncol 2016; 18 (Suppl. 6): vi23-

3.
[http://dx.doi.org/10.1093/neuonc/now212.091]

[19]   Omuro A, Vlahovic G, Baehring J, *et al.* Atim-16. Nivolumab combined with radiotherapy with or without temozolomide in patients with newly diagnosed glioblastoma: results from phase 1 safety cohorts in checkmate 143. Neuro-oncol 2016; 18 (Suppl. 6): vi21-1.
[http://dx.doi.org/10.1093/neuonc/now212.081]

[20]   Carter T, Brown N, Shaw H, Chester K, Cohn-Brown D, Mulholland P. ATIM-08. Survival and efficacy data for patients with glioblastoma treated with ipilimumab in combination with bevacizumab. Neuro-oncol 2016; 18 (Suppl. 6): vi19-9.
[http://dx.doi.org/10.1093/neuonc/now212.073]

[21]   Preusser M, Lim M, Hafler DA, Reardon DA, Sampson JH. Prospects of immune checkpoint modulators in the treatment of glioblastoma. Nat Rev Neurol 2015; 11(9): 504-14.
[http://dx.doi.org/10.1038/nrneurol.2015.139] [PMID: 26260659]

[22]   Bielamowicz K, Khawja S, Ahmed N. Adoptive cell therapies for glioblastoma. Front Oncol 2013; 3: 275.
[http://dx.doi.org/10.3389/fonc.2013.00275] [PMID: 24273748]

[23]   Fesnak AD, June CH, Levine BL. Engineered T cells: the promise and challenges of cancer immunotherapy. Nat Rev Cancer 2016; 16(9): 566-81.
[http://dx.doi.org/10.1038/nrc.2016.97] [PMID: 27550819]

[24]   Kong DS, Nam DH, Kang SH, *et al.* Phase III randomized trial of autologous cytokine-induced killer cell immunotherapy for newly diagnosed glioblastoma in korea. Oncotarget 2017; 8(4): 7003-13.
[http://dx.doi.org/10.18632/oncotarget.12273] [PMID: 27690294]

[25]   Brown C, Alizadeh D, Starr R, *et al.* ATIM-13. Phase I study of chimeric antigen receptor-engineered t cells targeting il13rα2 for the treatment of glioblastoma. Neuro-oncol 2016; 18 (Suppl. 6): vi20-0.
[http://dx.doi.org/10.1093/neuonc/now212.078]

[26]   Melero I, Gaudernack G, Gerritsen W, *et al.* Therapeutic vaccines for cancer: an overview of clinical trials. Nat Rev Clin Oncol 2014; 11(9): 509-24.
[http://dx.doi.org/10.1038/nrclinonc.2014.111] [PMID: 25001465]

[27]   Jeanbart L, Swartz MA. Engineering opportunities in cancer immunotherapy. Proc Natl Acad Sci USA 2015; 112(47): 14467-72.
[http://dx.doi.org/10.1073/pnas.1508516112] [PMID: 26598681]

[28]   Wong KK, Li WA, Mooney DJ, Dranoff G. Advances in therapeutic cancer vaccines, Elsevier Inc 2016; 130: 191-249.

[29]   Xu LW, Chow KKH, Lim M, Li G. Current vaccine trials in glioblastoma: a review. J Immunol Res 2014; 2014: 1-10.
[http://dx.doi.org/10.1155/2014/796856] [PMID: 24804271]

[30]   Comber JD, Philip R. MHC class I antigen presentation and implications for developing a new generation of therapeutic vaccines. Ther Adv Vaccines 2014; 2(3): 77-89.
[http://dx.doi.org/10.1177/2051013614525375] [PMID: 24790732]

[31]   Li W, Joshi M, Singhania S, Ramsey K, Murthy A. Peptide vaccine: progress and challenges. Vaccines (Basel) 2014; 2(3): 515-36.
[http://dx.doi.org/10.3390/vaccines2030515] [PMID: 26344743]

[32]   Akiyama Y, Komiyama M, Miyata H, *et al.* Novel cancer-testis antigen expression on glioma cell lines derived from high-grade glioma patients. Oncol Rep 2014; 31(4): 1683-90.
[http://dx.doi.org/10.3892/or.2014.3049] [PMID: 24573400]

[33]   Swartz AM, Batich KA, Fecci PE, Sampson JH. Peptide vaccines for the treatment of glioblastoma. J Neurooncol 2015; 123(3): 433-40.
[http://dx.doi.org/10.1007/s11060-014-1676-y] [PMID: 25491947]

[34] Weller M, Butowski N, Tran D, *et al.* ATIM-03. ACT IV: An international, double-blind, phase 3 trial of rindopepimut in newly diagnosed, EGFRvIII-expressing glioblastoma. Neuro-oncol 2016; 18 (Suppl. 6): vi17-8.
[http://dx.doi.org/10.1093/neuonc/now212.068]

[35] Sampson JH, Heimberger AB, Archer GE, *et al.* Immunologic escape after prolonged progression-free survival with epidermal growth factor receptor variant III peptide vaccination in patients with newly diagnosed glioblastoma. J Clin Oncol 2010; 28(31): 4722-9.
[http://dx.doi.org/10.1200/JCO.2010.28.6963] [PMID: 20921459]

[36] Verhaak RGW, Hoadley KA, Purdom E, *et al.* Integrated genomic analysis identifies clinically relevant subtypes of glioblastoma characterized by abnormalities in PDGFRA, IDH1, EGFR, and NF1. Cancer Cell 2010; 17(1): 98-110.
[http://dx.doi.org/10.1016/j.ccr.2009.12.020] [PMID: 20129251]

[37] Patel AP, Tirosh I, Trombetta JJ, Shalek AK, Gillespie SM, Wakimoto H. Single-cell RNA-seq highlights intratumoral heterogeneity in primary glioblastoma. Science (1979) 2014;344(6160):1396–401.

[38] Francis JM, Zhang CZ, Maire CL, *et al. EGFR* variant heterogeneity in glioblastoma resolved through single-nucleus sequencing. Cancer Discov 2014; 4(8): 956-71.
[http://dx.doi.org/10.1158/2159-8290.CD-13-0879] [PMID: 24893890]

[39] Sabado RL, Bhardwaj N. Dendritic-cell vaccines on the move. Nature 2015; 519(7543): 300-1.
[http://dx.doi.org/10.1038/nature14211] [PMID: 25762139]

[40] Kim W, Liau LM. Dendritic cell vaccines for brain tumors. Neurosurg Clin N Am 2010; 21(1): 139-57.
[http://dx.doi.org/10.1016/j.nec.2009.09.005] [PMID: 19944973]

[41] Jeanbart L, Swartz MA. Engineering opportunities in cancer immunotherapy. Proc Natl Acad Sci USA 2015; 112(47): 14467-72.
[http://dx.doi.org/10.1073/pnas.1508516112] [PMID: 26598681]

[42] Bregy A, Wong TM, Shah AH, Goldberg JM, Komotar RJ. Active immunotherapy using dendritic cells in the treatment of glioblastoma multiforme. Cancer Treat Rev 2013; 39(8): 891-907.
[http://dx.doi.org/10.1016/j.ctrv.2013.05.007] [PMID: 23790634]

[43] Mitchell DA, Batich KA, Gunn MD, *et al.* Tetanus toxoid and CCL3 improve dendritic cell vaccines in mice and glioblastoma patients. Nature 2015; 519(7543): 366-9.
[http://dx.doi.org/10.1038/nature14320] [PMID: 25762141]

[44] Weiner LM, Surana R, Wang S. Monoclonal antibodies: versatile platforms for cancer immunotherapy. Nat Rev Immunol 2010; 10(5): 317-27.
[http://dx.doi.org/10.1038/nri2744] [PMID: 20414205]

[45] Van den Bent M, Gan H, Lassman A, *et al.* OS07.4 Efficacy of a novel antibody-drug conjugate (ADC), ABT-414, as monotherapy in epidermal growth factor receptor (EGFR) amplified (EGFRamp), recurrent glioblastoma (rGBM). Neuro-oncol 2017; 19 (Suppl. 3): iii13-4.
[http://dx.doi.org/10.1093/neuonc/nox036.045]

[46] Reardon DA, Lassman AB, van den Bent M, Kumthekar P, Merrell R, Scott AM. Efficacy and safety results of ABT-414 in combination with radiation and temozolomide in newly diagnosed glioblastoma. Neuro Oncol 2016;now257: 1-11

[47] Sirachainan N, Pakakasama S, Anurathapan U, Hansasuta A, Boonkerd A, Dhanachai M. Hg-14outcomes of children with newly diagnosed high grade gliomas treated with nimotuzumab ® and irinotecan. Neuro Oncol 2016;18:iii50.4-iii50.

[48] Lyon JG, Mokarram N, Saxena T, Carroll SL, Bellamkonda RV. Engineering challenges for brain tumor immunotherapy. Adv Drug Deliv Rev 2017; 114: 19-32.
[http://dx.doi.org/10.1016/j.addr.2017.06.006] [PMID: 28625831]

[49] Han SJ, Zygourakis C, Lim M, Parsa AT. Immunotherapy for Glioma. Neurosurg Clin N Am 2012; 23(3): 357-70.
[http://dx.doi.org/10.1016/j.nec.2012.05.001] [PMID: 22748649]

[50] Majc B, Novak M, Kopitar-Jerala N, Jewett A, Breznik B. Immunotherapy of glioblastoma: Current strategies and challenges in tumor model development. Cells 2021; 10(2): 265.
[http://dx.doi.org/10.3390/cells10020265] [PMID: 33572835]

[51] Sampson JH, Maus MV, June CH. Immunotherapy for brain tumors. J Clin Oncol 2017; 35(21): 2450-6.
[http://dx.doi.org/10.1200/JCO.2017.72.8089] [PMID: 28640704]

[52] Pardridge WM. Drug transport across the blood-brain barrier. J Cereb Blood Flow Metab 2012; 32(11): 1959-72.
[http://dx.doi.org/10.1038/jcbfm.2012.126] [PMID: 22929442]

[53] Abbott NJ, Rönnbäck L, Hansson E. Astrocyte–endothelial interactions at the blood–brain barrier. Nat Rev Neurosci 2006; 7(1): 41-53.
[http://dx.doi.org/10.1038/nrn1824] [PMID: 16371949]

[54] Sarkaria JN, Hu LS, Parney IF, *et al.* Is the blood–brain barrier really disrupted in all glioblastomas? A critical assessment of existing clinical data. Neuro-oncol 2018; 20(2): 184-91.
[http://dx.doi.org/10.1093/neuonc/nox175] [PMID: 29016900]

[55] Sindhwani S, Syed AM, Ngai J, *et al.* The entry of nanoparticles into solid tumours. Nat Mater 2020; 19(5): 566-75.
[http://dx.doi.org/10.1038/s41563-019-0566-2] [PMID: 31932672]

[56] Jain RK, Stylianopoulos T. Delivering nanomedicine to solid tumors. Nat Rev Clin Oncol 2010; 7(11): 653-64.
[http://dx.doi.org/10.1038/nrclinonc.2010.139] [PMID: 20838415]

[57] Jackson C, Ruzevick J, Phallen J, Belcaid Z, Lim M. Challenges in immunotherapy presented by the glioblastoma multiforme microenvironment. Clin Dev Immunol 2011; 2011: 1-20.
[http://dx.doi.org/10.1155/2011/732413] [PMID: 22190972]

[58] Parney I, Petruk K, Hao C, Roa W, Turner J, Ramsay D. Cytokine and cytokine receptor mRNA expression in human glioblastomas: evidence of Th1, Th2 and Th3 cytokine dysregulation. Acta Neuropathol 2002; 103(2): 171-8.
[http://dx.doi.org/10.1007/s004010100448] [PMID: 11810184]

[59] Razavi SM, Lee KE, Jin BE, Aujla PS, Gholamin S, Li G. Immune evasion strategies of glioblastoma. Front Surg 2016; 3: 11.
[http://dx.doi.org/10.3389/fsurg.2016.00011] [PMID: 26973839]

[60] Chahlavi A, Rayman P, Richmond AL, *et al.* Glioblastomas induce T-lymphocyte death by two distinct pathways involving gangliosides and CD70. Cancer Res 2005; 65(12): 5428-38.
[http://dx.doi.org/10.1158/0008-5472.CAN-04-4395] [PMID: 15958592]

[61] Zhang L, Liu W, Alizadeh D, *et al.* S100B attenuates microglia activation in gliomas: Possible role of STAT3 pathway. Glia 2011; 59(3): 486-98.
[http://dx.doi.org/10.1002/glia.21118] [PMID: 21264954]

[62] Darnell JE, Kerr lan M, Stark GR. Jak-STAT pathways and transcriptional activation in response to IFNs and other extracellular signaling proteins. Science (1979) 1994; 264(5164): 1415–21.

[63] Sherry MM, Reeves A, Wu JK, Cochran BH. STAT3 is required for proliferation and maintenance of multipotency in glioblastoma stem cells. Stem Cells 2009; 27(10): 2383-92.
[http://dx.doi.org/10.1002/stem.185] [PMID: 19658181]

[64] Kang SH, Yu MO, Park KJ, Chi SG, Park DH, Chung YG. Activated STAT3 regulates hypoxia-induced angiogenesis and cell migration in human glioblastoma. Neurosurgery 2010; 67(5): 1386-95.

[http://dx.doi.org/10.1227/NEU.0b013e3181f1c0cd] [PMID: 20871442]

[65] Wei J, Wu A, Kong LY, *et al.* Hypoxia potentiates glioma-mediated immunosuppression. PLoS One 2011; 6(1): e16195.
[http://dx.doi.org/10.1371/journal.pone.0016195] [PMID: 21283755]

[66] Kortylewski M, Kujawski M, Wang T, *et al.* Inhibiting Stat3 signaling in the hematopoietic system elicits multicomponent antitumor immunity. Nat Med 2005; 11(12): 1314-21.
[http://dx.doi.org/10.1038/nm1325] [PMID: 16288283]

[67] Zhang L, Alizadeh D, Van Handel M, Kortylewski M, Yu H, Badie B. Stat3 inhibition activates tumor macrophages and abrogates glioma growth in mice. Glia 2009; 57(13): 1458-67.
[http://dx.doi.org/10.1002/glia.20863] [PMID: 19306372]

[68] Lin EY, Li JF, Gnatovskiy L, *et al.* Macrophages regulate the angiogenic switch in a mouse model of breast cancer. Cancer Res 2006; 66(23): 11238-46.
[http://dx.doi.org/10.1158/0008-5472.CAN-06-1278] [PMID: 17114237]

[69] Herrero Herrero JI, Ruiz Beltrán R, Sanchez AMM. Naturally acquired immunity: a means of resistance to Mediterranean spotted fever? J Infect Dis 1991; 164(3): 618-9.
[http://dx.doi.org/10.1093/infdis/164.3.618] [PMID: 1869855]

[70] Mittal SK, Roche PA. Suppression of antigen presentation by IL-10. Curr Opin Immunol 2015; 34: 22-7.
[http://dx.doi.org/10.1016/j.coi.2014.12.009] [PMID: 25597442]

[71] Parsa AT, Waldron JS, Panner A, *et al.* Loss of tumor suppressor PTEN function increases B7-H1 expression and immunoresistance in glioma. Nat Med 2007; 13(1): 84-8.
[http://dx.doi.org/10.1038/nm1517] [PMID: 17159987]

[72] Singh K, Hotchkiss KM, Patel KK, *et al.* Enhancing T cell chemotaxis and infiltration in glioblastoma. Cancers (Basel) 2021; 13(21): 5367.
[http://dx.doi.org/10.3390/cancers13215367] [PMID: 34771532]

[73] Daneman R, Prat A. The blood-brain barrier. Cold Spring Harb Perspect Biol 2015; 7(1): a020412.
[http://dx.doi.org/10.1101/cshperspect.a020412] [PMID: 25561720]

[74] Engelhardt B. Immune cell entry into the central nervous system: Involvement of adhesion molecules and chemokines. J Neurol Sci 2008; 274(1-2): 23-6.
[http://dx.doi.org/10.1016/j.jns.2008.05.019] [PMID: 18573502]

[75] Ott M, Prins RM, Heimberger AB. The immune landscape of common CNS malignancies: implications for immunotherapy. Nat Rev Clin Oncol 2021; 18(11): 729-44.
[http://dx.doi.org/10.1038/s41571-021-00518-9] [PMID: 34117475]

[76] Wei J, Barr J, Kong LY, *et al.* Glioblastoma cancer-initiating cells inhibit T-cell proliferation and effector responses by the signal transducers and activators of transcription 3 pathway. Mol Cancer Ther 2010; 9(1): 67-78.
[http://dx.doi.org/10.1158/1535-7163.MCT-09-0734] [PMID: 20053772]

[77] Muftuoglu Y, Pajonk F. Targeting glioma stem cells. Neurosurg Clin N Am 2021; 32(2): 283-9.
[http://dx.doi.org/10.1016/j.nec.2021.01.002] [PMID: 33781508]

[78] Galli R, Binda E, Orfanelli U, *et al.* Isolation and characterization of tumorigenic, stem-like neural precursors from human glioblastoma. Cancer Res 2004; 64(19): 7011-21.
[http://dx.doi.org/10.1158/0008-5472.CAN-04-1364] [PMID: 15466194]

[79] Wei J, Barr J, Kong LY, *et al.* Glioblastoma cancer-initiating cells inhibit T-cell proliferation and effector responses by the signal transducers and activators of transcription 3 pathway. Mol Cancer Ther 2010; 9(1): 67-78.
[http://dx.doi.org/10.1158/1535-7163.MCT-09-0734] [PMID: 20053772]

[80] Shnaper S, Desbaillets I, Brown DA, *et al.* Elevated levels of MIC-1/GDF15 in the cerebrospinal fluid

of patients are associated with glioblastoma and worse outcome. Int J Cancer 2009; 125(11): 2624-30.
[http://dx.doi.org/10.1002/ijc.24639] [PMID: 19521960]

[81]   Wu A, Wei J, Kong LY, *et al.* Glioma cancer stem cells induce immunosuppressive macrophages/microglia. Neuro-oncol 2010; 12(11): 1113-25.
[http://dx.doi.org/10.1093/neuonc/noq082] [PMID: 20667896]

[82]   Sharifzad F, Ghavami S, Verdi J, *et al.* Glioblastoma cancer stem cell biology: Potential theranostic targets. Drug Resist Updat 2019; 42: 35-45.
[http://dx.doi.org/10.1016/j.drup.2018.03.003] [PMID: 30877905]

[83]   Bao S, Wu Q, McLendon RE, *et al.* Glioma stem cells promote radioresistance by preferential activation of the DNA damage response. Nature 2006; 444(7120): 756-60.
[http://dx.doi.org/10.1038/nature05236] [PMID: 17051156]

[84]   Calabrese C, Poppleton H, Kocak M, *et al.* A perivascular niche for brain tumor stem cells. Cancer Cell 2007; 11(1): 69-82.
[http://dx.doi.org/10.1016/j.ccr.2006.11.020] [PMID: 17222791]

[85]   Jhaveri N, Chen TC, Hofman FM. Tumor vasculature and glioma stem cells: Contributions to glioma progression. Cancer Lett 2016; 380(2): 545-51.
[http://dx.doi.org/10.1016/j.canlet.2014.12.028] [PMID: 25527451]

[86]   Tang X, Zuo C, Fang P, *et al.* Targeting glioblastoma stem cells: A review on biomarkers, signal pathways and targeted therapy. Front Oncol 2021; 11: 701291.
[http://dx.doi.org/10.3389/fonc.2021.701291] [PMID: 34307170]

[87]   Leone RD, Powell JD. Metabolism of immune cells in cancer. Nat Rev Cancer 2020; 20(9): 516-31.
[http://dx.doi.org/10.1038/s41568-020-0273-y] [PMID: 32632251]

[88]   Nusblat LM, Tanna S, Roth CM. Gene silencing of HIF-2α disrupts glioblastoma stem cell phenotype. Cancer Drug Resist 2020; 3(2): 199-208.
[http://dx.doi.org/10.20517/cdr.2019.96] [PMID: 32566921]

[89]   Löfstedt T, Fredlund E, Holmquist-Mengelbier L, *et al.* Hypoxia inducible factor-2α in cancer. Cell Cycle 2007; 6(8): 919-26.
[http://dx.doi.org/10.4161/cc.6.8.4133] [PMID: 17404509]

[90]   Bergles DE, Roberts JDB, Somogyi P, Jahr CE. Glutamatergic synapses on oligodendrocyte precursor cells in the hippocampus. Nature 2000; 405(6783): 187-91.
[http://dx.doi.org/10.1038/35012083] [PMID: 10821275]

[91]   Venkatesh HS, Morishita W, Geraghty AC, *et al.* Electrical and synaptic integration of glioma into neural circuits. Nature 2019; 573(7775): 539-45.
[http://dx.doi.org/10.1038/s41586-019-1563-y] [PMID: 31534222]

[92]   Venkatesh HS, Morishita W, Geraghty AC, *et al.* Electrical and synaptic integration of glioma into neural circuits. Nature 2019; 573(7775): 539-45.
[http://dx.doi.org/10.1038/s41586-019-1563-y] [PMID: 31534222]

[93]   Buckingham SC, Campbell SL, Haas BR, *et al.* Glutamate release by primary brain tumors induces epileptic activity. Nat Med 2011; 17(10): 1269-74.
[http://dx.doi.org/10.1038/nm.2453] [PMID: 21909104]

[94]   Zhao R, Hirano M, Tanaka S, Sato K. Vocal fold epithelial hyperplasia. Vibratory behavior *vs* extent of lesion. Arch Otolaryngol Head Neck Surg 1991; 117(9): 1015-8.
[http://dx.doi.org/10.1001/archotol.1991.01870210087017] [PMID: 1910717]

[95]   Krishna S, Lee AT, Kakaizada S, Seo K, Raleigh DR, Almeida N. Glioma-Neuronal Interactions. Neurosurgery 2020;67(Supplemnet_1).
[http://dx.doi.org/10.1093/neuros/nyaa447_865]

[96]   Moore WS, Mohr JP, Najafi H, Robertson JT, Stoney RJ, Toole JF. Carotid endarterectomy: Practice

guidelines. Report of the Ad Hoc committee to the joint council of the society for vascular surgery and the north american chapter of the international society for cardiovascular surgery. J Vasc Surg 1992; 15(3): 469-79.
[http://dx.doi.org/10.1016/0741-5214(92)90185-B] [PMID: 1538503]

[97]   Hotchkiss KM, Sampson JH. Temozolomide treatment outcomes and immunotherapy efficacy in brain tumor. J Neurooncol 2021; 151(1): 55-62.
[http://dx.doi.org/10.1007/s11060-020-03598-2] [PMID: 32813186]

[98]   Suryadevara CM, Desai R, Abel ML, *et al.* Temozolomide lymphodepletion enhances CAR abundance and correlates with antitumor efficacy against established glioblastoma. OncoImmunology 2018; 7(6): e1434464.
[http://dx.doi.org/10.1080/2162402X.2018.1434464] [PMID: 29872570]

[99]   Karachi A, Yang C, Dastmalchi F, *et al.* Modulation of temozolomide dose differentially affects T-cell response to immune checkpoint inhibition. Neuro-oncol 2019; 21(6): 730-41.
[http://dx.doi.org/10.1093/neuonc/noz015] [PMID: 30668768]

[100]   Heynckes S, Daka K, Franco P, *et al.* Crosslink between Temozolomide and PD-L1 immune-checkpoint inhibition in glioblastoma multiforme. BMC Cancer 2019; 19(1): 117.
[http://dx.doi.org/10.1186/s12885-019-5308-y] [PMID: 30709339]

[101]   Kroschinsky F, Stölzel F, von Bonin S, *et al.* New drugs, new toxicities: severe side effects of modern targeted and immunotherapy of cancer and their management. Crit Care 2017; 21(1): 89.
[http://dx.doi.org/10.1186/s13054-017-1678-1] [PMID: 28407743]

[102]   Kwon ED, Drake CG, Scher HI, *et al.* Ipilimumab *versus* placebo after radiotherapy in patients with metastatic castration-resistant prostate cancer that had progressed after docetaxel chemotherapy (CA184-043): a multicentre, randomised, double-blind, phase 3 trial. Lancet Oncol 2014; 15(7): 700-12.
[http://dx.doi.org/10.1016/S1470-2045(14)70189-5] [PMID: 24831977]

[103]   Michot JM, Bigenwald C, Champiat S, *et al.* Immune-related adverse events with immune checkpoint blockade: a comprehensive review. Eur J Cancer 2016; 54: 139-48.
[http://dx.doi.org/10.1016/j.ejca.2015.11.016] [PMID: 26765102]

[104]   Horvat TZ, Adel NG, Dang TO, *et al.* Immune-related adverse events, need for systemic immunosuppression, and effects on survival and time to treatment failure in patients with melanoma treated with ipilimumab at memorial sloan kettering cancer center. J Clin Oncol 2015; 33(28): 3193-8.
[http://dx.doi.org/10.1200/JCO.2015.60.8448] [PMID: 26282644]

[105]   Topp MS, Gökbuget N, Stein AS, *et al.* Safety and activity of blinatumomab for adult patients with relapsed or refractory B-precursor acute lymphoblastic leukaemia: a multicentre, single-arm, phase 2 study. Lancet Oncol 2015; 16(1): 57-66.
[http://dx.doi.org/10.1016/S1470-2045(14)71170-2] [PMID: 25524800]

[106]   Lee DW, Gardner R, Porter DL, *et al.* Current concepts in the diagnosis and management of cytokine release syndrome. Blood 2014; 124(2): 188-95.
[http://dx.doi.org/10.1182/blood-2014-05-552729] [PMID: 24876563]

[107]   Cukier P, Santini FC, Scaranti M, Hoff AO. Endocrine side effects of cancer immunotherapy. Endocr Relat Cancer 2017; 24(12): T331-47.
[http://dx.doi.org/10.1530/ERC-17-0358] [PMID: 29025857]

[108]   Ranjan S, Quezado M, Garren N, *et al.* Clinical decision making in the era of immunotherapy for high grade-glioma: report of four cases. BMC Cancer 2018; 18(1): 239.
[http://dx.doi.org/10.1186/s12885-018-4131-1] [PMID: 29490632]

[109]   Colaco RJ, Martin P, Kluger HM, Yu JB, Chiang VL. Does immunotherapy increase the rate of radiation necrosis after radiosurgical treatment of brain metastases? J Neurosurg 2016; 125(1): 17-23.
[http://dx.doi.org/10.3171/2015.6.JNS142763] [PMID: 26544782]

[110] Martin AM, Cagney DN, Catalano PJ, *et al.* Immunotherapy and symptomatic radiation necrosis in patients with brain metastases treated with stereotactic radiation. JAMA Oncol 2018; 4(8): 1123-4.
[http://dx.doi.org/10.1001/jamaoncol.2017.3993] [PMID: 29327059]

[111] Wang JH, Wahid KA, van Dijk LV, Farahani K, Thompson RF, Fuller CD. Radiomic biomarkers of tumor immune biology and immunotherapy response. Clin Transl Radiat Oncol 2021; 28: 97-115.
[http://dx.doi.org/10.1016/j.ctro.2021.03.006] [PMID: 33937530]

[112] Xie T, Chen X, Fang J, *et al.* Non-invasive monitoring of the kinetic infiltration and therapeutic efficacy of nanoparticle-labeled chimeric antigen receptor T cells in glioblastoma *via* 7.0-Tesla magnetic resonance imaging. Cytotherapy 2021; 23(3): 211-22.
[http://dx.doi.org/10.1016/j.jcyt.2020.10.006] [PMID: 33334686]

[113] Maxwell R, Jackson CM, Lim M. Clinical trials investigating immune checkpoint blockade in glioblastoma. Curr Treat Options Oncol 2017; 18(8): 51.
[http://dx.doi.org/10.1007/s11864-017-0492-y] [PMID: 28785997]

[114] Rong L, Li N, Zhang Z. Emerging therapies for glioblastoma: current state and future directions. J Exp Clin Cancer Res 2022; 41(1): 142.
[http://dx.doi.org/10.1186/s13046-022-02349-7] [PMID: 35428347]

[115] McGranahan T, Therkelsen KE, Ahmad S, Nagpal S. Current state of immunotherapy for treatment of glioblastoma. Curr Treat Options Oncol 2019; 20(3): 24.
[http://dx.doi.org/10.1007/s11864-019-0619-4] [PMID: 30790064]

[116] McCracken MN, Cha AC, Weissman IL. Molecular pathways: Activating T cells after cancer ccll phagocytosis from blockade of CD47 "Don't Eat Me" signals. Clin Cancer Res 2015; 21(16): 3597-601.
[http://dx.doi.org/10.1158/1078-0432.CCR-14-2520] [PMID: 26116271]

[117] Casey SC, Tong L, Li Y, Do R, Walz S, Fitzgerald KN. MYC regulates the antitumor immune response through CD47 and PD-L1. Science (1979) 2016;352(6282):227–31.

[118] Vijayan D, Young A, Teng MWL, Smyth MJ. Erratum: Targeting immunosuppressive adenosine in cancer. Nat Rev Cancer 2017; 17(12): 765-5.
[http://dx.doi.org/10.1038/nrc.2017.110] [PMID: 29162946]

[119] Tang L, Zhang M, Liu C. Advances in nanotechnology-based immunotherapy for glioblastoma. Front Immunol 2022; 13: 882257.
[http://dx.doi.org/10.3389/fimmu.2022.882257] [PMID: 35651605]

[120] Zhao M, van Straten D, Broekman MLD, Préat V, Schiffelers RM. Nanocarrier-based drug combination therapy for glioblastoma. Theranostics 2020; 10(3): 1355-72.
[http://dx.doi.org/10.7150/thno.38147] [PMID: 31938069]

[121] Li Y, Teng X, Wang Y, Yang C, Yan X, Li J. Neutrophil delivered hollow titania covered persistent luminescent nanosensitizer for ultrosound augmented chemo/immuno glioblastoma therapy. Adv Sci (Weinh) 2021; 8(17): 2004381.
[http://dx.doi.org/10.1002/advs.202004381] [PMID: 34196474]

[122] Wang J, Huang CH, Echeagaray OH, *et al.* Microshell enhanced acoustic adjuvants for immunotherapy in glioblastoma. Adv Ther (Weinh) 2019; 2(10): 1900066.
[http://dx.doi.org/10.1002/adtp.201900066]

[123] Kadiyala P, Li D, Nuñez FM, *et al.* High-density lipoprotein-mimicking nanodiscs for chemo-immunotherapy against glioblastoma multiforme. ACS Nano 2019; 13(2): acsnano.8b06842.
[http://dx.doi.org/10.1021/acsnano.8b06842] [PMID: 30721028]

[124] Zhou J, Wang G, Chen Y, Wang H, Hua Y, Cai Z. Immunogenic cell death in cancer therapy: Present and emerging inducers. J Cell Mol Med 2019; 23(8): 4854-65.
[http://dx.doi.org/10.1111/jcmm.14356] [PMID: 31210425]

[125] Fucikova J, Kepp O, Kasikova L, *et al.* Detection of immunogenic cell death and its relevance for cancer therapy. Cell Death Dis 2020; 11(11): 1013.
[http://dx.doi.org/10.1038/s41419-020-03221-2] [PMID: 33243969]

[126] Li X. The inducers of immunogenic cell death for tumor immunotherapy. Tumori 2018; 104(1): 1-8.
[http://dx.doi.org/10.5301/tj.5000675] [PMID: 28967094]

[127] Zhang M, Wang W, Mohammadniaei M, *et al.* Upregulating aggregation-induced-emission nanoparticles with blood–tumor-barrier permeability for precise photothermal eradication of brain tumors and induction of local immune responses. Adv Mater 2021; 33(22): 2008802.
[http://dx.doi.org/10.1002/adma.202008802] [PMID: 33893670]

[128] Tang L, Zhang M, Liu C. Advances in nanotechnology-based immunotherapy for glioblastoma. Front Immunol 2022; 13: 882257.
[http://dx.doi.org/10.3389/fimmu.2022.882257] [PMID: 35651605]

[129] Segura-Collar B, Hiller-Vallina S, de Dios O, *et al.* Advanced immunotherapies for glioblastoma: tumor neoantigen vaccines in combination with immunomodulators. Acta Neuropathol Commun 2023; 11(1): 79.
[http://dx.doi.org/10.1186/s40478-023-01569-y] [PMID: 37165457]

[130] Pombo Antunes AR, Scheyltjens I, Duerinck J, Neyns B, Movahedi K, Van Ginderachter JA. Understanding the glioblastoma immune microenvironment as basis for the development of new immunotherapeutic strategies. eLife 2020; 9: e52176.
[http://dx.doi.org/10.7554/eLife.52176] [PMID: 32014107]

[131] Hegde PS, Karanikas V, Evers S. The where, the when, and the how of immune monitoring for cancer immunotherapies in the era of checkpoint inhibition. Clin Cancer Res 2016; 22(8): 1865-74.
[http://dx.doi.org/10.1158/1078-0432.CCR-15-1507] [PMID: 27084740]

[132] Thorsson V, Gibbs DL, Brown SD, *et al.* The immune landscape of cancer. Immunity 2018; 48(4): 812-830.e14.
[http://dx.doi.org/10.1016/j.immuni.2018.03.023] [PMID: 29628290]

[133] Agosti E, Zeppieri M, De Maria L, *et al.* Glioblastoma immunotherapy: A systematic review of the present strategies and prospects for advancements. Int J Mol Sci 2023; 24(20): 15037.
[http://dx.doi.org/10.3390/ijms242015037] [PMID: 37894718]

[134] Bregy A, Wong TM, Shah AH, Goldberg JM, Komotar RJ. Active immunotherapy using dendritic cells in the treatment of glioblastoma multiforme. Cancer Treat Rev 2013; 39(8): 891-907.
[http://dx.doi.org/10.1016/j.ctrv.2013.05.007] [PMID: 23790634]

[135] Pimplikar SW, Simons K. Role of heterotrimeric G proteins in polarized membrane transport. J Cell Sci 1993; 1993 (Suppl. 17): 27-32.
[http://dx.doi.org/10.1242/jcs.1993.Supplement_17.5] [PMID: 8144702]

[136] Wang X, Zhao HY, Zhang FC, Sun Y, Xiong ZY, Jiang XB. Dendritic cell-based vaccine for the treatment of malignant glioma: a systematic review. Cancer Invest 2014; 32(9): 451-7.
[http://dx.doi.org/10.3109/07357907.2014.958234] [PMID: 25259676]

[137] Vitanza NA, Johnson AJ, Wilson AL, *et al.* Locoregional infusion of HER2-specific CAR T cells in children and young adults with recurrent or refractory CNS tumors: an interim analysis. Nat Med 2021; 27(9): 1544-52.
[http://dx.doi.org/10.1038/s41591-021-01404-8] [PMID: 34253928]

[138] Benitez-Ribas D, Cabezón R, Flórez-Grau G, *et al.* Immune response generated with the administration of autologous dendritic cells pulsed with an allogenic tumoral cell-lines lysate in patients with newly diagnosed diffuse intrinsic pontine glioma. Front Oncol 2018; 8: 127.
[http://dx.doi.org/10.3389/fonc.2018.00127]

[139] Pollack IF, Jakacki RI, Butterfield LH, *et al.* Antigen-specific immunoreactivity and clinical outcome following vaccination with glioma-associated antigen peptides in children with recurrent high-grade

gliomas: results of a pilot study. J Neurooncol 2016; 130(3): 517-27.
[http://dx.doi.org/10.1007/s11060-016-2245-3] [PMID: 27624914]

[140] Pollack IF, Jakacki RI, Butterfield LH, *et al.* Immune responses and outcome after vaccination with glioma-associated antigen peptides and poly-ICLC in a pilot study for pediatric recurrent low-grade gliomas. Neuro-oncol 2016; 18(8): 1157-68.
[http://dx.doi.org/10.1093/neuonc/now026] [PMID: 26984745]

[141] Friedman GK, Johnston JM, Bag AK, *et al.* Oncolytic HSV-1 G207 immunovirotherapy for pediatric high-grade gliomas. N Engl J Med 2021; 384(17): 1613-22.
[http://dx.doi.org/10.1056/NEJMoa2024947] [PMID: 33838625]

[142] Kieran MW, Goumnerova L, Manley P, *et al.* Phase I study of gene-mediated cytotoxic immunotherapy with AdV-tk as adjuvant to surgery and radiation for pediatric malignant glioma and recurrent ependymoma. Neuro-oncol 2019; 21(4): 537-46.
[http://dx.doi.org/10.1093/neuonc/noy202] [PMID: 30883662]

[143] Fangusaro J, Mitchell DA, Kocak M, *et al.* Phase 1 study of pomalidomide in children with recurrent, refractory, and progressive central nervous system tumors: A Pediatric Brain Tumor Consortium trial. Pediatr Blood Cancer 2021; 68(2): e28756.
[http://dx.doi.org/10.1002/pbc.28756] [PMID: 33025730]

[144] Fangusaro J, Cefalo MG, Garré ML, *et al.* Phase 2 study of pomalidomide (CC-4047) monotherapy for children and young adults with recurrent or progressive primary brain tumors. Front Oncol 2021; 11: 660892.
[http://dx.doi.org/10.3389/fonc.2021.660892] [PMID: 34168987]

[145] Khatua S, Cooper LJN, Sandberg DI, *et al.* Phase I study of intraventricular infusions of autologous *ex vivo* expanded NK cells in children with recurrent medulloblastoma and ependymoma. Neuro-oncol 2020; 22(8): 1214-25.
[http://dx.doi.org/10.1093/neuonc/noaa047] [PMID: 32152626]

[146] Van Gool S, Makalowski J, Bonner E, *et al.* Addition of multimodal immunotherapy to combination treatment strategies for children with DIPG: A single institution experience. Medicines (Basel) 2020; 7(5): 29.
[http://dx.doi.org/10.3390/medicines7050029] [PMID: 32438648]

[147] Lynes JP, Nwankwo AK, Sur HP, *et al.* Biomarkers for immunotherapy for treatment of glioblastoma. J Immunother Cancer 2020; 8(1): e000348.
[http://dx.doi.org/10.1136/jitc-2019-000348] [PMID: 32474411]

[148] Brown CE, Bucktrout S, Butterfield LH, *et al.* The future of cancer immunotherapy for brain tumors: a collaborative workshop. J Transl Med 2022; 20(1): 236.
[http://dx.doi.org/10.1186/s12967-022-03438-z] [PMID: 35606815]

[149] Rong L, Li N, Zhang Z. Emerging therapies for glioblastoma: current state and future directions. J Exp Clin Cancer Res 2022; 41(1): 142.
[http://dx.doi.org/10.1186/s13046-022-02349-7] [PMID: 35428347]

CHAPTER 8

Pharmacological Modulation of Brain Tumors: Therapeutic Opportunities and Persistent Challenges

Thippeswamy Mallamma[1,*], **Nagaraj Sreeharsha**[2] and **Prakash Goudanavar**[1]

[1] *Department of Pharmaceutics, Sri Adichucnhanagiri College of Pharmacy, Adichunchanagiri University-571448, Karnataka, India*

[2] *Department of Pharmaceutics, Vidya Siri College of Pharmacy, Bengaluru-560 035, Karnataka, India*

Abstract: Brain tumours are an aggressive and rapidly progressing class of cancers, whose complexity limits effective treatment options. This chapter examines the complexities and hurdles associated with employing pharmacological modulations as a therapeutic approach. Pharmacological modulations are an emerging requirement in tumour management, which calls for improved and innovative strategies. Pharmacological modulations promise to alter the biological environment of tumors, sensitizing, potentiating, and overcoming drug resistance. Sensitisation to increase tumour vulnerability to drugs, potentiation to increase drug efficacy, and overcoming resistance by targeting pathways that stem the tumour proliferation are the main approaches in pharmacological modulation. Targeted therapies, like tyrosine kinase inhibitors, also play a crucial role. This chapter provides a concise overview of primary and metastatic tumors, highlighting the molecular and cellular interactions that influence drug response and the hindrance posed by the heterogeneous barrier, the blood-brain barrier (BBB). Various therapeutic approaches have been discussed, including small-molecule inhibitors, monoclonal antibodies, and innovative ones such as RNA-targeting therapies and nano-oncology. Case studies have been cited to prove that modulation strategies successfully overcome biological hurdles. Prevailing challenges, which almost seem unbeatable, such as the BBB, are discussed in detail, along with approaches to overcome them and enhance drug delivery, including nanoparticle formulations and combination therapies. Pharmacological modulations proved promising results in treating brain tumours despite challenges like the BBB. Continued research and the development of innovative approaches are essential for further progress in brain tumor treatment. Personalized therapies and improved drug delivery systems offer hope for more effective treatment options in the future.

* **Corresponding author Thippeswamy Mallamma:** Department of Pharmaceutics, Sri Adichucnhanagiri College of Pharmacy, Adichunchanagiri University-571448, Karnataka, India; E-mail: mallammareddy89@gmail.com

Prashant Tiwari, Pankaj Kumar Singh & Sunil Kumar Kadiri (Eds.)
All rights reserved-© 2025 Bentham Science Publishers

Keywords: Blood-brain barrier (BBB), Brain tumour, Biomarkers, Monoclonal antibodies and antibody-drug conjugates, Nano-oncology, Pharmacological modulation, PD-1/PD-L1 pathways, RNA-targeting therapies, Tyrosine kinase inhibitors.

INTRODUCTION

Brain tumors are a diverse, complex group of neoplasms with a broad spectrum of over 100 different types that affect the central nervous system [1]. Characterized by their aggressive nature, high-grade gliomas exhibit rapid growth and infiltrative tendencies, posing significant challenges [2]. Despite advancements in surgical techniques and adjuvant therapies, the prognosis for brain tumour patients remains poor, illuminating the pressing demand for a more efficient treatment approach.

A limited understanding of the pathobiology of these neoplasms remains a major challenge in brain tumor research. Although advancements in imaging have enhanced diagnostic accuracy, the development of targeted therapies continues to be hindered by the heterogeneous nature of brain tumours and the protective function of the blood-brain barrier [2, 3]. Furthermore, the tumour microenvironment significantly influences disease progression and treatment resistance, involving a complex interplay between tumour cells and surrounding neural tissues that affects therapeutic outcomes [4].

Treatment strategies are typically determined by tumor type, grade, and location; however, available options remain limited. Surgery is the primary modality; however, complete resection is often unachievable due to the infiltrative nature of most brain tumours. Adjuvant therapies, such as radiation and chemotherapy, are commonly employed; however, their effectiveness is limited by the blood-brain barrier and the development of therapeutic resistance [5].

Current advancements in nanoparticle-mediated drug delivery and immunotherapy are promising, but they face numerous hurdles and are still in their early stages of development. Recent advances in nanoparticle-mediated drug delivery and immunotherapy hold promise, but these approaches are still in the early stages of development and face numerous challenges [6, 7].

Brain tumours are a complex, aggressive, and diverse spectrum of neoplasms that pose significant hurdles due to their infiltrative nature, the blood-brain barrier, and the limited understanding of the brain tumour microenvironment. Although advancements in imaging and the development of novel therapeutics are promising, there is a pressing need for research to improve outcomes for affected patients. An overview of brain tumor classifications is depicted in Fig. (**1**).

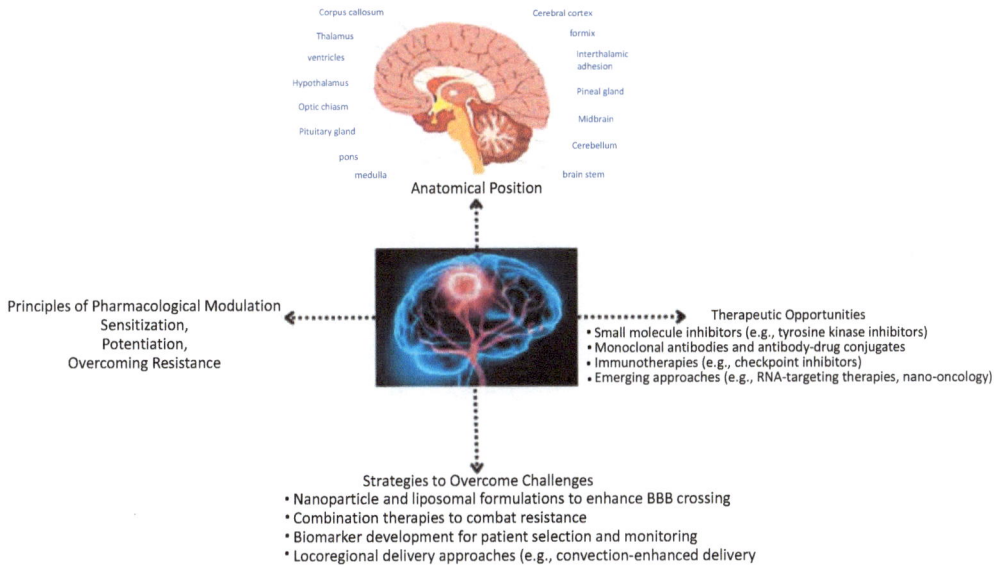

Fig. (1). Summary of brain tumour.

This literature review aims to process the current state of knowledge on pharmacological approaches to brain tumour treatment, highlighting key themes, methodologies, findings, and gaps in the existing research to guide further research in filling those gaps.

Concept of Pharmacological Modulation as a Promising Therapeutic Approach

Pharmacological modulation offers a therapeutic approach for a range of diseases. This strategy involves using drugs to modulate specific biological pathways or processes, thereby restoring normal functioning and improving patient outcomes.

Pharmacological modulation of Fibrinogen levels has been proposed as a therapeutic strategy for cardiovascular disease. Fibrinogen is a protein characterised by its crucial role in blood clotting, whose elevated levels pose an increased cardiovascular risk. A new advancement of oral or long-lasting parental therapies to lower fibrinogen could act as a risk reducer [8]. Autophagy, the process by which cells restore damaged or dysfunctional components, has emerged as a target for pharmacological modulation. Both pharmacological and nutritional autophagy modulators showcase their therapeutic potential in preclinical studies. Hurdles remain in the clinical translation of these findings, including the lack of understanding of how to optimally modulate autophagy in the context of various diseases [9].

An innovative therapeutic approach involving redox regulation systems in the brain has been proposed in the context of psychiatric disorders. These conditions are increasingly associated with oxidative stress and disruptions in redox homeostasis, suggesting that modulation of redox systems may offer significant therapeutic benefits [10].

Allosteric modulator agents that bind to sites on a protein distinct from its active site represent a novel pharmacological strategy. These modulators offer enhanced selectivity by targeting specific receptor isoforms or conformational states, thereby increasing therapeutic efficacy while minimizing adverse effects compared to traditional orthosteric ligands. For instance, the allosteric modulation of neural receptors presents a promising avenue for treating neuropsychiatric disorders [11]. Genetic modulation has also emerged as a critical component in pharmacological approaches to pain management. Genetic variability in drug targets and metabolizing enzymes contributes to the diverse responses observed among patients receiving analgesics. Incorporating personalized genetic profiles into treatment planning could enable more precise and effective pharmacological interventions for pain therapy.

Overall, pharmacological modulation provides a versatile therapeutic strategy for a wide range of diseases. By targeting specific biological pathways and processes, drugs can potentially restore normal function and improve clinical outcomes. Despite existing challenges in translating these findings into clinical practice, ongoing advancements in human biology and disease pathophysiology provide a robust foundation for the development of more refined and effective modulatory therapies.

Challenges Persist in Effectively Treating Brain Tumours

While pharmacological modulation presents promising therapeutic opportunities, several challenges persist in the effective treatment of brain tumours. One of the most significant barriers is the blood-brain barrier (BBB), which tightly regulates the passage of molecules into the brain, limiting drug delivery. Strategies to modulate the BBB have been explored to enhance drug penetration into tumour sites. For instance, certain compounds have demonstrated the ability to alter markers on brain tumor capillaries, enabling the selective delivery of drugs of varying sizes to tumor tissues [12]. Additionally, pharmacological modulation has been shown to increase the permeability of chemotherapeutic agents, such as doxorubicin, in rat models, indicating potential for treating refractory brain tumors [13].

Beyond drug delivery, tumour heterogeneity poses a substantial challenge to pharmacological intervention. Glioblastomas, the most common and aggressive primary brain tumours, exhibit extensive genetic, cellular, and microenvironmental heterogeneity. This diversity contributes to variable therapeutic responses and the emergence of resistance. Genetic heterogeneity arises from subclonal populations with distinct mutations, such as EGFR amplification, PTEN loss, and other oncogenic alterations, which influence sensitivity to targeted therapies. Single-cell sequencing studies have revealed that different regions within the same tumour may activate distinct oncogenic pathways, complicating the implementation of uniform treatment strategies. Moreover, glioblastomas comprise a mixture of differentiated tumor cells and glioma stem-like cells (GSCs), the latter possessing a high self-renewal capacity and resistance to conventional therapies, such as temozolomide and radiotherapy. GSCs contribute to tumour recurrence and poor prognosis by repopulating the tumour following initial treatment.

The tumour microenvironment further exacerbates therapy resistance. Tumour cells interact with immune cells, endothelial cells, and extracellular matrix components to form an immunosuppressive niche. Regulatory T cells (Tregs), tumor-associated macrophages (TAMs), and microglia contribute to immune evasion, thereby diminishing the efficacy of immune checkpoint inhibitors, such as PD-1 and CTLA-4. To counteract this, combination therapies such as immune checkpoint inhibitors with radiotherapy or chemotherapy are being evaluated to enhance response rates. Additionally, strategies targeting the tumour microenvironment aim to reshape immune dynamics and improve drug penetrance.

These challenges underscore the necessity of a multifaceted approach to overcome tumour heterogeneity. Molecular profiling of tumours enables the design of individualized therapies based on specific genetic mutations. Combination therapies, such as dual inhibition of PI3K/mTOR and receptor tyrosine kinases (RTKs), are under investigation to simultaneously target multiple resistance mechanisms. Furthermore, novel drug delivery systems, including nanoparticle-based platforms, have shown promise in bypassing the BBB and achieving more uniform drug distribution across heterogeneous tumour regions. Despite these advances, long-term clinical success remains elusive, highlighting the need for continued innovation and research [14].

Repurposing existing drugs has emerged as a viable strategy for pharmacological modulation of brain tumours. For example, the antiepileptic drug valproate has demonstrated the ability to modulate the tumour microenvironment and inhibit glioma invasion, suggesting its potential as a therapeutic agent [15]. Similarly, the mood stabiliser lithium has shown anti-tumour effects in preclinical models of brain cancer [16].

Neuroinflammation is another critical component of the brain tumour microenvironment targeted for pharmacological intervention. Chronic inflammation contributes to tumour development and progression, and anti-inflammatory agents have shown promise in preclinical studies. However, further investigation is required to elucidate the complex interplay between inflammation and tumour biology and to identify optimal therapeutic strategies [17]. In summary, while pharmacological modulation holds significant promise for treating brain tumors, several key challenges must be addressed. These include enhancing drug delivery across the BBB, targeting heterogeneous tumour populations, and modulating the tumour microenvironment. Approaches such as drug repurposing and targeting neuroinflammation offer additional therapeutic avenues. Continued research is essential to overcome these barriers and fully realize the potential of pharmacological modulation in brain tumour therapy.

BIOLOGY OF BRAIN TUMORS

Brain tumours are broadly classified into two categories: primary and metastatic. Primary brain tumours originate within the brain or from closely associated structures such as the meninges, pituitary gland, or pineal gland. These tumours result from abnormal cell proliferation, forming masses that can compress and damage surrounding brain tissues [18, 19]. Among primary brain tumours, glioblastoma multiforme is the most common and aggressive type in adults, often associated with poor prognosis. Other primary tumours include astrocytomas, oligodendrogliomas, ependymomas, and medulloblastomas, each exhibiting distinct histopathological features and clinical behaviours [18]. Fig. (2) provides an overview of the biological processes involved in brain tumour development.

In contrast, metastatic brain tumours arise from the spread of cancer cells from other parts of the body to the brain. These are significantly more prevalent, occurring approximately ten times more frequently than primary brain tumours. Common primary cancers that metastasise to the brain include those of the lung, breast, skin (melanoma), kidney, and colon [19, 20]. Metastatic brain tumours may present as single or multiple lesions and can occur in various regions of the brain. Clinical manifestations depend on the size and location of the lesions and typically include headaches, seizures, cognitive disturbances, and focal neurological deficits such as weakness or numbness [21, 22].

The most common primary brain tumors

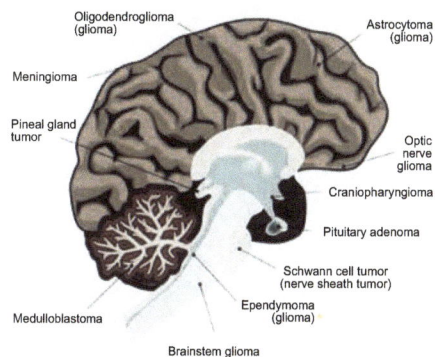

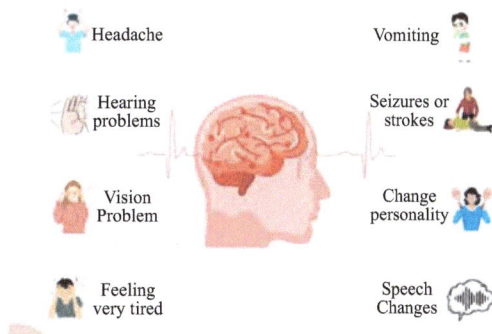

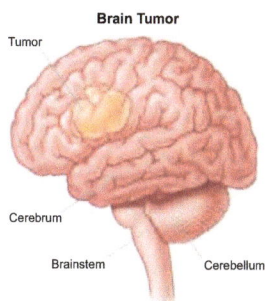

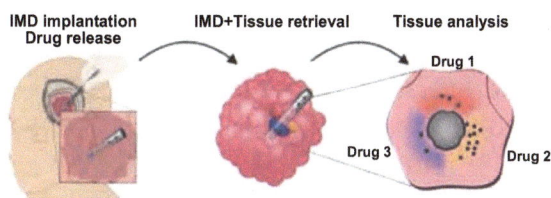

Brain Cancer

Fig. (2). Biology of brain tumour.

Management of both primary and metastatic brain tumours requires a multidisciplinary approach, incorporating surgery, radiation therapy, chemotherapy, and supportive care. Treatment strategies are tailored based on tumour type, location, the extent of disease, and the patient's overall health and functional status [18, 23]. In summary, primary brain tumours originate within the brain and exhibit diverse histopathological and clinical characteristics. Metastatic brain tumors, which are more common, arise from the spread of systemic malignancies to the brain. Effective management of both primary and metastatic tumors requires a comprehensive and individualized treatment plan developed through multidisciplinary collaboration.

Key Molecular and Cellular Characteristics Influencing Drug Response

A complex interplay of molecular and cellular factors governs the response of brain tumours to drug therapy. At the molecular level, the expression of specific genes and proteins significantly influences the efficacy of chemotherapy and other pharmacological agents. For example, the enzyme glutathione S-transferase P1

(GSTP1) has been implicated in drug resistance in malignant gliomas. Additionally, molecular markers involved in DNA repair and apoptosis pathways may modulate therapeutic response [24]. Pharmacogenomic studies have further elucidated the role of inherited genetic traits in drug response. Variations in genes encoding drug-metabolizing enzymes, transporters, and target proteins can affect both the efficacy and toxicity of cancer therapies. Notably, polymorphisms in the MDR1 gene, which encodes the P-glycoprotein efflux pump, have been associated with resistance to chemotherapy [25]. Beyond molecular factors, cellular interactions within the tumour microenvironment play a critical role in shaping drug response. The brain tumour microenvironment comprises diverse cell types, including glioma stem cells, endothelial cells, pericytes, immune cells, and components of the extracellular matrix (Faisal *et al.*, 2022) [26]. These cellular constituents participate in intricate signaling networks that influence tumor growth, angiogenesis, invasion, and resistance to therapy [26, 27].

Glioma stem cells (GSCs) are particularly notable for their contribution to chemoresistance and tumour recurrence in glioblastoma. GSCs possess enhanced self-renewal capacity and can generate heterogeneous progeny with increased resistance to chemotherapy and radiation. Targeting GSCs and their specialized microenvironmental niches holds promise for improving therapeutic outcomes [28]. Drug delivery to brain tumours is further complicated by the presence of the blood-brain barrier (BBB) and the blood-tumour barrier (BTB). These physiological barriers restrict the penetration of systemically administered drugs, thereby reducing therapeutic efficacy [29, 30]. Emerging technologies such as nanoparticle-based drug formulations and focused ultrasound are being actively investigated to enhance drug permeability across the BBB and BTB, offering new avenues for effective treatment [30].

The Blood-Brain Barrier (BBB) and its Impact on Drug Delivery

The blood-brain barrier (BBB) is a highly specialized and protective structure that safeguards the central nervous system (CNS) from harmful neurotoxins while permitting the passage of essential nutrients. It constitutes a neurovascular unit composed of tightly joined endothelial cells, pericytes, and astrocytes. This unique architecture renders the BBB nearly impermeable to most molecules and pharmacological agents, posing significant challenges in the treatment of CNS disorders, including brain tumours [31 - 33]. Several systemic mechanisms within the BBB actively oppose drug delivery: Tight junctions between endothelial cells restrict paracellular transport. The lipophilic nature of the BBB limits the diffusion of hydrophilic compounds. Efflux transporters, such as P-glycoprotein and breast cancer resistance protein (BCRP), actively expel drugs back into the bloodstream, thereby reducing their effectiveness. Enzymatic activity within the BBB metabolizes drugs, reducing their bioavailability in the brain [31, 32].

In the context of brain tumours, the BBB becomes more complex and heterogeneous. The tumour microenvironment can alter BBB integrity, leading to localized increases in permeability. However, this permeability is often confined to specific regions, resulting in non-uniform drug distribution and inadequate exposure in certain tumour areas. Moreover, cancer stem cells may remain protected behind an intact BBB, contributing to treatment resistance [34, 35]. To address these challenges, several enhanced drug delivery strategies have been developed, including molecular modification of drugs.

Altering lipophilicity, charge, and molecular weight to improve passive diffusion. Designing prodrugs that become pharmacologically active within the brain, thereby increasing bioavailability [31, 36]. Receptor-mediated transcytosis: Drugs conjugated to peptides or antibodies that bind to BBB receptors (*e.g.*, transferrin or insulin receptors) can be transported across the BBB into the brain. Adsorptive-mediated transcytosis: Cationic drugs and nanoparticles utilize electrostatic interactions for uptake across the BBB [31, 36]. Surface modification with ligands (*e.g.*, polysorbate 80 or apolipoprotein E) targets receptors such as the low-density lipoprotein receptor, facilitating receptor-mediated uptake. The enhanced permeability and retention (EPR) effect allows nanoparticles to accumulate in tumor regions with leaky vasculature [37 - 40]. Physiological and physical disruption techniques have also been explored to facilitate drug delivery by transiently disrupting the blood-brain barrier (BBB). For instance, hyperosmotic agents, such as mannitol, can temporarily open tight junctions, thereby enhancing drug permeability.

Focused ultrasound and bradykinin can reversibly increase BBB permeability. These methods, often combined with intra-arterial chemotherapy, have shown promise in treating brain tumours and other CNS disorders [31, 41].

PRINCIPLES OF PHARMACOLOGICAL MODULATION

Pharmacological modulation refers to the strategic use of drugs or pharmacological agents to alter the behaviour of brain tumours and improve therapeutic outcomes. This approach involves targeting molecular pathways that support tumour growth and survival, modulating the tumour microenvironment, and enhancing the body's immune response to malignancies. The primary objectives of pharmacological modulation are to inhibit cancer cell proliferation and viability, induce apoptosis, and prevent tumor invasion and metastasis [14, 42, 43]. Several pharmacological agents have been investigated for their potential to modulate brain tumors. Repurposed therapeutics, such as antiepileptic drugs, have demonstrated antiproliferative and proapoptotic effects on glioblastoma cells. These effects are believed to stem from the inhibition of histone deacetylases, enzymes that regulate gene expression and contribute to tumour progression. Lithium, traditionally used as a mood stabilizer, has shown potential in inhibiting glycogen synthase kinase-3 beta (GSK-3β), a kinase implicated in the survival and proliferation of cancer cells [44, 45]. Advanced Drug Delivery Systems Nanoparticles loaded with therapeutic agents are being explored for targeted delivery to tumour sites. These systems aim to enhance drug penetration across the blood-brain barrier (BBB) and concentrate antitumour effects within the tumour microenvironment. Surface modifications and ligand conjugation strategies are being employed to improve specificity and uptake, thereby increasing therapeutic efficacy [39]. Checkpoint inhibitors, such as PD-1 and CTLA-4 antagonists, are under investigation for their ability to stimulate the immune system's response to brain tumours. These agents aim to overcome the immunosuppressive tumour microenvironment and promote tumour clearance [46]. Pharmacological modulation represents a rapidly evolving frontier in brain tumour therapy. By repurposing existing drugs, developing advanced delivery systems, and integrating immunotherapeutic strategies, researchers are working to overcome the formidable challenges posed by brain tumours. Continued investigation is essential to validate these approaches and translate them into effective clinical interventions, especially given the persistent complexity and poor prognosis associated with many brain tumour types.

Modulation Effect on Sensitization, Potentiation, and Overcoming Resistance in Brain Tumour Treatment

The blood-brain barrier, inherent and acquired drug resistance, and the need to protect healthy brain tissue pose a significant hurdle in treating brain tumors. Pharmacological modulation strategies hold promise for enhancing the efficacy of existing therapies and improving patient outcomes. Three key types of modulation —sensitization, potentiation, and overcoming resistance —have been discussed in this chapter.

Sensitisation

Enhancing the sensitivity of brain tumour cells to cytotoxic agents and radiation therapy is a critical strategy in improving treatment efficacy. Several pharmacological agents have been investigated for their potential to increase tumour cell susceptibility to therapeutic interventions. Bromodeoxyuridine (BrdUrd), a thymidine analogue, has been explored as a radiosensitiser due to its incorporation into DNA in place of thymidine. This substitution increases the vulnerability of tumour cells to radiation-induced damage by disrupting DNA replication and repair mechanisms [47, 48].

Histone deacetylase inhibitors (HDACis) have shown promise in enhancing the response of brain tumour cells to both radiation and chemotherapy. HDACs modulate gene expression by altering chromatin structure, thereby influencing DNA accessibility and repair. Their radiosensitizing effects are attributed to the Inhibition of DNA repair pathways, making glioblastoma cells more susceptible to radiation. Downregulation of DNA repair proteins, such as MGMT and RAD51, which are associated with resistance to therapy. Potentiation of chemotherapeutic efficacy, as chromatin structure plays a role in the action of various chemotherapeutic agents. HDAC inhibitors have demonstrated additive effects when combined with standard chemotherapy and radiation therapy, offering a synergistic approach to brain tumour treatment [49]. The antiepileptic drug valproate has been identified as a potential chemosensitiser for glioblastoma cells. Its mechanism involves HDAC inhibition, which enhances the cytotoxic effects of chemotherapeutic agents and may overcome resistance. This dual functionality positions valproate as a promising adjunct in the treatment of glioblastoma [15].

Potentiation

Pharmacological potentiation aims to enhance the effect of a drug without increasing its dose. Ligands of purinergic receptors, such as ATP and BzATP, have been shown to enhance the cytotoxicity of temozolomide (TMZ), a chemotherapeutic drug widely used in the treatment of glioblastoma. The ligands act as epigenetic modifiers that are important in treating brain tumors when combined with histone deacetylase (HDAC) inhibitors. HDAC inhibitors in COL treatment enhance the susceptibility of glioma cells to existing therapies (*i.e.*, TMZ or radiation) by reprogramming the epigenetic profile of glioma cells and increasing their susceptibility to chemotherapy. Furthermore, they can inhibit critical molecular pathways, such as the PI3K/AKT/mTOR axis, thereby augmenting the effectiveness of other therapeutic agents and leading to an improved overall treatment response in glioblastoma patients [50]. Specific antiepileptic drugs possess antineoplastic properties, with the potential to be reused as adjuvants in brain tumour therapy [44].

Overcoming Resistance

Improving treatment to overcome brain tumours can be achieved by overcoming drug resistance. By exporting chemotherapy drugs from brain and tumour tissue, the BBB and blood-tumour barrier (BTB) play crucial roles in building multidrug resistance [51, 52]. Osmotic BBB disruption, bradykinin-mediated BTB opening, and inhibition of multidrug resistance proteins are a few strategies to overcome these barriers [53]. The combination of HDACs overcomes resistance, restoring the epigenetic frequency of brain tumor cells to both chemotherapy and targeted therapy. They modulate chromatin architecture and transcription, inducing sensitivity in tumor cells to cytotoxic drugs. This is especially important for glioblastomas, where many standard treatments are ineffective due to cellular resistance. HDAC inhibitors are typically used in combination therapies in conjunction with traditional chemotherapy or stable targeted molecular agents to overcome resistance pathways that tumors acquire over time. HDAC inhibitors reprogram resistant cancer cells by modulating gene expression, thereby resensitizing cancer cells to tumor progression. They can also negate oncogenic pathways at play in resistance, thereby enhancing therapeutic efficacy. Nanomedicine strategies can further optimize the blood-brain barrier penetration of HDAC inhibitors, adding another dimension of synergy to their combination with brain tumor therapy. Consequently, HDAC inhibitors can be used as a pharmacological approach to overcome resistance, leading to improved patient outcomes in the treatment of brain tumors [54]

Rationale & Strategies of Targeted Therapies in Brain Tumor Modulation

Targeted therapies represent a pivotal advancement in the modulation of brain tumours. These therapies aim to selectively inhibit specific molecular targets that drive tumour growth and progression, offering the potential for more effective and less toxic treatment compared to conventional cytotoxic chemotherapy. The importance of targeted therapies is underscored by the unique biology of brain malignancies, particularly the molecular heterogeneity and genetic complexity exhibited by gliomas, the most common type of brain tumour [55]. Glioblastomas, the most aggressive glioma subtype, are composed of multiple subclones with distinct genetic and epigenetic alterations. This intratumoural heterogeneity presents significant therapeutic challenges and highlights the need for precision-targeted strategies. Additionally, physiological barriers such as the blood-brain barrier (BBB) and blood-tumor barrier (BTB) further complicate effective drug delivery, underscoring the importance of targeted approaches [54].

Several targeted therapeutic strategies have been explored to disrupt signaling pathways essential for tumor cell survival and proliferation. EGFR-targeted therapy: The epidermal growth factor receptor (EGFR) pathway is frequently altered in glioblastoma. Targeting EGFR has shown promise in inhibiting tumour growth and modulating the tumour microenvironment (TME) [56]. MET pathway inhibition: The mesenchymal-epithelial transition (MET) pathway is implicated in glioma progression. Potent MET inhibitors have demonstrated anti-tumour activity in preclinical models and are currently being evaluated in clinical trials for safety and efficacy in glioblastoma patients [57]. Targeted therapies also play a role in enhancing the effectiveness of immunotherapy by reshaping the immunosuppressive TME: EGFR-targeted agents have been shown to reduce immunosuppressive cell populations and increase cytotoxic T cell activity, thereby improving immune-mediated tumour clearance [58]. Despite the promise of targeted therapies, several challenges remain: Heterogeneous target expression within tumours can limit therapeutic efficacy. Development of resistance to targeted agents is a common obstacle. Lack of predictive biomarkers hinders patient stratification and treatment personalization.

THERAPEUTIC OPPORTUNITIES: SMALL MOLECULE INHIBITORS IN BRAIN TUMORS

The class of small molecule inhibitors promises an effective treatment for brain tumors, including highly aggressive glioblastoma. They target specific molecular pathways that drive the proliferation and survival of cancer cells, promising more selective and enhanced treatment compared to conventional ones.

EGFR, Tyrosine Kinases, PI3K/AKT/mTOR, FAK and Pyk2 Pathway Inhibitors

The epidermal growth factor receptor (EGFR) pathway is the most targeted research in glioblastoma. Preclinical efficacy was demonstrated by small-molecule inhibitors like gefitinib and erlotinib in altering EGFR, which is commonly found in glioblastomas [59]. The limited effectiveness of these inhibitors is primarily attributed to poor penetration across the BBB [60]. New EGFR inhibitors are being developed to enhance their ability to penetrate the BBB. An example is a potent brain-penetrant EGFR inhibitor that demonstrated activity in preclinical glioblastoma models [61].

Other receptor tyrosine kinases, in addition to EGFR, are being targeted in glioblastoma. Inhibitors of MET, a receptor crucial for glioblastoma proliferation and invasion, are under investigation. Targeting angiogenesis through the inhibition of vascular endothelial growth factor receptor (VEGFR) is also being explored [59]. Inhibition of the PI3K/AKT/mTOR pathway has demonstrated a notable increase in treatment efficacy in glioblastoma [59].

The FAK and Pyk2 pathways, which play a significant role in the cellular mechanics of glioblastoma, are suitable targets for treating glioblastoma and are being studied [59].

Challenges and Future Directions

Glioblastoma is characterized by profound heterogeneity, comprising multiple subclonal populations with distinct genetic and epigenetic alterations. This complexity contributes to the rapid development of therapeutic resistance, presenting a formidable challenge to effective treatment. Additionally, the blood-brain barrier (BBB) remains a significant obstacle to drug delivery, thereby limiting the efficacy of many promising small-molecule therapies. Despite these challenges, innovative strategies such as nanoparticle-based drug delivery systems and combination therapies have demonstrated potential in overcoming these barriers. Nanoparticles can enhance drug penetration across the BBB and facilitate targeted delivery to tumour sites, while combination approaches may synergistically address resistance mechanisms and improve therapeutic outcomes [62].

While small molecule inhibitors offer substantial promise and renewed hope in the treatment of glioblastoma, several critical challenges must still be addressed. These include overcoming tumor heterogeneity and resistance, enhancing drug delivery across the blood-brain barrier (BBB), and identifying patients who are most likely to benefit from targeted therapies.

Continued research is essential to confront these obstacles and unlock the full potential of small molecule inhibitors. Their further investigation may pave the way for more personalized and effective treatment strategies for this devastating disease.

Monoclonal Antibodies and Antibody-Drug Conjugates in Brain Tumors

Involving monoclonal antibodies (mAbs) and antibody-drug conjugates (ADCs) is revolutionary for various cancers, including brain tumours. Targeted therapies may help in treating brain tumors, potentially leading to reduced systemic toxicity. This chapter delves into the therapeutic opportunities and challenges associated with mAbs and ADCs in treating brain tumours.

Monoclonal Antibodies in Brain Tumors

Monoclonal antibodies (mAbs) are engineered molecules designed to function as substitute antibodies that restore, enhance, or mimic the immune system's response against cancer cells. They have demonstrated significant potential in both the diagnosis and treatment of brain tumours. Monoclonal antibodies (mAbs) diagnose and treat primary brain tumours by specifically targeting tumour-associated antigens. These antibodies effectively bind to neuroectodermal tumour-associated antigens, facilitating precise localization and targeted treatment of tumours [63]. Tumour-specific, unarmed mAbs targeting the EGFRvIII mutation have shown efficacy in managing malignant brain tumours [64]. Recent advancements include the development of monoclonal antibodies (mAbs) that inhibit the interaction between growth factors and their receptors. For instance, mAbs targeting insulin-like growth factor receptors (IGF-IR) have shown promise in preclinical brain tumour models [65]. Additionally, mAbs encapsulated within nanoparticles are being explored for their potential to traverse the blood–brain barrier (BBB), thereby enhancing drug delivery and therapeutic efficacy in brain tumour treatment [66].

Antibody-drug conjugates (ADCs) represent a novel class of targeted cancer therapies that combine the specificity of monoclonal antibodies (mAbs) with the cytotoxic potency of chemotherapeutic agents. This design enables the selective delivery of cytotoxic compounds to tumour cells, thereby minimizing systemic toxicity. The therapeutic efficacy of ADCs in brain tumours is significantly influenced by the homogeneity of the conjugates and their ability to cross the blood–brain barrier (BBB). Interestingly, heterogeneous conjugates have demonstrated superior efficacy in brain tumour models [67]. ADCs targeting both wild-type EGFR and the EGFR variant III are currently under investigation for their potential to treat brain metastases originating from HER2-positive breast cancer [68]. A major challenge in ADC development is the BBB itself—its structural and functional heterogeneity poses a significant barrier to effective drug delivery in glioblastoma treatment [69]. Ongoing research aims to develop ADCs with enhanced permeability, leveraging accumulated insights to overcome these limitations [70].

Immunotherapies in Brain Tumors

Immunotherapies, comprising checkpoint inhibitors, are innovative strategies that promise enhanced management of brain tumors. This chapter provides a concise overview of prevailing immunotherapies in brain tumors, emphasizing their mechanisms, key findings, and future directions.

Unique hurdles are posed by glioblastomas, in particular, due to their strategic location and the BBB, limiting drug delivery. Promising approaches to overcome these challenges, like immunotherapies, including checkpoint inhibitors, have emerged. The current state and potential of immunotherapies in treating brain tumours have been discussed in this section.

IMMUNOTHERAPY APPROACHES

- **Checkpoint Inhibitors:** A significant advancement in immunotherapy is the development of immune checkpoint inhibitors targeting proteins such as PD-1,

PD-L1, and CTLA-4, which evade the immune system. Checkpoint inhibitors block these proteins, enabling the immune system to recognise and attack tumour cells more efficiently.

Clinical trials have proven the potential of inhibitors in treating brain metastasis from cancers like melanoma and non-small cell lung cancer (NSCLC) [71]. Table **1** presents an overview of clinical trials and therapeutic agents evaluated for brain cancer between 2015 and 2024, highlighting the number of trials conducted during this timeframe. Achieving efficacy in primary brain tumours like glioblastoma is a more challenging aspect. Significant hurdles are posed by the heterogeneity of glioblastoma and the immunosuppressive tumour microenvironment [72].

- **Combinatorial Approaches:** The combination of checkpoint inhibitors with other treatments, such as radiation therapy, has shown promise. When checkpoint inhibitors are used in conjunction with radiation, local treatment of tumors leads to anti-tumor effects, and the abscopal effect is enhanced [73]. Additionally, combined strategies involving checkpoint inhibitors and other immunotherapies, such as CAR-T cells or oncolytic viruses, are being investigated to improve patient outcomes [74].
- **Overcoming the Blood-Brain Barrier:** The BBB is a significant obstacle against drug delivery to brain tumours. Nanoparticles, focused ultrasound, and modified molecular structure of drugs to increase permeability are a few approaches to enhance drug delivery across the BBB [75].

KEY FINDINGS AND INSIGHTS

- **Efficacy in Brain Metastases:**

Checkpoint inhibitors have demonstrated efficacy in treating brain metastases from various cancers. The efficacy of pembrolizumab and nivolumab in treating melanoma and NSCLC brain metastases can be considered an example [71].

- **Challenges in Glioblastoma:**
- Success of checkpoint inhibitors in glioblastoma is yet limited. The presence of BBB and its highly immunosuppressive microenvironment are significant hurdles. Biomarkers that predict response to therapies have yet to be identified, and research to develop combination strategies that overcome all hurdles is still ongoing [72].
- **Future Directions:** Future research aspirations include the enhanced delivery of immunotherapies to the brain, identifying effective combinations of therapies, and understanding the mechanisms of resistance. The success of immunotherapies in brain tumors will likely be attributed to personalized approaches based on the tumors' molecular and genetic profiles [76].

RNA-Targeting Therapies and Nano-Oncology

The strategic location and protective nature of the blood-brain barrier (BBB) present significant therapeutic challenges, particularly in the case of glioblastomas. Emerging approaches, such as RNA-targeting therapies and nano-oncology, promise new avenues for treatment. Innovative strategies and their potential impact on brain tumour therapy are explored in this section.

RNA-Targeting Therapies: Mechanism, Challenges and Solutions, Clinical Potential

RNA-targeting therapies utilize oligonucleotides such as small interfering RNA (siRNA), antisense oligonucleotides, and microRNA (miRNA) mimics to modulate gene expression. These approaches aim to silence oncogenes or restore tumour suppressor genes, thereby inhibiting tumour progression [77]. The blood–brain barrier (BBB) presents a significant obstacle to the effective delivery of drugs within the central nervous system (CNS). However, advancements in delivery systems, including lipid nanoparticles and polymer-based carriers, have markedly improved RNA transport to the brain [78]. Additionally, focused ultrasound techniques that transiently disrupt the BBB have shown promise in enhancing drug penetration. Preclinical models of brain tumours have demonstrated encouraging outcomes with RNA-targeting therapies. Targeting the mTOR pathway using RNA-based strategies has shown potential in promoting axon regeneration and inhibiting tumour growth [79]. Furthermore, RNA-mediated immunotherapies are being actively investigated for their ability to modulate the tumour immune microenvironment, potentially enhancing the efficacy of other therapeutic modalities [80].

Nano-Oncology: Principles and Techniques

Nano-oncology involves the application of engineered nanoparticles for the diagnosis and treatment of cancer. These nanoparticles can be designed to deliver chemotherapeutic agents, RNA-based therapeutics, or imaging compounds directly to targeted tumour sites, thereby enhancing therapeutic efficacy while minimizing systemic side effects [81]. Due to their small size, nanoparticles exhibit a higher rate of drug delivery across the blood–brain barrier (BBB) compared to larger molecules, making them particularly suitable for treating brain tumours. For instance, metallic nanoparticles have been employed in glioblastoma models to improve the delivery of chemotherapeutic agents [82]. Furthermore, personalized nanobiotechnology-assisted therapies can be developed by tailoring nanoparticle formulations to the unique molecular characteristics of an individual's brain tumour [82]. Integrated nano-oncology has achieved notable progress in the controlled synthesis of multifunctional nanoparticles capable of simultaneous drug delivery and imaging. This integration enables real-time monitoring of treatment responses and dynamic adjustments to therapeutic strategies [83]. Enhancing BBB permeability remains a key focus in nano-oncology, with innovative approaches being developed to improve drug delivery to brain tumours [84].

Table 1. Clinical trials and drugs on brain cancer, the number of clinical trials carried out from 2015 to 2024.

Sl No	Study Title	Conditions	Interventions	Phases
1.	Scalp Nerve Block and Opioid Consumption in Brain Surgery	Brain Tumour	DRUG: Scalp block with 0.5% plain Marcaine DRUG: Scalp block with 0.9% normal saline	PHASE4
2.	Study of Immunotoxin, MR1-1	Supratentorial Malignant Brain Tumor	BIOLOGICAL: MR1-1	PHASE1
3.	Treatment of Recurrent Brain Tumors: Metabolic Manipulation Combined With Radiotherapy	Brain Neoplasms	RADIATION: Partial brain re-irradiation. DRUG: Metformin BEHAVIORAL: low carbohydrate diet	PHASE1
4.	[F-18]Fluoro-DOPA PET Imaging of Brain Tumors in Children	Brain Tumors	DRUG: [F-18]Fluoro-DOPA PET, a diagnostic radiopharmaceutical	-------
5.	Ferumoxytol in Magnetic Resonance Imaging of Pediatric Patients With Brain Tumors	Brain Neoplasm	DRUG: Ferumoxytol PROCEDURE: Magnetic Resonance Imaging	PHASE2
6.	Dexanabinol in Patients With Brain Cancer	Brain Cancer	DRUG: Dexanabinol	PHASE1
7.	A Study of the Specificity and Sensitivity of 5-ALA Fluorescence in Malignant Brain Tumors	Brain Neoplasms	DRUG: 5-aminolevulinic acid	PHASE1|PHASE2
8.	Safety Study of Intracerebral Topotecan for Recurrent Brain Tumors	Brain Neoplasms, Primary Malignant	PROCEDURE: Convection-Enhanced Delivery DRUG: Topotecan	PHASE1|PHASE2

CASE STUDIES: SUCCESSFUL MODULATION STRATEGIES IN BRAIN TUMOR THERAPY

Examples of Approved Therapies Demonstrating Successful Modulation

- **Gliadel Wafer (Carmustine Implant):** Gliadel Wafer was FDA-approved in 1996, marking significant advancements in targeted brain tumour therapies. After surgical removal of the tumour, this wafer is implanted. It eventually starts releasing carmustine (BCNU), a chemotherapeutic agent, achieving localised high concentrations over time and parallelly minimising systemic toxicity [85, 86]. This targeted delivery method effectively circumvents the blood-brain barrier (BBB), a significant challenge in brain tumour therapies.
- **Temozolomide:** Temozolomide (TMZ) is a widely used oral alkylating agent for the treatment of glioblastoma multiforme (GBM). It is administered alongside radiotherapy and maintenance therapy. The capacity to penetrate the BBB and its favourable safety profile stems from TMZ's efficacy. Its action involves methylating guanine residues within DNA to induce apoptosis of tumour cells [87].
- **Bevacizumab (Avastin;)** Bevacizumab, an anti-VEGF monoclonal antibody, gained approval in 2009 for recurrent GBM—its mechanism of action inhibits angiogenesis, reducing blood supply to the tumor and impeding its growth. Although not curative, bevacizumab enhances progression-free survival and improves the quality of life for patients with recurrent GBM [88].

Overcome Specific Biological Challenges

• Overcoming the Blood-Brain Barrier:

The blood-brain barrier poses a significant challenge in treating brain tumours by limiting access to many therapeutic agents. Gliadel Wafer overcomes this limitation by targeting the delivery of therapeutics and circumventing the BBB (Lesniak & Brem, 2004). The ability of Temozolomide to bypass the BBB is credited for its effectiveness in drug delivery [87].

• Targeting Tumor Microenvironment:

Bevacizumab's mechanism targets the tumour microenvironment by inhibiting angiogenesis, which blocks VEGF and diminishes the formation of new blood vessels that nourish tumours, thereby cutting off their supply and potentially saving them. These approaches underscore the importance of modulating the tumor microenvironment.

- **Localized Drug Delivery:**

Localised drug delivery systems, such as the Gliadel Wafer, deliver concentrated chemotherapeutics directly to tumour sites, minimising systemic toxicity and its side effects. While enhancing drug efficacy, it also plays a crucial role in reducing the risk of systemic toxicity [86].

- **Enhancing Drug Penetration:** Diverse methods have improved drug delivery. For example, nanoparticles and liposomes enhance chemotherapeutic drug delivery across the BBB. They can be engineered to release their cargo, increasing the therapeutic concentration at tumour sites [2].

PERSISTING CHALLENGES IN BRAIN TUMOR THERAPY: THE BLOOD-BRAIN BARRIER

Strategies to Overcome the Blood-Brain Barrier (BBB) in Brain Tumor Therapy

Therapeutic delivery and its efficacy are limited due to the BBB. Different approaches have been devised to overcome the BBB, although each has specific drawbacks. This section explores current strategies while discussing the room for developed and enhanced approaches.

Current Strategies to Overcome the BBB

- **Nanotechnology-Based Approaches:**

Nanotechnology utilises nanoparticles to circumvent the BBB. They are designed to deliver therapeutics and precisely release them at tumour sites. Additionally, they can be designed to bind to specific receptors on the BBB to aid BBB penetration [89]. However, the BBB's complexity and variable permittivity at different regions in the brain are significant hurdles to the effectiveness of the nanoparticle [90].

- **Focused Ultrasound (FUS):**

Focused ultrasound (FUS) coupled with microbubbles is a non-invasive method that transiently opens the BBB and enables drugs to reach the tumour sites. This approach has demonstrated promising results in preclinical and clinical models [91]. Despite its benefits, FUS may induce tissue damage and inflammation; it necessitates precisely timed drug administration since its impact on the BBB is temporary [92].

• Chemical Modulation:

Chemical agents, such as hyperosmotic solutions, may temporarily disrupt the BBB by shrinking endothelial cells and loosening tight junctions. This approach improves drug delivery efficiency to brain tumours [93]. Nonetheless, the crude disruption may allow harmful substances to enter the brain, potentially leading to a neurotoxic effect [94].

• Receptor-Mediated Transcytosis:

This approach utilizes ligands or antibodies that selectively bind to receptors on the BBB, triggering endocytosis and facilitating drug delivery into the brain. This targeted delivery approach enhances precision in drug delivery [41]. Nonetheless, a potential risk is the saturation of the receptors, which restricts the quantity of drug that can traverse the BBB.

• Viral Vectors:

Viral vectors have been modified to circumvent the BBB to transport genetic material to the brain cells regarding gene therapy for brain tumours (Helms *et al.*, 2020). However, they pose potential risks, such as immune responses and insertional mutagenesis, which limit their safety and efficiency [95].

Limitations and Potential Solutions

• Heterogeneity of the BBB:

The variable permittivity of the blood–brain barrier (BBB) across different brain regions leads to inconsistent drug delivery. To address this challenge, ongoing investigations are focused on developing personalized medicine strategies that tailor treatments based on an individual's BBB characteristics and tumour profile. Brain tumour heterogeneity remains a significant obstacle to effective treatment, underscoring the need for innovative therapeutic approaches. Precision medicine offers a promising solution by enabling the identification of specific molecular subtypes of tumours, which facilitates the customization of targeted therapies. For example, patients with glioblastomas that exhibit amplification of the epidermal growth factor receptor (EGFR) may respond favorably to EGFR inhibitors, making this a viable targeted strategy for improving efficacy.

Combination therapies that target multiple pathways involved in tumour progression have shown potential in overcoming resistance mechanisms driven by signaling activation. By eliminating tumour cells that may be resistant to single-agent therapies, such approaches reduce the likelihood of recurrence through resistant subpopulations.

A central strategy in this context involves targeting the tumour microenvironment, which plays a critical role in tumour survival and therapeutic resistance. Modulating immune responses and disrupting tumour-supportive stromal interactions can significantly diminish the tumour's ability to withstand treatment. Advancements in nanotechnology have further propelled innovation in drug delivery systems, offering enhanced penetration of the BBB and precise targeting of heterogeneous tumour subpopulations. Nanoparticles can be engineered to deliver chemotherapeutic agents, genetic material, or immunotherapeutics directly to tumour sites, thereby minimizing systemic toxicity and improving therapeutic outcomes. Despite these advancements, tumour heterogeneity continues to pose significant challenges. This necessitates further exploration of novel therapeutic agents, biomarker-guided therapy selection, and advanced drug delivery platforms to fully realize the potential of personalized and effective treatments for brain tumour patients [96].

- **Transient Nature of BBB Disruption:**

Methods like FUS and clinical modulation transiently disrupt the BBB, emphasizing the crucial requirement for sustained and controlled modulation techniques to optimize drug delivery safety and effectiveness [97] precisely.

- **Non-Specific Delivery:**

Numerous existing strategies are deficient in specificity, resulting in indiscriminate BB opening and potential neurological harm. Enhancing delivery systems with advanced targeting capabilities through dual-targeting nanoparticles could minimize off-target effects and improve specificity [98].

- **Immune Response:**

Viral vectors and biological agents often provoke immune responses, compromising efficacy and safety. Addressing this issue involves engineering vectors that evade immune surveillance and employ immunosuppressive strategies to alleviate these challenges and enhance clinical utility [29].

• Scalability and Clinical Translation:

Despite their promises, numerous innovative strategies are still in the early developmental phases, and obstacles have been encountered in scaling up clinical trials. Effective collaborations among researchers, clinicians, and industry stakeholders are crucial for accelerating the transition of these innovations to clinical utility [99].

STRATEGIES TO OVERCOME CHALLENGES IN BRAIN TUMOR TREATMENT

Brain tumours present formidable therapeutic challenges due to their location, the protective role of the BBB, and the emergence of treatment resistance. This section explores diverse strategies to address these obstacles, including nanoparticle and liposomal technologies to enhance BBB permeability, combination therapies to combat resistance, biomarker advances for patient stratification and monitoring, and localised drug delivery.

Nanoparticle and Liposomal Formulations to Enhance BBB Crossing

• Nanoparticles:

Nanoparticles are intricately designed to ferry therapeutics across the BBB, employing processes like receptor-mediated transcytosis and the enhanced permeability and retention effect (EPR). Specifically, polymeric nanoparticles can be engineered to equip ligands tailored to bind to the BBB receptors, facilitating delivery into the brain [100].

• Liposomal Formulations:

Liposomes, spherical vesicles structured with a phospholipid bilayer, are effective carriers for facilitating drug delivery across the BBB. Liposomes embedded with transferrin have been shown to cross the BBB by delivering drugs directly to tumor sites, thereby enhancing treatment efficacy while mitigating systemic side effects [101]. Furthermore, engineered liposomal formulations can circumvent efflux transporters like p-glycoprotein, thereby optimising drug delivery [102].

• Combination with Magnetic Nanoparticles:

Integrating magnetic nanoparticles into liposomal formulations enables precise delivery and controlled release of therapeutics. This method enhances BBB penetration and facilitates dynamic drug distribution imaging and monitoring [103].

Combination Therapies to Combat Resistance

• Gene Therapy and Chemotherapy:

The synergy of gene therapy and conventional chemotherapy has proven potential in combating drug-resistant brain tumours. Gene therapy sensitises tumour cells to chemotherapy agents, effectively disrupting resistance mechanisms to improve patient outcomes.Gene therapy for brain cancer can be significantly impacted when combined with chemotherapy, the study said. Common approaches include the delivery of constructs containing therapeutic genes (such as p53 or pro-apoptotic genes) through viral vectors (either oncolytic viruses or genetically modified viral vectors) or non-viral vectors. Gene therapy can also increase drug sensitivity, decrease tumor resistance, and improve the efficacy of treatment in combination with chemotherapeutics. The synergistic effect was characterized by an enhanced apoptotic destruction of tumor cells, lower toxicity, and increased transmembrane capacity. However, research on improving the efficacy of gene delivery, enhanced targeting with specific vectors, and the use of rational combinations with traditional/random body therapies6 suggests that gene therapy represents a potential avenue in multimodal therapies for controlling brain cancer [104].

• Epigenetic Modulation:

Histone deacetylase inhibitors (HDACis) and other epigenetic drugs are increasingly used in combination with chemotherapy to counteract resistance mechanisms by modulating gene expression and restoring treatment sensitivity. Epigenetic alterations including DNA methylation, histone modifications, and non-coding RNA activity play a pivotal role in brain tumorigenesis, drug resistance, and therapeutic response.

This body of research underscores the potential of targeting epigenetic pathways to overcome drug resistance and enhance tumour sensitization to chemotherapy and radiotherapy. Given the reversible nature of epigenetic modifications, agents such as HDAC inhibitors and DNA methyltransferase inhibitors are being actively investigated for their ability to reverse aberrant epigenetic states and restore normal gene expression in tumour cells. Moreover, combination therapies incorporating epigenetic modulators have shown promise in improving treatment efficacy, particularly in brain tumours that exhibit resistance to conventional therapies [105]. These strategies represent a critical advancement in the pursuit of personalized and adaptive treatment approaches for patients with brain tumors.

• Targeted Therapies:

Synergistic approaches combining targeted therapies, such as tyrosine kinase inhibitors, with traditional treatments offer promising avenues for overcoming resistance by addressing specific mutations and signaling pathways. Notably, the combination of MEK and BRAF inhibitors has demonstrated efficacy in bypassing resistance mechanisms in glioblastoma, particularly in tumors with defined molecular and genetic profiles.

These targeted therapies encompass a range of modalities, including small-molecule inhibitors, monoclonal antibodies, and receptor-specific strategies aimed at suppressing tumor growth and angiogenesis. Key agents in this domain include tyrosine kinase inhibitors, epidermal growth factor receptor (EGFR) inhibitors, and vascular endothelial growth factor (VEGF) inhibitors, all of which play a critical role in brain tumor therapy.

Innovations in nanotechnology and drug delivery systems, such as nanoparticles and liposomal formulations, are being actively explored to enhance drug permeability across the blood–brain barrier and improve selectivity for tumor cells. These advancements are essential for optimizing therapeutic efficacy while minimizing systemic toxicity. Equally important is the personalization of treatment strategies based on the molecular and biological profile of each tumour. Tailoring therapies to individual tumour characteristics enhances susceptibility to targeted interventions and supports the broader goals of precision oncology [106].

Biomarker Development for Patient Selection and Monitoring

• Circulating Biomarkers:

Innovating circulating biomarker panels could help identify patients who would benefit from therapies. Real-time monitoring of treatment response and tumor progression can be achieved using biomarkers such as circulating tumor DNA (ctDNA) [107].

• Imaging Biomarkers:

Dynamic susceptibility contrast (DSC) imaging is an advanced MRI technology that detects resistance signs as early as possible. Biomarkers are valuable informants on the tumour microenvironment and work as guides for therapeutic decisions [108].

- **Molecular Biomarkers:** Biomarkers associated with identifying resistance mechanisms can guide the tailoring of therapies according to the patient's profile. An example is the response to temozolomide in glioblastoma, which is tracked with the help of MGMT methylation [109].

Locoregional Delivery Approaches

- **Convection-Enhanced Delivery (CED):**

Convection-enhanced delivery (CED) enables the direct administration of therapeutics into the brain *via* implanted catheters. This method bypasses the blood–brain barrier (BBB), resulting in a localized increase in drug concentration while minimizing systemic toxicity [110].

- **Polymer-Based Drug Depots:**

Implantable polymer-drug deposits can deliver precise medication to the tumor site. It has shown promising results in glioblastoma, ensuring long-lasting drug delivery [111].

- **Radioisotope Conjugates:**

Radioisotope conjugates, such as Y-90-DOTATOC, when delivered under targeted conditions, lead to minimized damage to systemic tissues. This procedure has yielded promising results [112].

The synergistic integration of emerging drug delivery systems, targeted therapies, and advanced biomarkers presents a broad and promising spectrum of possibilities for treating brain tumors. Continued innovation, deeper translational research, and rigorous clinical trials are crucial to advancing these therapeutic strategies toward enhanced patient outcomes and personalized care.

CONCLUSION

Pharmacological modulations have sparked renewed hope in enhancing drug delivery across the blood–brain barrier (BBB) and overcoming other protective mechanisms of the central nervous system. Strategies aimed at modulating or transiently disrupting the BBB have proven pivotal in facilitating efficient drug delivery and targeting molecular pathways responsible for tumour progression. Promising results have demonstrated that pharmacological agents, such as doxorubicin, can circumvent the BBB to improve therapeutic efficacy. Additionally, repurposing antiepileptic drugs for their potential antineoplastic effects is being explored, offering dual benefits of tumour growth inhibition and seizure control. However, several challenges persist, including the complex anatomical structure and location of brain tumours, the impermeability of the BBB, and the emergence of drug resistance. Misleading phenomena such as pseudoprogression and pseudo-response continue to complicate clinical decision-making, hindering accurate assessment of treatment outcomes. Achieving localized and optimal concentrations of chemotherapeutic agents without inducing systemic toxicity remains a significant hurdle. Immunotherapy, despite its promise, faces challenges related to precise delivery, immune system engineering, and managing adverse immune responses. Looking ahead, nanotechnology platforms such as magnetic nanoparticles are being actively investigated for their dual capabilities in imaging and therapy. These innovations enable real-time monitoring and targeted treatment of brain tumours, marking a significant advancement in precision oncology. Personalized treatment strategies, guided by biomarker profiling, are forming the backbone of future therapeutic development. Synergistic approaches that combine chemotherapy, targeted therapy, and immunotherapy are being evaluated for their potential to overcome tumour resistance and improve patient outcomes. With continued innovation, rigorous research, and strategic integration of emerging technologies and personalized medicine, the remaining challenges in brain tumour therapy can be effectively addressed. This convergence of pharmacological modulation and advanced therapeutic platforms holds immense promise for the future of brain tumour treatment.

AUTHORS' CONTRIBUTIONS

Mallamma T: Writing – Original draft, Resources Formal analysis. Nagaraj Sreeharsha: Formulated chapter headings and drafted or revised the chapter critically. Prakash Goudanavar: Conceptualization, Supervision Resources, Visualization.

REFERENCES

[1]   Dubey S, Suraj MR, Goni T, *et al.* 3D QSAR studies of 3, 16 and 17 position modifications in steroidal derivatives for CNS anticancer activity. Curr Res Chem. 2023;15(1):1–4.
[http://dx.doi.org/10.3923/crc.2023.1.4]

[2]   Singh D, Tiwari P, Nagdev S. Particulate vaccine dispersions emerge as a novel carrier for deep pulmonary immunization. Curr Nanomedicine. 2023;13(2):71–4.
[http://dx.doi.org/10.2174/2468187313666230714124009]

[3]   Mabray MC, Barajas RF Jr, Cha S. Modern brain tumor imaging. Brain Tumor Res Treat 2015; 3(1): 8-23.
[http://dx.doi.org/10.14791/btrt.2015.3.1.8] [PMID: 25977902]

[4]   Charles NA, Holland EC, Gilbertson R, Glass R, Kettenmann H. The brain tumor microenvironment. Glia 2011; 59(8): 1169-80.
[http://dx.doi.org/10.1002/glia.21136] [PMID: 21446047]

[5]   Graham C, Cloughesy T. Brain tumor treatment: Chemotherapy and other new developments. Semin Oncol Nurs 2004; 20(4): 260-72.
[http://dx.doi.org/10.1016/S0749-2081(04)00090-7] [PMID: 15612602]

[6]   Liu Y, Lu W. Recent advances in brain tumor-targeted nano-drug delivery systems. Expert Opin Drug Deliv 2012; 9(6): 671-86.
[http://dx.doi.org/10.1517/17425247.2012.682726] [PMID: 22607535]

[7]   Lyon JG, Mokarram N, Saxena T, Carroll SL, Bellamkonda RV. Engineering challenges for brain tumor immunotherapy. Adv Drug Deliv Rev 2017; 114: 19-32.
[http://dx.doi.org/10.1016/j.addr.2017.06.006] [PMID: 28625831]

[8]   Handley DA, Hughes TE. Pharmacological approaches and strategies for therapeutic modulation of fibrinogen. Thromb Res 1997; 87(1): 1-36.
[http://dx.doi.org/10.1016/S0049-3848(97)00091-1] [PMID: 9253797]

[9]   Galluzzi L, Bravo-San Pedro JM, Levine B, Green DR, Kroemer G. Pharmacological modulation of autophagy: therapeutic potential and persisting obstacles. Nat Rev Drug Discov 2017; 16(7): 487-511.
[http://dx.doi.org/10.1038/nrd.2017.22] [PMID: 28529316]

[10]   Schiavone S, Trabace L. Pharmacological targeting of redox regulation systems as new therapeutic approach for psychiatric disorders: A literature overview. Pharmacol Res 2016; 107: 195-204.
[http://dx.doi.org/10.1016/j.phrs.2016.03.019] [PMID: 26995306]

[11]   Marcoli M, Agnati LF, Franco R, *et al.* Modulating brain integrative actions as a new perspective on pharmacological approaches to neuropsychiatric diseases. Front Endocrinol (Lausanne) 2023; 13: 1038874.
[http://dx.doi.org/10.3389/fendo.2022.1038874] [PMID: 36699033]

[12]   Black KL, Ningaraj NS. Modulation of brain tumor capillaries for enhanced drug delivery selectively to brain tumor. Cancer Contr 2004; 11(3): 165-73.
[http://dx.doi.org/10.1177/107327480401100304] [PMID: 15153840]

[13]   Sardi I, la Marca G, Cardellicchio S, *et al.* Pharmacological modulation of blood-brain barrier increases permeability of doxorubicin into the rat brain. Am J Cancer Res 2013; 3(4): 424-32.
[PMID: 23977451]

[14]   Alghamri MS, McClellan BL, Hartlage CS, *et al.* Targeting neuroinflammation in brain cancer: Uncovering mechanisms, pharmacological targets, and neuropharmaceutical developments. Front Pharmacol 2021; 12: 680021.
[http://dx.doi.org/10.3389/fphar.2021.680021] [PMID: 34084145]

[15]   Bianchi L, Zhou Y, Bocci G, Scarselli M. Repurposing mood stabilizers for glioblastoma: mechanistic insights and translational advances. Trends Pharmacol Sci.

[16]   Zhou Y, Cucchiara F, Bocci G, Danesi R. Antiepileptic drugs as dual modulators of neuroexcitation and tumor progression: emerging evidence in glioma therapy. Neurotherapeutics. 2024;21(2):312–28.
[http://dx.doi.org/10.1007/s13311-024-01345-2]

[17]   Chandana SR, Movva S, Arora M, Singh T, Ferris SP. Primary brain tumors in adults. Am Fam Physician 2008; 77(10): 1423-30.
[PMID: 18533376]

[18]   Perkins A, Liu G. Primary brain tumors in adults: Diagnosis and treatment. Am Fam Physician 2016; 93(3): 211-7.
[PMID: 26926614]

[19]   McFaline-Figueroa JR, Lee EQ. Brain tumors. Am J Med 2018; 131(8): 874-82.
[http://dx.doi.org/10.1016/j.amjmed.2017.12.039] [PMID: 29371158]

[20]   Fox BD, Cheung VJ, Patel AJ, Suki D, Rao G. Epidemiology of metastatic brain tumors. Neurosurg Clin N Am 2011; 22(1): 1-6, v.

[http://dx.doi.org/10.1016/j.nec.2010.08.007] [PMID: 21109143]

[21]   Sawaya R, Ligon BL, Bindal RK. Management of metastatic brain tumors. Ann Surg Oncol 1994; 1(2): 169-78.
[http://dx.doi.org/10.1007/BF02303562] [PMID: 7834443]

[22]   Pérez-Larraya JG, Hildebrand J. Brain Metastases. Handb Clin Neurol 2014; 119: 115-32.

[23]   Fabi A, Felici A, Metro G, *et al.* Brain metastases from solid tumors: disease outcome according to type of treatment and therapeutic resources of the treating center. J Exp Clin Cancer Res 2011; 30(1): 10.
[http://dx.doi.org/10.1186/1756-9966-30-10] [PMID: 21244695]

[24]   Fruehauf JP, Brem H, Brem S, *et al. In vitro* drug response and molecular markers associated with drug resistance in malignant gliomas. Clin Cancer Res 2006; 12(15): 4523-32.
[http://dx.doi.org/10.1158/1078-0432.CCR-05-1830] [PMID: 16899598]

[25]   Mehrian Shai R, Reichardt JKV, Chen TC. Pharmacogenomics of brain cancer and personalized medicine in malignant gliomas. Future Oncol 2008; 4(4): 525-34.
[http://dx.doi.org/10.2217/14796694.4.4.525] [PMID: 18684063]

[26]   Faisal SM, Comba A, Varela ML, *et al.* The complex interactions between the cellular and non-cellular components of the brain tumor microenvironmental landscape and their therapeutic implications. Front Oncol 2022; 12: 1005069.
[http://dx.doi.org/10.3389/fonc.2022.1005069] [PMID: 36276147]

[27]   Quail DF, Joyce JA. The microenvironmental landscape of brain tumors. Cancer Cell 2017; 31(3): 326-41.
[http://dx.doi.org/10.1016/j.ccell.2017.02.009] [PMID: 28292436]

[28]   Pasqualini R, Kozaki T, Bruschi M, Nguyen T H & Borggren U. Modeling the interaction between the microenvironment and tumor cells in brain tumors. Neuron.2020;106(6):1039-1053.E5.
[http://dx.doi.org/10.1016/j.neuron.2020.09.018]

[29]   Fortin D. The blood-brain barrier: its influence in the treatment of brain tumors metastases. Curr Cancer Drug Targets 2012; 12(3): 247-59.
[http://dx.doi.org/10.2174/156800912799277511] [PMID: 22229251]

[30]   Haumann R, Videira JC, Kaspers GJL, van Vuurden DG, Hulleman E. Overview of current drug delivery methods across the blood–brain barrier for the treatment of primary brain tumors. CNS Drugs 2020; 34(11): 1121-31.
[http://dx.doi.org/10.1007/s40263-020-00766-w] [PMID: 32965590]

[31]   Pandit R, Chen L, Götz J, Götz J. The blood-brain barrier: Physiology and strategies for drug delivery. Adv Drug Deliv Rev 2020; 165-166: 1-14.
[http://dx.doi.org/10.1016/j.addr.2019.11.009] [PMID: 31790711]

[32]   Chen Y, Dalwadi G, Benson H. Drug delivery across the blood-brain barrier. Curr Drug Deliv 2004; 1(4): 361-76.
[http://dx.doi.org/10.2174/1567201043334542] [PMID: 16305398]

[33]   Omidi Y, Barar J. Impacts of blood-brain barrier in drug delivery and targeting of brain tumors. Bioimpacts 2012; 2(1): 5-22.
[PMID: 23678437]

[34]   Parrish KE, Sarkaria JN, Elmquist WF, Sarkaria JN. Improving drug delivery to primary and metastatic brain tumors: Strategies to overcome the blood–brain barrier. Clin Pharmacol Ther 2015; 97(4): 336-46.
[http://dx.doi.org/10.1002/cpt.71] [PMID: 25669487]

[35]   Omidi Y, Kianinejad N, Kwon Y, Omidian H. Drug delivery and targeting to brain tumors: considerations for crossing the blood-brain barrier. Expert Rev Clin Pharmacol 2021; 14(3): 357-81.
[http://dx.doi.org/10.1080/17512433.2021.1887729] [PMID: 33554678]

[36]   Scherrmann JM. Drug delivery to brain *via* the blood–brain barrier. Vascul Pharmacol 2002; 38(6): 349-54.
[http://dx.doi.org/10.1016/S1537-1891(02)00202-1] [PMID: 12529929]

[37]   Tsou YH, Zhang XQ, Zhu H, Syed S, Xu X, Makhoul G. Drug delivery to the brain across the blood–brain barrier using nanomaterials. Small 2017; 13(43): 1701921.
[http://dx.doi.org/10.1002/smll.201701921] [PMID: 29045030]

[38]   Ding S, Khan AI, Cai X, *et al.* Overcoming blood–brain barrier transport: Advances in nanoparticle-based drug delivery strategies. Mater Today 2020; 37(3): 112-25.
[http://dx.doi.org/10.1016/j.mattod.2020.02.001] [PMID: 33093794]

[39]   Kumar PB, Kadiri SK, Khobragade DS, *et al.* Synthesis, characterization and biological investigations of some new Oxadiazoles: *In-vitro* and *In-Silico* approach. Results in Chemistry. 2024; 7:101241.
[http://dx.doi.org/10.1016/j.rechem.2023.101241]

[40]   Ferraris C, Cavalli R, Panciani PP, Battaglia L. Overcoming the blood–brain barrier: successes and challenges in developing nanoparticle-mediated drug delivery systems for the treatment of brain tumours. Int J Nanomedicine 2020; 15: 2999-3022.
[http://dx.doi.org/10.2147/IJN.S231479] [PMID: 32431498]

[41]   Hersh DS, Wadajkar AS, Roberts N, *et al.* Evolving drug delivery strategies to overcome the blood brain barrier. Curr Pharm Des 2016; 22(9): 1177-93.
[http://dx.doi.org/10.2174/1381612822666151221150733] [PMID: 26685681]

[42]   Sathornsumetee S, Rich JN. New approaches to primary brain tumor treatment. Anticancer Drugs 2006; 17(9): 1003-16.
[http://dx.doi.org/10.1097/01.cad.0000231473.00030.1f] [PMID: 17001172]

[43]   Shah V, Kochar P. Brain cancer: Implication to disease, therapeutic strategies and tumor targeted drug delivery approaches. Recent Patents Anticancer Drug Discov 2018; 13(1): 70-85.
[PMID: 29189177]

[44]   Cucchiara F, Pasqualetti F, Giorgi FS, Danesi R, Bocci G. Epileptogenesis and oncogenesis: An antineoplastic role for antiepileptic drugs in brain tumours? Pharmacol Res 2020; 156: 104786.
[http://dx.doi.org/10.1016/j.phrs.2020.104786] [PMID: 32278037]

[45]   Natale G, Fini E, Calabrò PF, Carli M, Scarselli M, Bocci G. Valproate and lithium: Old drugs for new pharmacological approaches in brain tumors? Cancer Lett 2023; 560: 216125.
[http://dx.doi.org/10.1016/j.canlet.2023.216125] [PMID: 36914086]

[46]   Del Burgo L S, Hernández R M, Orive G. J Control Release 2014; 195: 32-46.

[47]   Russo A, Gianni L, Kinsella TJ, *et al.* Pharmacological evaluation of intravenous delivery of 5-bromodeoxyuridine to patients with brain tumors. Cancer Res 1984; 44(4): 1702-5.
[PMID: 6704976]

[48]   Sano K, Hoshino T, Nagai M. Radiosensitization of brain tumor cells with a thymidine analogue (bromouridine). J Neurosurg 1968; 28(6): 530-8.
[http://dx.doi.org/10.3171/jns.1968.28.6.0530] [PMID: 4233585]

[49]   Beg U, Snyder B M, Madhani S I, Hamidi N. Current landscape and future prospects of radiation sensitizers for malignant brain tumors: A systematic review. World Neurosurg. 2021;147:E1378–E1387.E2.
[http://dx.doi.org/10.1016/j.wneu.2021.04.134]

[50]   D'Alimonte I, Nargi E, Zuccarini M, *et al.* Potentiation of temozolomide antitumor effect by purine receptor ligands able to restrain the *in vitro* growth of human glioblastoma stem cells. Purinergic Signal 2015; 11(3): 331-46.
[http://dx.doi.org/10.1007/s11302-015-9454-7] [PMID: 25976165]

[51]   Singh R, Khaitan D, Ningaraj N. Nanomedicine strategies to overcome the blood–brain barrier in

glioblastoma: recent advances and translational outlook. Adv Drug Deliv Rev. 2024;205:114589.
[http://dx.doi.org/10.1016/j.addr.2023.114589]

[52]   Régina A, Poirier J, Duquette M, *et al.* ABCA1 is a target of PPARγ in human brain microvascular endothelial cells and is a cerebrovascular inflammation regulator. Mol Cell Biol 2001; 21(16): 5466-78.

[53]   Van Tellingen O, Yetkin-Arik B, de Gooijer MC, Wesseling P, Würdinger T, de Vries HE. Overcoming the blood–brain tumor barrier for effective glioblastoma treatment. Drug Resist Updat 2015; 19: 1-12.
[http://dx.doi.org/10.1016/j.drup.2015.02.002] [PMID: 25791797]

[54]   Khaitan D, Reddy PL, Ningaraj N. Targeting brain tumors with nanomedicines: Overcoming blood brain barrier challenges. Curr Clin Pharmacol 2018; 13(2): 110-9.
[http://dx.doi.org/10.2174/1574884713666180412150153] [PMID: 29651960]

[55]   Yang K, Wu Z, Zhang H, *et al.* Glioma targeted therapy: insight into future of molecular approaches. Mol Cancer 2022; 21(1): 39.
[http://dx.doi.org/10.1186/s12943-022-01513-z] [PMID: 35135556]

[56]   Huang TT, Sarkaria SM, Cloughesy TF, Mischel PS. Targeted therapy for malignant glioma patients: lessons learned and the road ahead. Neurotherapeutics 2009; 6(3): 500-12.
[http://dx.doi.org/10.1016/j.nurt.2009.04.008] [PMID: 19560740]

[57]   Cheng F, Guo D. MET in glioma: signaling pathways and targeted therapies. J Exp Clin Cancer Res 2019; 38(1): 270.
[http://dx.doi.org/10.1186/s13046-019-1269-x] [PMID: 31221203]

[58]   Barnestein R, Galland L, Kalfeist L, Ghiringhelli F, Ladoire S, Limagne E. Immunosuppressive tumor microenvironment modulation by chemotherapies and targeted therapies to enhance immunotherapy effectiveness. OncoImmunology 2022; 11(1): 2120676.
[http://dx.doi.org/10.1080/2162402X.2022.2120676] [PMID: 36117524]

[59]   Heffron TP. Small molecule kinase inhibitors for the treatment of brain cancer. J Med Chem 2016; 59(22): 10030-66.
[http://dx.doi.org/10.1021/acs.jmedchem.6b00618] [PMID: 27414067]

[60]   Heffron TP. Challenges of developing small-molecule kinase inhibitors for brain tumors and the need for emphasis on free drug levels. Neuro-oncol 2018; 20(3): 307-12.
[http://dx.doi.org/10.1093/neuonc/nox179] [PMID: 29016919]

[61]   Tsang JE, Urner LM, Kim G, *et al.* Development of a potent brain-penetrant EGFR tyrosine kinase inhibitor against malignant brain tumors. ACS Med Chem Lett 2020; 11(10): 1799-809.
[http://dx.doi.org/10.1021/acsmedchemlett.9b00599] [PMID: 33062157]

[62]   Wang D, Wang C, Wang L, Chen Y. A comprehensive review in improving delivery of small-molecule chemotherapeutic agents overcoming the blood-brain/brain tumor barriers for glioblastoma treatment. Drug Deliv 2019; 26(1): 551-65.
[http://dx.doi.org/10.1080/10717544.2019.1616235] [PMID: 31928355]

[63]   Bullard DE, Bigner DD. Applications of monoclonal antibodies in the diagnosis and treatment of primary brain tumors. J Neurosurg 1985; 63(1): 2-16.
[http://dx.doi.org/10.3171/jns.1985.63.1.0002] [PMID: 2409248]

[64]   Sampson JH, Crotty LE, Lee S, *et al.* Unarmed, tumor-specific monoclonal antibody effectively treats brain tumors. Proc Natl Acad Sci USA 2000; 97(13): 7503-8.
[http://dx.doi.org/10.1073/pnas.130166597] [PMID: 10852962]

[65]   Ashley DM, Batra SK, Bigner DD. Monoclonal antibodies to growth factors and growth factor receptors: their diagnostic and therapeutic potential in brain tumors. J Neurooncol 1997; 35(3): 259-73.
[http://dx.doi.org/10.1023/A:1005812417638] [PMID: 9440024]

[66] Han L, Liu C, Qi H, *et al.* Systemic delivery of monoclonal antibodies to the central nervous system for brain tumor therapy. Adv Mater 2019; 31(19): 1805697.
[http://dx.doi.org/10.1002/adma.201805697] [PMID: 30773720]

[67] Anami Y, Otani Y, Xiong W, *et al.* Homogeneity of antibody-drug conjugates critically impacts the therapeutic efficacy in brain tumors. Cell Rep 2022; 39(8): 110839.
[http://dx.doi.org/10.1016/j.celrep.2022.110839] [PMID: 35613589]

[68] Mair MJ, Bartsch R, Le Rhun E, *et al.* Understanding the activity of antibody–drug conjugates in primary and secondary brain tumours. Nat Rev Clin Oncol 2023; 20(6): 372-89.
[http://dx.doi.org/10.1038/s41571-023-00756-z] [PMID: 37085569]

[69] Marin BM, Porath KA, Jain S, *et al.* Heterogeneous delivery across the blood-brain barrier limits the efficacy of an EGFR-targeting antibody drug conjugate in glioblastoma. Neuro-oncol 2021; 23(12): 2042-53.
[http://dx.doi.org/10.1093/neuonc/noab133] [PMID: 34050676]

[70] Epaillard N, Bassil J, Pistilli B. Current indications and future perspectives for antibody-drug conjugates in brain metastases of breast cancer. Cancer Treat Rev 2023; 119: 102597.
[http://dx.doi.org/10.1016/j.ctrv.2023.102597] [PMID: 37454577]

[71] Johanns T, Waqar SN, Morgensztern D. Immune checkpoint inhibition in patients with brain metastases. Ann Transl Med 2016; 4(S1) (Suppl. 1): S9.
[http://dx.doi.org/10.21037/atm.2016.09.40] [PMID: 27867977]

[72] Tan AC, Heimberger AB, Menzies AM, Pavlakis N, Khasraw M. Immune checkpoint inhibitors for brain metastases. Curr Oncol Rep 2017; 19(6): 38.
[http://dx.doi.org/10.1007/s11912-017-0596-3] [PMID: 28417311]

[73] Berghoff AS, Venur VA, Preusser M, Ahluwalia MS. Immune checkpoint inhibitors in brain metastases: From biology to treatment. Am Soc Clin Oncol Educ Book 2016; 35(36): e116-22.
[http://dx.doi.org/10.1200/EDBK_100005] [PMID: 27249713]

[74] Sampson JH, Maus MV, June CH. Immunotherapy for brain tumors. J Clin Oncol 2017; 35(21): 2450-6.
[http://dx.doi.org/10.1200/JCO.2017.72.8089] [PMID: 28640704]

[75] Mitchell DA, Fecci PE, Sampson JH. Immunotherapy of malignant brain tumors. Immunol Rev 2008; 222(1): 70-100.
[http://dx.doi.org/10.1111/j.1600-065X.2008.00603.x] [PMID: 18363995]

[76] Sampson JH, Gunn MD, Fecci PE, Ashley DM. Brain immunology and immunotherapy in brain tumours. Nat Rev Cancer 2020; 20(1): 12-25.
[http://dx.doi.org/10.1038/s41568-019-0224-7] [PMID: 31806885]

[77] Krichevsky AM, Uhlmann EJ. Oligonucleotide therapeutics as a new class of drugs for malignant brain tumors: targeting MRNAS, regulatory rnas, mutations, combinations, and beyond. Neurotherapeutics 2019; 16(2): 319-47.
[http://dx.doi.org/10.1007/s13311-018-00702-3] [PMID: 30644073]

[78] Wang S, Huang R. Non-viral nucleic acid delivery to the central nervous system and brain tumors. J Gene Med 2019; 21(7): e3091.
[http://dx.doi.org/10.1002/jgm.3091] [PMID: 30980444]

[79] Ho ES, Chow SF, Tsang CK, Li M-X, Weng J-W. Brain delivering RNA-based therapeutic strategies by targeting mTOR pathway for axon regeneration after central nervous system injury. Neural Regen Res 2022; 17(10): 2157-65.
[http://dx.doi.org/10.4103/1673-5374.335830] [PMID: 35259823]

[80] Pandey PR, Young KH, Kumar D, Jain N. RNA-mediated immunotherapy regulating tumor immune microenvironment: next wave of cancer therapeutics. Mol Cancer 2022; 21(1): 58.
[http://dx.doi.org/10.1186/s12943-022-01528-6] [PMID: 35189921]

[81]   Alphandéry E. Natural metallic nanoparticles for application in nano-oncology. Int J Mol Sci 2020; 21(12): 4412.
[http://dx.doi.org/10.3390/ijms21124412] [PMID: 32575884]

[82]   Nehra M, Uthappa UT, Kumar V, *et al.* Nanobiotechnology-assisted therapies to manage brain cancer in personalized manner. J Control Release 2021; 338: 224-43.
[http://dx.doi.org/10.1016/j.jconrel.2021.08.027] [PMID: 34418523]

[83]   Jiang J, Cui X, Huang Y, *et al.* Advances and prospects in integrated nano-oncology. Nano Biomed Eng 2024; 16(1): 45-60.

[84]   L Fymat A. Nanooncology: Perspective on promising anti-tumor therapies. Journal of Tumor Medicine & Prevention 2017; 1(1): 1-10.
[http://dx.doi.org/10.19080/JTMP.2017.01.555555]

[85]   Lesniak MS, Brem H. Targeted therapy for brain tumours. Nat Rev Drug Discov 2004; 3(6): 499-508.
[http://dx.doi.org/10.1038/nrd1414] [PMID: 15173839]

[86]   Guerin C, Olivi A, Weingart JD, Lawson HC, Brem H. Recent advances in brain tumor therapy: local intracerebral drug delivery by polymers. Invest New Drugs 2004; 22(1): 27-37.
[http://dx.doi.org/10.1023/B:DRUG.0000006172.65135.3e] [PMID: 14707492]

[87]   Fisher JP, Adamson DC. Current FDA-approved therapies for high-grade malignant gliomas. Biomedicines 2021; 9(3): 324.
[http://dx.doi.org/10.3390/biomedicines9030324] [PMID: 33810154]

[88]   Castro MG, Cowen R, Williamson IK, *et al.* Current and future strategies for the treatment of malignant brain tumors. Pharmacol Ther 2003; 98(1): 71-108.
[http://dx.doi.org/10.1016/S0163-7258(03)00014-7] [PMID: 12667889]

[89]   De Rosa G, Salzano G, Caraglia M, Abbruzzese A. Nanotechnologies: a strategy to overcome blood-brain barrier. Curr Drug Metab 2012; 13(1): 61-9.
[http://dx.doi.org/10.2174/138920012798356943] [PMID: 22292810]

[90]   Jain KK. Nanobiotechnology-based strategies for crossing the blood-brain barrier. Nanomedicine (Lond) 2012; 7(8): 1225-33.
[http://dx.doi.org/10.2217/nnm.12.86] [PMID: 22931448]

[91]   Vykhodtseva N, McDannold N, Hynynen K. Progress and problems in the application of focused ultrasound for blood–brain barrier disruption. Ultrasonics 2008; 48(4): 279-96.
[http://dx.doi.org/10.1016/j.ultras.2008.04.004] [PMID: 18511095]

[92]   Bellavance MA, Blanchette M, Fortin D. Recent advances in blood-brain barrier disruption as a CNS delivery strategy. AAPS J 2008; 10(1): 166-77.
[http://dx.doi.org/10.1208/s12248-008-9018-7] [PMID: 18446517]

[93]   Patel MM, Patel BM. Crossing the blood–brain barrier: recent advances in drug delivery to the brain. CNS Drugs 2017; 31(2): 109-33.
[http://dx.doi.org/10.1007/s40263-016-0405-9] [PMID: 28101766]

[94]   Azad TD, Pan J, Connolly ID, Remington A, Wilson CM, Grant GA. Therapeutic strategies to improve drug delivery across the blood-brain barrier. Neurosurg Focus 2015; 38(3): E9.
[http://dx.doi.org/10.3171/2014.12.FOCUS14758] [PMID: 25727231]

[95]   Griffith JI, Rathi S, Zhang W, *et al.* Addressing BBB heterogeneity: A new paradigm for drug delivery to brain tumors. Pharmaceutics 2020; 12(12): 1205.
[http://dx.doi.org/10.3390/pharmaceutics12121205] [PMID: 33322488]

[96]   Kassner A, Merali Z. Assessment of blood–brain barrier disruption in stroke. Stroke 2015; 46(11): 3310-5.
[http://dx.doi.org/10.1161/STROKEAHA.115.008861] [PMID: 26463696]

[97]   Pinkiewicz M, Pinkiewicz M, Walecki J, Zaczyński A, Zawadzki M. Breaking barriers in neuro-

oncology: A scoping literature review on invasive and non-invasive techniques for blood–brain barrier disruption. Cancers (Basel) 2024; 16(1): 236.
[http://dx.doi.org/10.3390/cancers16010236] [PMID: 38201663]

[98]   Karmur BS, Philteos J, Abbasian A, *et al.* Blood-brain barrier disruption in neuro-oncology: Strategies, failures, and challenges to overcome. Front Oncol 2020; 10: 563840.
[http://dx.doi.org/10.3389/fonc.2020.563840] [PMID: 33072591]

[99]   Hempel C, Johnsen KB, Kostrikov S, Hamerlik P, Andresen TL. Brain tumor vessels—a barrier for drug delivery. Cancer Metastasis Rev 2020; 39(3): 959-68.
[http://dx.doi.org/10.1007/s10555-020-09877-8] [PMID: 32488404]

[100] Lombardo SM, Schneider M, Türeli AE, Günday Türeli N. Key for crossing the BBB with nanoparticles: the rational design. Beilstein J Nanotechnol 2020; 11: 866-83.
[http://dx.doi.org/10.3762/bjnano.11.72] [PMID: 32551212]

[101] Ding H, Sagar V, Agudelo M, *et al.* Enhanced blood–brain barrier transmigration using a novel transferrin embedded fluorescent magneto-liposome nanoformulation. Nanotechnology 2014; 25(5): 055101.
[http://dx.doi.org/10.1088/0957-4484/25/5/055101] [PMID: 24406534]

[102] Correia AC, Monteiro AR, Silva R, Moreira JN, Sousa Lobo JM, Silva AC. Lipid nanoparticles strategies to modify pharmacokinetics of central nervous system targeting drugs: Crossing or circumventing the blood–brain barrier (BBB) to manage neurological disorders. Adv Drug Deliv Rev 2022; 189: 114485.
[http://dx.doi.org/10.1016/j.addr.2022.114485] [PMID: 35970274]

[103] Pinzón-Daza M, Campia I, Kopecka J, Garzón R, Ghigo D, Rigant C. Nanoparticle- and liposome-carried drugs: new strategies for active targeting and drug delivery across blood-brain barrier. Curr Drug Metab 2013; 14(6): 625-40.
[http://dx.doi.org/10.2174/1389200211314060001] [PMID: 23869808]

[104] Candolfi M, Kroeger K, Muhammad A, *et al.* Gene therapy for brain cancer: combination therapies provide enhanced efficacy and safety. Curr Gene Ther 2009; 9(5): 409-21.
[http://dx.doi.org/10.2174/156652309789753301] [PMID: 19860655]

[105] Sarkar S, Deyoung T, Ressler H, Chandler W. Brain tumors: Development, drug resistance, and sensitization – an epigenetic approach. Epigenetics 2023; 18(1): 2237761.
[http://dx.doi.org/10.1080/15592294.2023.2237761] [PMID: 37499114]

[106] El-Habashy SE, Nazief AM, Adkins CE, *et al.* Novel treatment strategies for brain tumors and metastases. Pharm Pat Anal 2014; 3(3): 279-96.
[http://dx.doi.org/10.4155/ppa.14.19] [PMID: 24998288]

[107] Tanase C, Albulescu R, Codrici E, *et al.* Circulating biomarker panels for targeted therapy in brain tumors. Future Oncol 2015; 11(3): 511-24.
[http://dx.doi.org/10.2217/fon.14.238] [PMID: 25241806]

[108] Henriksen OM, del Mar Álvarez-Torres M, Figueiredo P, *et al.* High-grade glioma treatment response monitoring biomarkers: A position statement on the evidence supporting the use of advanced MRI techniques in the clinic, and the latest bench-to-bedside developments. Part 1: Perfusion and diffusion techniques. Front Oncol 2022; 12: 810263.
[http://dx.doi.org/10.3389/fonc.2022.810263] [PMID: 35359414]

[109] Hafeez U, Cher LM. Biomarkers and smart intracranial devices for the diagnosis, treatment, and monitoring of high-grade gliomas: a review of the literature and future prospects. Neurooncol Adv 2019; 1(1): vdz013.
[http://dx.doi.org/10.1093/noajnl/vdz013] [PMID: 32642651]

[110] Chaichana KL, Pinheiro L, Brem H. Delivery of local therapeutics to the brain: working toward advancing treatment for malignant gliomas. Ther Deliv 2015; 6(3): 353-69.
[http://dx.doi.org/10.4155/tde.14.114] [PMID: 25853310]

[111] Ramazani F, van Nostrum CF, Storm G, *et al.* Locoregional cancer therapy using polymer-based drug depots. Drug Discov Today 2016; 21(4): 640-7.
[http://dx.doi.org/10.1016/j.drudis.2016.02.014] [PMID: 26969576]

[112] Ferrari M, Cremonesi M, Bartolomei M, *et al.* Dosimetric model for locoregional treatments of brain tumors with 90Y-conjugates: clinical application with 90Y-DOTATOC. J Nucl Med 2006; 47(1): 105-12.
[PMID: 16391194]

SUBJECT INDEX

A

Abnormalities 1, 9, 110, 140, 143, 147, 150, 153, 154, 157

Advanced 181, 186, 192, 194, 196, 219, 231, 234, 235, 236

Availability 1

Acetyltransferases 154

Acetyl 28

Accumulation 5, 21, 22, 24, 55, 57, 62, 83, 104, 109, 146,

Administering 7, 26

Antitumor 6, 19, 26, 27, 83, 107, 108, 129, 179, 186

Anticancer 6, 11, 17, 18, 21, 22, 27, 84, 107, 108, 186

Antibody-drug conjugates 100, 108, 110, 183, 184, 211, 223

Arachnoid 76, 77, 140, 142, 177

Astrocytes 4, 55, 79, 103, 141, 158, 159, 217

Aggressive Materials 3

Autophagy 99, 212

B

Brain tumour 1, 98, 104, 105, 109, 111, 179, 180, 211, 212, 215, 216, 217, 219, 220, 221, 224, 226, 227, 230, 236

Brain cells 4, 5, 55, 57, 86, 103, 105, 139, 230

Blood-brain barrier 4, 7, 8, 9, 10, 11, 15, 16, 17, 18, 19, 21, 22, 25, 26, 30, 52, 53, 54, 55, 57, 58, 62, 64, 73, 74, 79, 80, 82, 84, 87, 88, 90, 98, 102, 103, 109, 111, 119, 145, 146, 175, 176, 184, 185, 199, 210, 211, , 213, 217, 218, 219, 221, 224, 225, 226, 228, 229, 230.234, 235

Biomarkers 106, 107, 109, 110, 121, 129, 175, 191, 198, 211, 222, 225, 234

Biosynthesis 58

Brain Protection 77

Biological barriers 11, 18, 102

C

Bevacizumab 6, 15, 23, 26, 27, 29, 99, 131, 178, 228,

Cerebrospinal fluid 4, 10, 12, 19, 73, 77, 185

Clinical trials 8, 14, 27, 29, 57, 130, 176, 184, 185, 192, 193, 195, 221, 225, 227, 235

Cellular 3, 4, 7, 8, 12, 16, 22, 23, 26, 28, 29, 55, 58, 60, 74, 78, 80, 81, 84, 88, 89, 101, 103, 106, 122, 123, 1, 118, 136, 146, 147, 149, 150, 192, 216, 221, 222

Cortex 178

Cerebrospinal 9, 10, 11, 17, 19, 73, 77, 87, 178

Clinical trials 1, 8, 14, 15, 25, 27, 28, 29, 30, 57, 76, 130, 148, 176.184, 185, 190, 192, 193, 191, 221, 225, 227, 232, 235

Clinical studies 10, 11, 12, 23, 26, 27, 28, 75, 82, 84, 89, 119, 120, 128, 149, 175, 180, 191, 212, 215

Chemotherapy 6, 7, 10, 53, 74, 99, 100, 101, 107, 108, 120, 128, 129, 131, 132, 136, 175, 176, 189, 190, 191, 912, 193, 195, 199, 211, 214, 216, 217, 218, 220, 221, 233, 236

Combination therapy 30, 61, 98, 123, 128, 136

Convection Enhanced Delivery 9, 19, 25, 29, 88, 227, 235

Cellular processes 28, 122, 123, 128, 142, 148, 156, 159

Combination 6, 9, 12, 15, 19, 23, 24, 26, 27, 28, 29, 30, 60, 61, 64, 84, 98, 107, 108, 109, 123, 124, 127, 128129, 1301, 131, 136, 148, 175, 181, 184, 185, 191, 192, 193, 199, 210, 214, 221, 223, 225, 230, 232, 233, 234

Cerebrovascular disease 2

Cognitive 6, 10, 74, 141, 144, 215

Convection-enhanced delivery 9, 19, 25, 29, 88, 227, 235

Prashant Tiwari, Pankaj Kumar Singh & Sunil Kumar Kadiri (Eds.)
All rights reserved-© 2025 Bentham Science Publishers

www.ingramcontent.com/pod-product-compliance
Lightning Source LLC
Chambersburg PA
CBHW050821220326
41598CB00006B/284